International Review of Cytology

A Survey of Cell Biology

MOLECULAR BIOLOGY OF RECEPTORS AND TRANSPORTERS

BACTERIAL AND GLUCOSE TRANSPORTERS

VOLUME 137A

International Review of Cytology

A Survey of Cell Biology

Guest Edited by

Martin Friedlander
Howard Hughes Medical Institute
Jules Stein Eye Institute
and Department of Physiology
UCLA School of Medicine
Los Angeles, California

Michael Mueckler
Department of Cell Biology
Washington University
School of Medicine
St. Louis, Missouri

MOLECULAR BIOLOGY OF RECEPTORS AND TRANSPORTERS

BACTERIAL AND GLUCOSE TRANSPORTERS

VOLUME 137A

Academic Press, Inc.
Harcourt Brace Jovanovich, Publishers
San Diego   New York   Boston   London   Sydney   Tokyo   Toronto

Academic Press, Inc.
1250 Sixth Avenue, San Diego, California 92101-4311

United Kingdom Edition published by
Academic Press Limited
24–28 Oval Road, London NW1 7DX

Library of Congress Catalog Number: 52-5203

International Standard Book Number: 0-12-364537-9

PRINTED IN THE UNITED STATES OF AMERICA
92 93 94 95 96 97   EB   9 8 7 6 5 4 3 2 1

CONTENTS

Bacterial Periplasmic Permeases as Model Systems for the Superfamily of Traffic ATPases Including the Multidrug Resistance Protein and the Cystic Fibrosis Transmembrane Conductance Regulator

Giovanna Ferro-Luzzi Ames

Amino Acid Transport in Bacteria

Steven A. Haney and Dale. L. Oxender

In and Out and Up and Down with Lac Permease

H. Ronald Kaback

Group Translocation of Glucose and Other Carbohydrates by the Bacterial Phosphotransferase System

Bernhard Erni

Sugar–Cation Symport Systems in Bacteria

Peter J. F. Henderson, Stephen A. Baldwin, Michael T. Cairns, Bambos M. Charalambous, H. Claire Dent, Frank Gunn, Wei-Jun Liang, Valerie A. Lucas, Giles E. Martin, Terry P. McDonald, Brian J. McKeown, Jennifer A. R. Muiry, Kathleen R. Petro, Paul E. Roberts, Karolyn P. Shatwell, Glenn Smith, and Christopher G. Tate

Molecular and Cellular Physiology of GLUT-2, a High-K_m Facilitated Diffusion Glucose Transporter

Bernard Thorens

The Insulin-Sensitive Glucose Transporter

Morris J. Birnbaum

Molecular Genetics of Yeast Ion Transport

Richard F. Gaber

CONTRIBUTORS

Numbers in parentheses indicate the pages on which the authors' contributions begin.

Giovanna Ferro-Luzzi Ames (1), *Department of Molecular and Cell Biology, Division of Biochemistry and Molecular Biology, University of California, Berkeley, Berkeley, California 94720*

Stephen A. Baldwin (149), *Department of Biochemistry, University of Cambridge, Cambridge CB2 1QW, England*

Morris J. Birnbaum (239), *Department of Cellular and Molecular Physiology, Harvard Medical School, Boston, Massachusetts 02115*

Michael T. Cairns (149), *Department of Biochemistry, University of Cambridge, Cambridge CB2 1QW, England*

Bambos M. Charalambous (149), *Department of Biochemistry, University of Cambridge, Cambridge CB2 1QW, England*

H. Claire Dent (149), *Department of Biochemistry, University of Cambridge, Cambridge CB2 1QW, England*

Bernhard Erni (127), *Institut für Biochemie, Universität Bern, CH-3012 Bern, Switzerland*

Richard F. Gaber (299), *Department of Biochemistry, Molecular Biology and Cell Biology, Northwestern University, Evanston, Illinois 60208*

Frank Gunn (149), *Department of Biochemistry, University of Cambridge, Cambridge CB2 1QW, England*

Steven A. Haney (37), *Department of Biological Chemistry, University of Michigan Medical School, Ann Arbor, Michigan 48109*

Peter J. F. Henderson (149), *Department of Biochemistry, University of Cambridge, Cambridge CB2 1QW, England*

H. Ronald Kaback (97), *Howard Hughes Medical Institute, Departments of Physiology and Microbiology and Molecular Genetics, Molecular Biology Institute, University of California, Los Angeles, Los Angeles, California 90024*

Wei-Jun Liang (149), *Department of Biochemistry, University of Cambridge, Cambridge CB2 1QW, England*

Valerie A. Lucas (149), *Department of Biochemistry, University of Cambridge, Cambridge CB2 1QW, England*

Giles E. Martin (149), *Department of Biochemistry, University of Cambridge, Cambridge CB2 1QW, England*

Terry P. McDonald (149), *Department of Biochemistry, University of Cambridge, Cambridge CB2 1QW, England*

Brian J. McKeown (149), *Department of Biochemistry, University of Cambridge, Cambridge CB2 1QW, England*

Jennifer A. R. Muiry (149), *Department of Biochemistry, University of Cambridge, Cambridge CB2 1QW, England*

Dale L. Oxender (37), *Department of Biological Chemistry, University of Michigan Medical School, Ann Arbor, Michigan 48109 and Warner-Lambert/ Parke-Davis, Ann Arbor, Michigan 48105*

Kathleen R. Petro (149), *Department of Biochemistry, University of Cambridge, Cambridge CB2 1QW, England*

Paul E. Roberts (149), *Department of Biochemistry, University of Cambridge, Cambridge CB2 1QW, England*

Karolyn P. Shatwell (149), *Department of Biochemistry, University of Cambridge, Cambridge CB2 1QW, England*

Glenn Smith (149), *Department of Biochemistry, University of Cambridge, Cambridge CB2 1QW, England*

Christopher G. Tate (149), *Department of Biochemistry, University of Cambridge, Cambridge CB2 1QW, England*

Bernard Thorens (209), *Institute of Pharmacology, University of Lausanne, CH-1005 Lausanne, Switzerland*

FOREWORD

It is clear from this volume of *International Review of Cytology: A Survey of Cell Biology* that over the past few years there has been a dramatic expansion of knowledge related to the nature of the proteins involved in the transport of various molecules through biological membranes. This knowledge has come primarily from the elucidation of the amino acid sequences of these proteins through the use of recombinant DNA technology and from the use of computer algorithms that allow secondary-structure predictions from primary sequence information. The information has led to the widespread belief that most membrane transport proteins, in analogy to bacteriorhodopsin, contain multiple hydrophobic domains in α-helical conformation that traverse the membrane in a zigzag manner connected by hydrophilic "loops." It has also become apparent that membrane transport proteins fall into at least two superfamilies that contain representatives from bacteria to man: (i) the 12-helix superfamily, reviewed by Henderson *et al.*, which contains a number of bacterial sugar-cation symporters as well as various mammalian glucose facilitated diffusion transporters and (ii) the traffic ATPases or ABC (ATP-binding cassette) transporters, reviewed by Ames, which include a number of multicomponent bacterial transport systems in addition to the multidrug resistance protein and the cystic fibrosis transport regulator which may have resulted from gene duplications followed by fusion events.

With the rapid development of technology in molecular biology, it has become possible to manipulate virtually any protein in a fashion that was almost unimaginable only a decade ago. Individual amino acid residues can be mutagenized selectively, portions of molecules can be deleted,and one molecule or a portion thereof can be fused to another to produce chimeras of any variety. Thus, in many ways, the modern molecular biologist is in a position to "play God." However, if the problem in question involves membrane transport proteins, "God" is blindfolded. That is, there is a large barrier yet to be overcome, particularly with the class of membrane proteins discussed in this volume, and that barrier is the difficulty inherent in obtaining high resolution structural information about hydrophobic membrane proteins. Thus far, only two membrane proteins have been crystallized in a form that yields a high resolution X-ray structure, photosynthetic reaction centers and porin, although a high resolution structure for bacteriorhodopsin has been elucidated from electron diffraction studies and sequence information. Since the techniques of molecular biology allow manipulation of proteins at the 2 Å level of resolution, it is clear that structural information at the same level of resolution is required to obtain mechanistically relevant information. This is not to say that a structure will yield the mechanism directly. Rather, it is not possible to obtain the mechanism without the structure. In summary, therefore, knowledge of the structure of membrane transport proteins is a necessary foundation for deciphering the molecular mechanisms underlying the phenomenon of membrane transport.

H. Ronald Kaback

PREFACE

It has been over 6 years since we first considered putting together a volume devoted to the molecular biology of membrane proteins. We thought a single volume would provide sufficient space for several authors to describe the analysis, through molecular and cell biological approaches, of membrane protein structure and function. That was an exciting time since the dozen or so membrane protein receptors that had been cloned and sequenced appeared to fall into several large superfamiles that were related either through sequence homology or putative structural similarities. In the year it took us to prepare an outline for the volume's contents, the number of receptors had increased logarithmically; by 1988 the cloning and sequencing of several dozen receptors, transporters, and channels had been reported in the literature and we both retreated to our laboratories to try and keep up with the flood of information emerging from these studies.

By early 1990, we again began to think about putting together a multivolume treatise that summarized our knowledge of membrane protein receptors, transporters, and channels to serve a useful function for both investigators already in the field as well as those "extramembranous" students and established investigators who wanted to familiarize themselves with a rapidly expanding area of membrane biology. We invited definitive reviews from active investigators in the field; we asked prospective contributors to include a summary of their knowledge of a particular membrane protein as well as to speculate on future directions. The response was very satisfying; most of the major classes and superfamilies of membrane proteins are represented and the chapters have been written by authors whose laboratories are very active in the field. Thus, the chapters are stimulating and authoritative even though we recognize that the speed with which the field is moving makes it difficult to be current by the time a volume is published. Nevertheless, we hope these volumes will provide a useful resource for those individuals interested in the field of membrane receptors, transporters, and channels.

Assembling nearly 30 chapters from as many laboratories has required the usual cajoling, pleading and, on occasion, threats. However, the credit for ultimate production is due to the authors themselves. We are grateful to the many scientists who contributed their efforts to writing the chapters in these volumes. For our part, the solicitation and editing of these volumes required extensive time that otherwise would have been spent with our families, and we are both very grateful to our wives, Sheila and Paula, and our children for their patience and understanding. Both of us would also like to thank Ron Kaback for his enthusiastic involvement in many of the chapters other than his own, as well as for his foreword to the volume.

Each of us has our own thanks to extend to individuals who have helped with various aspects of assembling these volumes. Martin Friedlander would like to acknowledge the support of the Heed Ophthalmic and Heed Knapp Foundations. My chairman, Bradley Straatsma, has been firmly supportive since my arrival at UCLA and has encouraged me throughout the course of this project. I am also grateful to Allan Kreiger, Bart Mondino, Gordon Grimes, and Joe Demer for their advice throughout the preparation of these volumes. Suraj Bhat and Dean Bok provided advice and encouragement during the early phases of this project and Eileen Fallon provided excellent editorial assistance. I am particularly grateful to Günter Blobel for introducing me to the field of membrane protein topogenesis and sharing his infectious enthusiasm for its study with me.

Michael Mueckler would like to acknowledge the support of the Juvenile Diabetes Foundation International. I am indebted to the past and current members of my laboratory for their patience during the preparation of these volumes and for making it a (usually) pleasant experience to come into the lab in the morning. I thank my colleagues in the Department of Cell Biology and Physiology, Robert Mercer, Edwin McCleskey, Philip Stahl, and Stephen Gluck, for their contributions to this series and for many stimulating discussions on membrane proteins over the past few years. I also thank my Ph.D. thesis advisor, Henry Pitot, and my postdoctoral mentor, Harvey Lodish, for sparking my interest in membrane proteins. Lastly, I am grateful to Alan Permutt for his continued friendship, support, and encouragement since my arrival in St. Louis.

Martin Friedlander

Michael Mueckler

Bacterial Periplasmic Permeases as Model Systems for the Superfamily of Traffic ATPases, Including the Multidrug Resistance Protein and the Cystic Fibrosis Transmembrane Conductance Regulator

Giovanna Ferro-Luzzi Ames

Department of Molecular and Cell Biology, Division of Biochemistry and Molecular Biology, University of California, Berkeley, Berkeley, CA 94720

I. Introduction

Traffic ATPases: Universality of the Overall Structure

Recent work has revealed that numerous proteins involved in the transport of a vast variety of molecules bear strong primary sequence and secondary structure similarity, suggesting that they have the same basic mechanism of action (Ames et al., 1990; Higgins et al., 1990). These proteins include numerous prokaryotic transporters, which transport both small and large substrates and several of which transport in the outward direction. Over the last few years numerous eukaryotic proteins have been discovered that are clearly members of this family, some of which are known to be transporters and others having unknown function; because of the high level of homology with known transporters it is likely that the latter are also transporters. We have proposed that members of this superfamily of transporters be called *traffic ATPases* (Ames et al., 1990), the rationale for this nomenclature being that (1) the similarity invariably includes a putative ATP binding site, which implies that they hydrolyze ATP, presumably for energization; (2) they translocate substrates that are extremely varied in nature and size; and (3) the term *traffic* implies that the translocation occurs in either direction and it helps distinguish this class from other transport ATPases, usually conducting ions. The overall picture is reminiscent of a busy and heterogeneous thoroughfare. Another name suggested for these proteins is ABC (ATP binding cassette) proteins (Hyde *et al.*, 1990). Universal features of traffic ATPases are the presence of a hydrophobic domain, which is usually postulated to be embedded in the membrane, and a hydrophilic domain characterized by a striking level of homology shared by all proteins that belong to this superfamily and that includes, in particular, an ATP binding consensus sequence.

The well-studied family of prokaryotic periplasmic transport systems (permeases) belongs to this superfamily. It is the purpose of this article to discuss the prokaryotic members of this family. A discussion of the eukaryotic proteins is also presented to put the knowledge acquired in the prokaryotic field in perspective relative to the eukaryotic traffic ATPases. Among the eukaryotic proteins assigned to this superfamily are several medically important eukaryotic transporters. Details on these transporters appear in other articles in a third volume of this series. They include the medically important P-glycoprotein, responsible for the phenomenon of multidrug resistance in tumor cells (MDR or P170; see Gros, this series, volume 137C), which acts by expelling cytoxic drugs from drug-resistant cancer cells using ATP as an energy source; the closely related *Plasmodium falciparum* protein involved in imparting chloroquine resistance to the malarial parasite; the cystic fibrosis gene product [referred to as CFTR (cystic fibrosis transmembrane conductance regulator); Kerem *et al.*, 1989; Riordan *et al.*, 1989; Rommens *et al.*, 1989; Riordan, this series]; a protein component of the major histocompatibility complex class I (Spies *et al.*, 1990; Parham, 1990; Deverson *et al.*, 1990; Monaco *et al.*, 1990; Trowsdale *et al.*, 1990); and the *STE6* gene product, responsible for the export of the **a** factor mating pheromone of *Saccharomyces cerevisiae*. The general structure of eukaryotic traffic ATPases is that of a single polypeptide composed of two homologous halves that are likely to have resulted from ancestral duplications followed by fusion; each half comprises a hydrophobic and a hydrophilic domain. This monocomponent structure is in contrast to that of the prokaryotic permeases, in which the translocating structure always includes distinct polypeptides that separately embody the hydrophilic and hydrophobic membrane components, and which also includes a soluble, substrate-binding receptor.

The general composition and basic features of periplasmic permeases of gram-negative bacteria are discussed in an article in this volume by Oxender and Haney, to which the reader is referred. Other articles also have reviewed this material (Ames and Lecar, 1992; Ames, 1986a; Shuman, 1987; Furlong, 1987). This article will stress those aspects of periplasmic transport that relate to their energy-coupling mechanism and to their structural organization. The functional and evolutionary relationships between these and the eukaryotic transporters will be discussed once the characteristics of the much better studied prokaryotic members of the superfamily have been described. The significance of the latest data for the mechanism of action of the eukaryotic transporters also will be discussed.

II. Periplasmic Permeases

A. General Features

In brief, periplasmic permeases are typically composed of a receptor (a periplasmic substrate binding protein) and two to three membrane bound components. The loss of the periplasmic component and subsequent loss of transport activity on osmotic shock has resulted in the alternative nomenclature for these systems of shock-sensitive permeases. Many periplasmic permeases acting on a wide variety of substrates have been discovered (Furlong, 1987). All periplasmic permeases have the same overall composition (Ames, 1986a), which is schematically represented in Fig. 1 for the histidine permease. The components of the histidine permease are the periplasmic histidine binding protein, HisJ, and the three membrane bound components, HisQ, HisM, and HisP. Two of the membrane bound components, HisQ and HisM, are very hydrophobic, while the third one, HisP, is hydrophilic, although also firmly bound to the membrane. The membrane bound components form a complex within the cytoplasmic membrane. Histidine enters the periplasm through pores in the outer membrane, is then bound by the substrate binding protein, and then translocated to the interior of the cell through the membrane bound complex. Genetic and biochemical evidence suggests an interaction between the periplasmic binding protein and the membrane bound proteins (Ames and Spudich, 1976; G.F.-L. Ames, unpublished results). Some periplasmic permeases have a single hydrophobic membrane component instead of the usual two, or two homologous hydrophilic membrane components instead of one. The genetic structure of all periplasmic permeases is also very similar. In all cases a single operon (or two divergent operons) contains all the genes coding for the known transport components. In some cases the operon includes additional genes whose function is either unknown at present or is involved in further catabolism of the transported substrates (Ames, 1986a). The similarity in the composition of these permeases indicates a conserved structural organization and mechanism of action despite the vast variety of substrates transported and a common evolutionary origin. Their rather complex composition and organization is to be contrasted with that of shock-resistant transporters, such as the β-galactoside permease, which is composed of a single integral membrane protein.

Periplasmic permeases typically concentrate substrates against a very

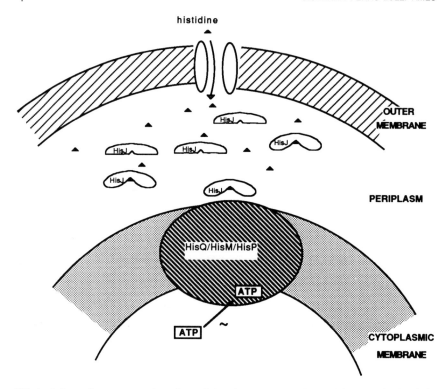

FIG. 1   Schematic representation of a periplasmic permease. The model system is the histidine permease. The outer membrane is represented as containing proteinaceous pores that allow entrance of substrate (histidine) into the periplasm, where it is bound by the periplasmic binding protein or a receptor (HisJ). The three membrane components (HisQ, HisM, and HisP) are represented as forming a complex within the cytoplasmic membrane. The receptor changes conformation on binding histidine and interacts with the membrane complex. The squiggle (~) indicates the involvement of ATP in energy coupling.

large (10,000-fold) concentration gradient. They are also able to scavenge solutes from very low concentrations, with apparent K_m values for uptake ranging from 0.01 to 1 μM. The high efficiency of transport and good concentrative ability might underlie the requirements leading to such a complex structure. The very high efficiency may also constitute an evolutionary advantage for these bacteria, at least where amino acid transport is conserved, since biosynthetically produced amino acid can be lost from the cell, but it is efficiently recaptured by these high-affinity permeases (Ames, 1972). This recapture may constitute a considerable advantage, thus justifying the existence of complex, multicomponent permeases, because amino acid biosynthesis is expensive: e.g., it has been calculated that 41

ATP molecules are being sacrificed for each histidine molecule synthesized (Brenner and Ames, 1971). An interesting aspect of some periplasmic systems is the fact that the membrane bound complex is multifunctional, i.e., it is used for transport of several unrelated substrates and for that purpose it is utilized by more than one binding protein. For example, in the case of the histidine permease, the membrane bound proteins are also essential for transport of arginine via the lysine–arginine–ornithine binding protein (LAO protein), coded for by the *argT* gene (Kustu and Ames, 1973; Ames, 1986a). The existence of alternative periplasmic binding proteins utilizing the same set of membrane bound components has also been shown for the branched chain amino acid permease (Landick and Oxender, 1985; Landick *et al.*, 1985).

A representation of a cycle of transport for the histidine permease is shown in Fig. 2, which in essence postulates a series of conformational changes that occur concomitantly with ATP hydrolysis and substrate translocation. The first step in transport is the liganding of the substrate by the substrate binding protein. The liganded binding protein, which is the actual substrate for translocation (Prossnitz *et al.*, 1989), interacts with the membrane bound complex, making direct contact at least with the hydrophobic components. This interaction has been demonstrated experimentally for the histidine permease by chemical cross-linking experiments, which yielded a cross-linked product containing HisJ and HisQ (Prossnitz *et al.*, 1988). The function of this initial interaction is presumably to signal the occupancy of the specific substrate binding site and to offer the substrate in a concentrated form to the next step in transport (Prossnitz *et al.*, 1989). In the next step the substrate dissociates from the periplasmic protein and traverses the membrane, either through a pore or by interaction with a substrate binding site(s) located on the membrane bound complex. This transport mechanism must utilize energy to achieve a concentration gradient. Indeed, energy is likely to be applied toward "forcing" the release of the substrate from the binding protein. It has been firmly established that ATP hydrolysis is the energy source, resolving a decade-long controversy; thus, Fig. 2 includes the hydrolysis of ATP as part of the transport cycle.

The hydrophilic membrane components of all periplasmic permeases share extensive amino acid sequence similarity and constitute a family of homologous proteins, also referred to as "conserved components." In particular, two regions of unusually high sequence homology correspond to sequence motifs that commonly occur in mononucleotide binding proteins and that have been implicated in the structure of the nucleotide binding pocket (Walker *et al.*, 1982; Mimura *et al.*, 1991; reviewed in Ames, 1986a; Higgins *et al.*, 1986). In agreement with this observation, the conserved components have been shown to bind ATP by the use of various

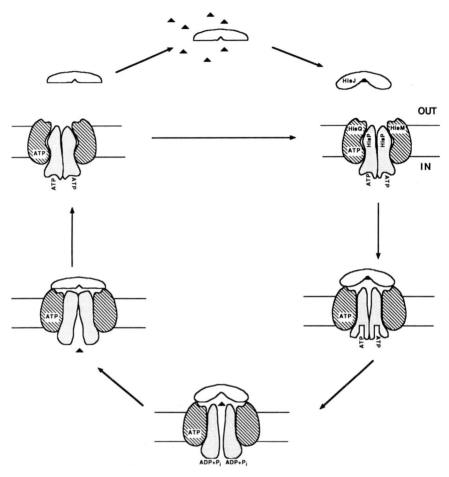

FIG. 2 A speculative model for periplasmic transport. The components undergo a series of conformational changes initiated by the binding of histidine to the periplasmic binding protein HisJ. The liganded HisJ binds to the membrane complex. This interaction occurs probably with both hydrophobic membrane components (HisQ and HisM) and elicits a conformational change in the ATP-binding membrane protein (HisP), thus causing ATP hydrolysis. Hydrolysis of ATP leads to the opening of a pore [possibly containing a substrate-specific binding site(s)] that allows the unidirectional diffusion of the substrate to the interior. After the substrate has been released to the interior, an additional conformation change (release of ADP?) closes the pore. The unliganded binding protein disassociates itself from the membrane complex, which binds ATP again, ready to start a new cycle. An ATP binding site is represented in HisQ; since the function of this site is at present unknown, the possibility should be left open that it may be the actual site of ATP hydrolysis and energy coupling. (Modified from Ames et al., 1990.)

affinity labeling ATP analogs (Hobson *et al.*, 1984; Higgins *et al.*, 1985). The conserved components have been proposed to be peripheral membrane proteins localized to the inner surface of the cytoplasmic membrane and held attached to it only by surface interactions with the hydrophobic components (Shuman, 1987; Gallagher *et al.*, 1989). We obtained data that are not consistent with this hypothesis, indicating that these ATP binding proteins span the membrane (Kerppola *et al.*, 1991; Baichwal *et al.*, 1992). Thus, our working model for the histidine permease postulates the organization shown in Fig. 2, in which two copies of HisP are embedded in the membrane and protected from contact with the hydrophobic bilayer by interactions with HisQ and HisM. The evidence leading to this model is summarized below in Section C.4.

B. Energy Coupling

1. The Source of Energy

After a long period of uncertainty, the energy-coupling mechanism of periplasmic permeases has now been clearly elucidated as being ATP hydrolysis (reviewed in Ames and Joshi, 1990; Ames 1990a,b). The notion that ATP is responsible for energy coupling received initial support from the finding that several members of the conserved family bind ATP and its analogs. These data, while not proving the involvement of ATP in energy coupling, gave a strong impetus into this aspect of research. Four different experimental systems, one *in vivo* and three *in vitro*, were used in several laboratories to prove this point.

1. The first one (Joshi *et al.*, 1989) used intact cells that lacked the proton-conducting ATPase (*unc* deletion mutants) and thus were unable to interconvert the two important pools of energy, ATP and the proton motive force. In these mutants it was possible to manipulate separately the levels of ATP and of the proton motive force. This work showed that the proton motive force is both unnecessary and insufficient as an energy source for either the histidine or the maltose permease. Importantly, this experimental system was also used to demonstrate that the use of valinomycin with K^+, or of TPP^+ (tetraphenylphosphorium bromide), to dissipate the proton motive force results in artifactual side effects that cause the inhibition of several of the periplasmic permeases. Thus, dissipation of the membrane potential by these reagents cannot be used for studying the energetics mechanism of periplasmic permeases. Similarly, the use of arsenate to lower the ATP pool in whole cells (Klein and Boyer, 1972), which unavoidably also lowers the pool of several other energy-rich molecules, was shown to result in side effects (Joshi *et al.*, 1989). An alterna-

tive method of lowering the ATP pool was used, which depends on creating an idle cycle of ATP consumption without addition of poisons (Galloway and Taylor, 1980). Interestingly, with this method there was essentially no change in histidine transport levels, despite a depletion of the ATP pool to less than 10% of the normal level; this indicates that a very low concentration of ATP is sufficient to energize transport.

2. The second system consisted of right-side-out membrane vesicles reconstituted with added binding protein and energized with ascorbate and phenazine methosulfate, or D-lactate (Prossnitz et al., 1989; Dean et al., 1989). Addition of ascorbate/phenazine methosulfate or D-lactate generated a proton motive force that was converted into ATP by the residual proton-conducting ATPase activity in vesicles prepared from wild-type strains. This conversion did not occur in vesicles prepared from unc mutants, which therefore could not transport. This result and the finding that various treatments lowering or raising the ATP pool correspondingly inhibited and increased transport led to the conclusion that ATP was the most likely energy source. However, this system is still metabolically too complicated to be useful for detailed energetics studies. Thus, the results obtained with the histidine permease (Prossnitz et al., 1988) and those with the maltose permease (Dean et al., 1989; Mimmack et al., 1989) could be taken as strongly suggestive that ATP is the energy source, but not conclusive.

3. Very strong support that ATP is indeed the direct energy source and that its hydrolysis is necessary was obtained for the histidine permease with a novel inside-out vesicles system that contained histidine and the histidine binding protein trapped internally (Ames et al., 1989). This system allows the controlled and direct presentation of the normally nonpermeating energy substrate to the energy-coupling aspect of the permease. Addition of ATP caused histidine translocation from inside to the outside, a movement that corresponds to inward transport in intact cells. A nonhydrolyzable analog of ATP was inactive, suggesting the need for ATP hydrolysis. Pretreatment of the vesicles with 8-azido-ATP and UV light, thus inactivating the nucleotide binding component of the histidine permease, HisP, eliminated transport, as expected (Hobson et al., 1984). With this system it was also possible to establish that the affinity of the permease for ATP is about 200 μM, which is in agreement with the demonstrated ability of intact cells to transport when the ATP level was lowered 10-fold (Joshi et al., 1989).

4. A reconstituted proteoliposome system finally resulted in the incontrovertible demonstration that ATP is the energy source. Once the important parameters of periplasmic permease function were known, such as the nature of the liganded binding protein as the true transport substrate

(Prossnitz *et al.*, 1989), the elimination of the proton motive force as energy source (Prossnitz *et al.*, 1989; Joshi *et al.*, 1989), and the knowledge of the concentrations of liganded binding protein and of ATP appropriate for active transport (Prosspitz *et al.*, 1989; Ames *et al.*, 1989), it became possible to develop a reconstituted proteoliposome system. Proteoliposomes were formed utilizing a solubilized, partially purified histidine permease membrane complex and *Escherichia coli* phospholipids (Bishop *et al.*, 1989; Davidson and Nikaido, 1990), and they contained internally trapped ATP. Active uptake occurred on addition of liganded binding protein. The reconstituted system was entirely dependent on all four permease proteins and ATP. Dissipators of the membrane potential had absolutely no effect on transport. The demonstration that ATP was hydrolyzed only concomitantly with histidine transport finally confirmed incontrovertibly that ATP hydrolysis drives solute transport in these permeases. GTP could replace ATP reasonably well.

The proteoliposome-reconstituted histidine permease functions with an efficiency that is comparable to that obtained in reconstituted right-side-out vesicles: 0.55 mol of histidine transported per minute per mole of HisP, as compared to 1 mol/min/mol of HisP in vesicles. Both values are comparable to those obtained for intact cells. Thus, the proteoliposome system must reflect reasonably well the conditions *in vivo*.

The stoichiometry between ATP hydrolysis and transport was determined initially in the proteoliposome system to be an average of five molecules of ATP per molecule of histidine transported. This value is most likely artifactual, possibly the result of damage inflicted on the membrane complex during purification and reconstitution, thus causing slipping (uncoupling). A stoichiometry of one or two ATP molecules per histidine is more likely to be correct. In particular, since periplasmic permeases have two ATP binding domains, either on the same polypeptide or in the form of two separate ATP binding proteins (Ames, 1986a; Bell *et al.*, 1986; Hiles *et al.*, 1987; Kerppola *et al.*, 1991; discussed in Section D.1. below), the value of two may be correct. Data obtained in intact cells inhibited with iodoacetate indicates a stoichiometry of one to two for the maltose permease (Mimmack *et al.*, 1989). This result should be interpreted with caution since intact cells may have complicated metabolic routes for recycling ATP, despite and possibly because of the poisoning procedure.

2. Mechanism of Energy Coupling

Relatively little is known of how ATP energy is converted into translocation. Presumably ATP binding and/or hydrolysis results in conformational changes that are transmitted to the rest of the permease, causing the opening of a pore or the creation of a substrate binding site in the membrane bound complex. In particular, conformational changes may be necessary as a "signal" for the receptor to release the substrate, which is deeply buried between its two lobes. Indeed, this relatively simple signaling system could be used as a model for studying means of communication across the thickness of the cell membrane. It is likely that there is a very tight coupling between transport and ATP hydrolysis, since in no case has hydrolysis of ATP by a periplasmic permease complex or by a partially purified conserved component been observed in the absence of reconstituted transport. It should also be mentioned that several efforts in various laboratories at detecting a phosphorylated derivative of conserved components have failed (Ames and Nikaido, 1981). Thus, it is likely, although not proven, that the formation of a phosphorylated intermediate is not part of the mechanism of these proteins.

It is important to note that while it is generally assumed that the conserved components, with their ATP binding site, are the subunits responsible for effecting the energy-coupling step, this hypothesis has not been proven and should be accepted only tentatively. In fact, they are not the only proteins of the membrane complex to interact with ATP. It has been shown in the case of the histidine permease that the hydrophobic component, HisQ, is also labeled with the ATP affinity analog, 8-azido-ATP, as efficiently as HisP (Shyamala and Ames, 1992a), leaving open the possibility that HisQ, rather than the conserved component, is directly responsible for coupling ATP energy to accumulation. It has been excluded that affinity labeling of HisQ is an artifact due to indirect labeling because of the proximity of HisQ to HisP within the membrane complex; in fact, HisQ reacts with 8-azido-ATP even in *hisP* deletion mutants. Thus, the histidine permease membrane complex carries at least three ATP binding sites: one in each of the two HisP molecules and one in HisQ. It should be noted that the sequence of HisQ shows no evidence of the ATP binding motifs that are clearly identifiable in the conserved components, suggesting that the binding and utilization of ATP by HisQ is entirely different than in the case of HisP and that it possibly has a regulatory function.

A better understanding of the process by which energy liberated in the hydrolysis of ATP is converted into accumulation of solute and/or into transmission of signals between the exterior liganded binding protein and the membrane bound complex requires the dissection and characterization of the various protein domain(s) involved. In this respect, a structure–function analysis of HisP has been initiated as discussed below.

C. Structural Considerations

1. The Periplasmic Substrate Binding Proteins

The state of the art on the structure of periplasmic binding proteins has been reviewed (Adams and Oxender, 1989; S. A. Haney and D. L. Oxender, this volume) and will only be summarized. Briefly, the three-dimensional analysis of several of these proteins indicates that in all cases they exhibit the shape of a kidney bean, with two domains separated by a cleft that contains the substrate binding site. The proteins are postulated to have a flexible structure that is either "open," with a widely accessible substrate binding site when the protein is unliganded, or "closed" when the substrate is bound and the cleft is closed by the two domains moving toward each other. Until recently, in no case yet had the same protein been crystallized in both the liganded and unliganded form and the structures of both forms resolved. The liganded histidine binding protein (HisJ) and the lysine–arginine–ornithine binding protein (LAO), in both the liganded and unliganded forms, have been crystallized and their structures are now being resolved (Kang *et al.*, 1989, 1991; Oh *et al.*, 1992, in preparation). The histidine binding protein has been shown to have a substrate binding domain that is separate from a domain(s) responsible for interacting with the membrane bound complex (Kustu and Ames, 1974; Ames and Spudich, 1976; Speiser and Ames, 1991). The latter domain has been identified by both genetic and biochemical techniques. The genetic evidence suggested an interaction between the binding protein and the conserved component, as indicated by the isolation of suppressor mutations in the conserved component that corrected binding protein defects causing an inability to interact with the membrane bound complex. The biochemical evidence demonstrated the existence of a physical interaction between the binding protein and both hydrophobic components (Prossnitz *et al.*, 1988; Ames *et al.*, 1992, unpublished results). Thus, it is likely that the overall interaction is complex and extensive. Structure–function studies are in progress for both HisJ and LAO that involve the characterization of numerous mutations affecting either the substrate binding site or the domain postulated to be involved in membrane interaction (Kang *et al.*, 1989, 1991; Oh *et al.*, 1992; Ames *et al.*, 1992). These studies, in combination with the X-ray analysis of crystals, will allow the precise definition of the molecular details relevant to the binding of substrates, interaction with the membrane, and the consequent conformational changes. In addition, the analysis of suppressor mutations in the conserved component provides useful information concerning its tertiary structure and topology.

2. The Hydrophobic Components

From studies on the topology of the various components of the histidine permease, a clear picture is now emerging (Kerppola *et al.*, 1991; Kerppola and Ames, 1992). HisQ and HisM are proteins of molecular weight around 25,000 and their sequences include several stretches that are quite hydrophobic (Higgins *et al.*, 1982; Ames, 1985). These two proteins (and equivalent ones from several of the other permeases) show significant homology to each other and they probably arose by duplication from an ancestral gene. Their sequence similarity suggests structural similarity, and they are likely to form a pseudodimer within the membrane complex (Ames, 1985). This view is compatible with and supported by the existence of some permeases containing only one hydrophobic component (such as the arabinose periplasmic permease), in which case the single component presumably forms a homodimer. The topology of HisQ and HisM has been investigated by the use of several impermeant reagents, such as proteolytic enzymes and antibodies (Kerppola *et al.*, 1991; Kerppola and Ames, 1992), using oriented vesicles, both inside out and right side out. As expected for integral membrane proteins, both HisQ and HisM were shown to span the membrane by being proteolytically digestible at either membrane surface and by requiring strong detergent action for solubilization. In addition, their behavior with respect to Triton X-114 solubilization and partition analysis (Bordier, 1981) was examined. This method relies on the low cloud point of Triton X-114, a property that causes the partitioning into water and detergent phases when the temperature is raised. Integral membrane proteins require bound detergent to remain soluble, and thus migrate to the detergent phase when the cloud point is reached. Both HisQ and HisM partition with the detergent phase, thus behaving like typical integral membrane proteins. Their carboxy termini were shown to be located on the cytoplasmic surface (i.e., facing the interior of the cell) and their amino termini to be on the periplasmic surface (Kerppola and Ames, 1992).

The actual number of membrane spanners in HisQ and HisM was investigated by the use of transposon Tn*phoA*, which distinguishes between the hydrophilic loops of integral membrane proteins that are located on the periplasmic versus cytoplasmic membrane surface (Manoil *et al.*, 1990). Such an analysis clearly indicated five membrane spanners (Kerppola and Ames, 1992). How general is this type of organization for the hydrophobic components? No such analysis has been performed for other hydrophobic periplasmic permease components except for MalF from the maltose permease (Boyd and Beckwith, 1989; Froshauer *et al.*, 1988). The topological model of HisQ and HisM has been compared with that of MalF. Even though the sequence homology is limited, topological homology is very extensive: the periodicity of loops and spanners present throughout the length of HisQ and HisM matches exactly the pattern at the carboxy-terminal end of MalF. Since MalF is considerably larger than HisQ or HisM, its amino-terminal third forms additional membrane spanners and hydrophilic loops. A comparison of the topology of the other 23 known hydrophobic subunits, as predicted by hydropathicity analysis, indicates a striking structural similarity in the carboxy-terminal ends of all of them. Thus, we propose that the minimum structure (as represented by HisQ and HisM) of these subunits consists of five hydrophobic helices spanning the membrane, with the amino and carboxy termini on the exterior and interior surfaces, respectively. The distribution of charges between inside and outside loops within this five-spanner arrangement was shown to adhere very well to the "rules" established for integral membrane proteins, with positive charges preferentially inside (Michel *et al.*, 1986; von Heijne, 1986; Boyd and Beckwith, 1989; Dalbey, 1990).

A conserved motif in all these proteins is a conserved region located in a large hydrophilic cytoplasmic loop that is always present between the third and fourth spanners of the minimum structure. A salient feature of this region is the triplet Glu-Ala-Ala (EAA) that is strongly conserved in all prokaryotic subunits (Dassa and Hofnung, 1985). At present, there is no information as to the function of this motif. A possible reason for its conservation is in the common purpose of interacting with a domain of the well-conserved hydrophilic membrane component for signal transmission.

3. The Hydrophilic (Conserved) Components

An extensive analysis of the membrane association of the conserved component from the histidine permease, HisP, has also been performed. HisP has a molecular weight of 28,000, a hydrophilic sequence, and no hydrophobic stretches of sufficient length to form a membrane spanner. It is, however, firmly bound to the membrane (Kerppola *et al.*, 1991; Ames and Nikaido, 1978). This tight membrane association, which was also shown for the analogous oligopeptide permease conserved component (Gallagher *et al.*, 1989), is puzzling considering the hydrophilicity of the HisP sequence and of all conserved components. Subcellular fractionation, proteolytic susceptibility, and solubilization studies have yielded important information that explains the peculiar behavior of this class of proteins (Kerppola *et al.*, 1991; Kerppola and Ames, 1992). HisP requires strong solubilization procedures to be released. Agents that readily release peripheral membrane proteins, such as urea (8 M), alkaline pH (Na_2CO_3, pH 11.5), and Triton X-114, only partially release HisP from the membrane. This indicates an attachment to the membrane firmer than that of peripheral membrane proteins, but not as firm as that displayed by classic integral membrane proteins. A sizeable proportion of HisP is still firmly associated with the membrane even in the absence of both HisQ and HisM. However, in the latter case the membrane association is clearly of a different nature than in the presence of HisQ and HisM, as shown by an easier release from the membrane by urea, alkaline pH, and Triton X-114, and by increased proteolytic susceptibility. Therefore, it seems that HisP folds into a specific complex with the integral membrane transport proteins when they are present and acts as a sort of template. Because of this behavior, the possibility that it spans the membrane was considered. HisP was shown to be proteolytically digestible in both inside-out and right-side-out vesicles, and accessible to an impermeant biotinylating reagent, while peripheral internally located proteins such as the α and β subunits of the proton-conducting F_0F_1 ATPase, are accessible only in inside-out vesicles (Baichwal, et al., 1992). These results suggest that HisP spans the membrane and they are in agreement with the genetic experiments described above, suggesting that HisP interacts directly with the periplasmic binding protein. However, since HisP is not digestible, or is biotinylated at the outer surface of the membrane in the absence of HisQ and HisM, it has been concluded that it spans the membrane only through an association with HisQ and HisM (see Fig. 2). This membrane-spanning arrangement for conserved components is contrary to all models of periplasmic permease organization published to date. Possible discrepancies between the model presented here and data on the maltose and oligopeptide permeases have been discussed (Kerppola et al., 1991).

4. The Membrane Bound Proteins Form a Complex

The architectural organization depicted in Fig. 2 implies that the membrane bound proteins of the histidine permease form a complex. The existence of such a complex has been demonstrated for the histidine permease by cross-linking and coimmunoprecipitation studies with antibodies raised against each of the individual proteins. The stoichiometry of the proteins within the complex was established to be 1 HisQ: 1 HisM: 2 HisP subunits (Kerppola et al., 1991). Similar results have been obtained for the maltose permease by the Nikaido laboratory (Davidson and Nikaido, 1991). This stoichiometry is likely to be true for all bacterial permeases and to take slightly different forms depending on whether a single or two separate hydrophobic components are present. In addition, three possible basic compositions are known to occur among periplasmic permeases, also with respect to the conserved component: (1) a single conserved component that is present in the membrane bound complex as a homodimer (Fig. 2), (2) two separate and homologous conserved components that presumably form a heterodimer within the complex (e.g., oligopeptide permease), and (3) a single protein encoding two homologous halves, each with a nucleotide binding site, which presumably forms an intramolecular heterodimer (e.g., ribose permease). The latter case can be taken as tentative evidence that the dimer assembles first and then interacts with the hydrophobic components to form the final complex.

In conclusion, with the definition of the architecture and localization of the membrane bound components of the histidine permease, the novel possibility has arisen that the conserved component spans the membrane and, therefore, that this protein carries the major burden of the transport process, directly receiving the substrate from the binding protein and translocating it while consuming energy. The hydrophobic components may have mostly a structural role. Given the high level of similarity between periplasmic permeases, the structure as deduced for the histidine permease is likely to apply to periplasmic permeases in general, with minor modifications. This topological model constitutes a framework to be extended by additional experimentation aiming at structure–function analysis. The implications that this structure has for the mechanism of action of eukaryotic transporters is discussed in Section III.D.

D. Structure—Function Analysis

1. Structural Modeling

An understanding of the molecular mechanism of action of these permeases requires a structure–function analysis of each of the component polypeptides. An extensive characterization is under way for the substrate binding protein, as described above. The conserved components have acquired a new light because of their close relationship to the eukaryotic traffic ATPases. Thus, efforts have been aimed also in that direction. An approach that has been useful for this purpose is chemical, genetic, and *in vitro* modification of residues suspected of being involved in specific functions. To understand the role played by individual residues it is necessary to relate the position of the residues to the tertiary structure of HisP. However, crystals are not available yet for any of the conserved components. Two laboratories have applied methods of computer-assisted molecular modeling to propose a structure for the conserved component and particularly for the ATP binding domain of the traffic ATPases (Mimura *et al.*, 1991; Hyde *et al.*, 1990). In one approach (Mimura *et al.*, 1991), the amino acid sequences of 17 prokaryotic conserved components were aligned and this alignment used to identify conserved motifs and to predict a consensus secondary structure. Such alignment is shown in Fig. 3. A three-dimensional model was then inferred by alignment of predicted secondary structural motifs with the corresponding features of several nucleotide binding proteins of known crystal structure as detailed below. The proteins of known structure used as the basis for the model were p21^{ras} (Milburn et al., 1990), the elongation factor EF-Tu (Woolley and Clark, 1989), and porcine adenylate kinase (Dreusicke et al., 1988). These proteins were chosen because they all bind mononucleotide triphosphates, all are known to moderately high resolution, and in the cases of p21^{ras} and EF-Tu the position where the nucleotide binds is also known. A computer-generated superposition of the C_α backbones of the three proteins was obtained illustrating the extent of structural similarity between them (Mimura et al., 1991). All three proteins possess a common framework structure, consisting of five β strands and several intervening α helices and turns. A very highly conserved secondary structural feature that also displays strong primary sequence conservation is the glycine-rich flexible loop, characterized by the sequence G(X)$_4$GK, which has been implicated as a region of large conformational change on ATP hydrolysis (Fry et al., 1986). Another highly conserved feature is an aspartic acid residue near the end of the third β strand, which appears to bind to a water molecule in the hydration shell of a magnesium ion chelated to the β and γ phosphates of the bound nucleotide (Milburn et al., 1990).

This common architecture among known structures of mononucleotide binding proteins was then used as a template to formulate a structural model for the conserved components. The corresponding features among the consensus secondary structures of the conserved components were identified and used as the basis for construction of a three-dimensional model of the ATP-binding domain. The glycine-rich loop and a conserved aspartate residue (D-178 in HisP) were easily identified as two of the key

---

FIG. 3   Alignment of HisP with the ATP binding domains of CFTR. The consensus sequences of both the prokaryotic and the eukaryotic proteins are shown for comparison. The eukaryotic consensus was derived from an alignment of all known eukaryotic components (Ames et al., 1991). Consensus secondary structure based on predictions made using only the prokaryotic proteins is indicated and corresponds to the structural model of Mimura et al., (1991). Regions of greater than 50% conservation among the prokaryotic proteins are indicated by the letter "b", regions conserved among the eukaryotic proteins by the letter "e", and for all proteins by the letter "c". A position at which a residue is 100% conserved over all prokaryotic and eukaryotic proteins is indicated by a capital letter corresponding to its one-letter amino acid code. If the position is completely conserved in either prokaryotic or eukaryotic sequences, but not in both, it is indicated by a lower-case letter. The first three lines indicate residues mutated in cystic fibrosis patients, as summarized by Davies (1990). Symbols are the same as in Fig. 3. An "X" among the CFTR replacements indicates a termination codon and an exclamation mark (!) indicates a splice mutation.

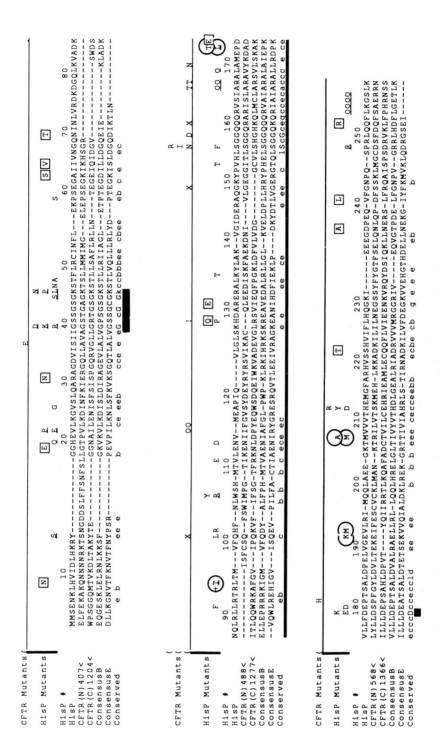

motifs. Additional topological considerations and partial sequence homology were used to align all the other key structural motifs. A structural model of HisP, which serves as a prototype for the conserved components, was generated (see also Ames *et al.*, 1991). A striking feature of the HisP structural model is a large helical domain that is relatively poorly conserved among the permeases and that bears no sequence or structural homology with the three proteins used for generating the model. This domain contains several predicted α helices, one moderately hydrophobic, H2, and a very long one, H3, and it may span the membrane (Kerppola *et al.*, 1991; Baichwal *et al.*, 1992). We postulated (Mimura *et al.*, 1991) that the helical domain interacts with the hydrophobic components in the formation of the complex. Such an interaction may be responsible for transmitting the conformational changes necessary for a cycle of transport. The helical domain would be expected to have diverged through evolution because of its postulated interaction with hydrophobic components, which are poorly conserved.

The second model for conserved components (Hyde *et al.*, 1990) used a similar computer-assisted approach, but used only adenylate kinase for structural modeling. It differs from the model shown here primarily due to alignment of the conserved aspartate in the transporters with a different aspartate in adenylate kinase. The reason for and the consequences of this difference have been discussed (Mimura *et al.*, 1991).

2. Chemical Modification

Initial experiments have been directed at identifying the HisP residues that form the nucleotide binding pocket. Modification of HisP with 8-azido-ATP resulted in derivatization of two peptides, which therefore presumably contribute to the structure of the ATP binding pocket (Mimura *et al.*, 1990). The specific residues modified by 8-azido-ATP within these two peptides are His-19 and Ser-41, respectively (Fig. 4). An inspection of the positions of these two residues within the structural model of Mimura *et al.* indicates that Ser-41 is located in the easily modeled glycine-rich flexible loop of the ATP-binding site, but it is not located immediately adjacent to the 8-position of adenine (Mimura *et al.*, 1990). Therefore, the movement of the glycine-rich loop (Fry *et al.*, 1986) may bring it into the vicinity of that position. The peptide in which His-19 is located cannot be accurately modeled, being outside the regions of homology. Because it is modified by 8-azido-ATP, it is likely that it is in the proximity of the 8-position of adenine (as depicted in Mimura *et al.*, 1991).

3. Mutant Analysis

A second approach to be used toward an understanding of the energy-coupling mechanisms, and of the transport mechanism in general, is the correlation between the genetic analysis of *hisP* mutants that do not function in transport and the ability of the HisP protein to bind ATP (Shyamala, *et al.*, 1992b). In particular, mutations can be introduced by site-directed mutagenesis into sites of special interest. Numerous residue replacements, obtained either as spontaneous mutations or by *in vitro* mutagenesis, and their effects on ATP binding are shown in Fig. 4. The following mutations resulted in loss of ATP binding activity concomitantly with loss of transport, indicating that those residues contribute to the structure of the ATP binding pocket: Gly-39 to aspartate, lysine, or arginine; Gly-44 to serine; and Lys-45 to asparagine, proline, or leucine. The substitution of the universally conserved aspartate at position 178 in HisP with lysine also resulted in the simultaneous elimination of transport and ATP binding; this is strong evidence for the importance of this residue for nucleotide binding. The substitutions of Ser-46 with asparagine, Thr-47 with alanine, and Glu-179 with aspartate do not cause a loss in ATP binding activity, even though they are located in sites that may contribute to the ATP binding pocket. These residues are probably facing away from the pocket, as also indicated by their predicted exposure to solvent; alternatively, the replacing residues are compatible with the structure of the binding pocket. In any case, they must be inactivating some important function distinct from ATP binding, such as hydrolysis. A few mutations eliminating ATP binding are in domains that cannot be modeled; either they are indeed close to the binding pocket, or they have a general disruptive effect. A more extensive discussion of the relationship between the effects of these mutations and the predicted structure appears in Shyamala *et al.* (1992b).

Additional information can be gathered by the creation of chimeric proteins fusing portions of HisP to conserved components that transport unrelated substrates. An example is a chimeric protein composed of 155 residues from the amino-terminal portion of HisP and of 281 residues from the carboxy-terminal portion of MalK (Schneider and Walter, 1991). This protein, which functions in maltose but not histidine transport, contains the B1, T1, A1, B2, and T2 motifs from HisP (Fig. 4). (which include the glycine-rich loop) and almost all of its helical domain. The carboxy-terminal end contains almost all of the helical domain from MalK plus all the remaining secondary structure motifs of the nucleotide binding site. The construction demonstrates that the two different motifs of the ATP binding site can be derived from different proteins. Less clear is the consequence of the presence of what amounts to two different helical domains. The data indicate that of the two helical domains, the one derived from MalK must have the preponderance in determining the specificity of transport in this chimeric protein. A possible basis for this behavior lies in the fact that the well-conserved "linker peptide," LSGGQQQRV, is derived from MalK and, therefore, acts as a linker to the MalK helical

```
8-AzidoATP
                    ___                 ___        ___                                          __        Y
HisP Mutants        |N|                 |E| P     |N|                          S           F (|I|) LR  |R|  E   D
                           Q E   G       D   N                                                    LR
                           B L           K   B                                            |S||V|
                                         B   B   SLNA                                     |T|
                    10        20        30        40        50      S    60        70        80        90       100       110       120
HisP #
HisP        MMSENKLHVIDLHKRY--------------GGHEVLKGVSLQARAGDVISIIGSSGSGKSTFLRCINFL----EKPSEGAIIVNGQNINLVRDKDGQLKVADKNQLRLLRTLTMVFQHF--NLWSHMTVLENV--MEAPIQ----VLG
GlnQ        ----MIEFKNVSKHF---------------GPTQVLHNIDLNIAQGEVVVIGPSGSGKSTLLRCINKL----EEITSGDLIV---------GDLKVNDPKVDERLIRQEAGMVFQQ--FYLFPHLTALEN--VMFGPL----RVR
CysA        ----MSIEIANIKKSF--------------GRTQVLNDISLDIPSGQMVALLGPSGSGKTTLLRIIAGL----EHQTSGHIRFHGTDVS-------RLHARDRKVGFVFQHY--ALFRHMTVFDNIAFGLIVLP----RRE
RbsA        APGDIRLKVDNLCGP----------------GVND-VSFTLRKGEILGVSGLMGAGRTEIMKVLYGA----LPRTSGVTLDGHEVVTRSPQDG--LANGIVYISEDRKRDGLVLGM--SVKENMSLTALRYFSRAGGS----LKH
AraC<       SYGEERLRLDAVKAP----------------GVRTPISLAVRSGEIVGLFGLVGAGRSELMKGMFBG----TQITAGQVYIDQQPIDIRKPSHA---IAAGMMLCPEDRKAEGIIPVH--SVRDNINISARRKHVLGGCV----INN
FhuC<       NHSDTTFALRNISFRV---------------PGRTLLHPLSLTFPAGKVTGLIGHNGSGKSTLLKMLGRH----QPPSEGEILLDAQPLE------SWSSKA----FARKVAYLPQQL--PPAEGMTVRELVAIGRYPWHGALGRFGA
SfuC<       ---MSTELHGIGKSY----------------NAIRVLEHIDLQVAAGSRTAIVGPSGSGKSTLLRIIAGF----EIPDGGILLQGQAMG------NGSGWVPAHLRGIGFVPQDG--ALFPHFTVAGNIGFGL-----------K
            EKGLSKEQILEKT-----------------GLSLGVKDASLAIEEGEIFVIMGLSGSGKSTMVRLLNRL----IEPTRGQVLIDGVDIA-------KISDAELREVRRKKIAMVFQSF--ALMPHMTVLDNTAFGM-----------ELAGIAA
ProV<       --MAGLKLQAVTKSW----------------DGKTQVIKPLTLDVADGEFIVMVGPSGCGKSTLLRMVAGL----ERVTEGDIWINDQRVT-------EMEPKD----RGIAMVFQNY--ALYPHMSVEENMAWGL------------KIRGMGK
UgpC        ETAPSKIQVRNLNFYY---------------GKFHALKNINLDIAKNQVTAFIGPSGCGKSTLLRTFNKMFELYPEQRAEGEILLDGDNI-------LTNSQDIALLRAKVGMVFQKP--TPFP-MSIYDNIAFGV----RLFKLSR
PstB<       --MEALLQLKGIDKAF---------------PG-VKALSGAALNVYPGRVMALVGENGAGKSTMKKVLTGI----YTRDAGTLWLGKETTFTGPKSSQEA------GIGIIHQEL--NLIPQLTIAENIFLGR-EFVNRFGKIDW
RbsA        QQSTPYLSFRGIGKTF---------------PG-VKALTDISFDCYAGQVHALMGENGAGKSTLLKILSGN----YAPTGSVVINGQEMSFSDTTAALNA------GVAIIYQEL--HLVPEMTVAENIYLG--QLPHKGIVNR
AraG<       QPANVLLEVNDLRVTFAI-------------PDGDVTAVNDLNFTLRAGETLGIVGESGSGKSQTAFALMGL--LATNGRIGGSATFNGREI-------LNLPERELNTRRAEQISMIFQDPMTSLNPYMRVGEQL----MEVLMLHKG
OppD<       EQRKVLLEIADLKVHFDIKEGKQWFWQPPKTLKAVDGVTLRLYEGETLGVVGESGCGKSTFARAIIGL----VKATDGKVAWLGKDL-------LGMKADEWREVRSD-IQMIFQDPLASLNPRMTIGEII----AEPLRTYHP
OppF<       ---MLELNFSQTLGNHCL-------------TINETLPANGITAIFGVSGAGKTSLINAISGL----TRPQKGRIVLNGRVL-------NDAEKGICLTPEKRRVGYVFQDA--RLFPHYKV----RGNLRYGMS
ChlD        -------MASVQLQNVTKAWGEVVVSKDINLDIHEGEFVVFVGPSGCGKSTLLRMIAGL----ETITSGDLFIGEKRMNDTPPAE------RGVGMVFQSY--ALYPHLSVAENMSFGLKPAGAK----K
MalK        ----------MSIVMQLQDVAESTRLGPLSGEVRAGEILHLVGPNGAGKSTLLARMAGM----TSGKGSIQFAGQPLEAWSATKLALHRA-----YLSQQTPPFATPVWHYLTLHQHD
BtuD        EQGESKELERNLKKSF---------------GEVKVLKDISLDIRAGEVLALVGPSGSGKSTLLRIIAGL----ETPEGDILLDGQEIE-------KIADKELLEPRRRKIGMVFQDY--ALFPHMTVAENIAFGL-PWP-KLRKIHR
Consensus
Conserved     o              o                o   o  oo  oooooooo  o                 o  o                  oo
Secondary     ++++++                ++++++    +++++++++++         +++++++++++++      ++++++++++++++++++     +++++++++++++     +++++++
Cons #      10        20        30        40        50        60        70        80        90       100       110       120
                                                                      ^^^^^--^^^^-^^^^                                 H1                H2        H3
Secondary             B1   T1    A1                                   B2   T2
```

```
HisP Mutants   Q E        T              T  F              Ω Ω Ω      TE                                              R        A M B        K A L R R QQQQ
               Q P                                          T                                    A M B        K ED        R Y D        T
                130        140       150       160       170       180       190       200       210       220       230       240       250

HisP #    LSKHDARERALKYLAK--VGIDERAQGKYPVHLSGGQQQRVSIA-RALAMEPD------VLLLDEPTSALDPELVGEVLLRIMQQLAEE-GKTMVVTHEMGFARHVSSHVIFLHQGKI------EEEGDPEQVFGNFQSPRLQQFLKGSLK
HisP      GANKEEAEKIARELLA--KVGLAERAHHYSPSELSGGQQQRVAIA-RALAVKPK---------MMLFDGPTSALDPELRHEVLKVMQDLAEE-GMTMVIVTHEIGFAEKVASRLIFIDKGRI------AEDGNPQVLIKNPFSQRLQEFLHQVS-
GlnQ      RPNAAAIKAKVTKLLE--MVQLAHLADRYPAHVSGGQQQRVALA-RALAVEPQ---------ILLLDEPFGALDAQVREKELRRWIRQLHEELKFTSVFVTHDQEEATEVADRVVMSQGNI------EQADAPDQVWREPATRFVLEPMGENPD
CysA      ADEQQAVSDFIRLFNV--K---TPSMEQAIGLLSGGQRWIA-TRPK-----------VILLDEPT-GVDVGAKKEIYQLINQFKAD-GLSIILVSSEMPEVLGMSDRIIVMHNGHLSGEFTREQA--------SLAMPKVS
RbsA      GWEENNADHHIRSLNI--K---TPGAEQLIMNLSGGNQQKAILG-RWLSEEMK------VILLDEPTRGIDVGAKHEIYNVIYALAAQ-GVAVLFASSDLPEVLGVADRIVMREGIAGELLHEQADERQAL-----QAGTPAEIMRGETLEMIYGIPMGILP
AraG      ADR-EKVEEAISLVGL--K---KPLAHRLVDSLSGGERQRAWIA-MLVAQDSR------CLLLDEPTSALDIAHQVDVLSLVHRLSQERGLTVIAVLHDINMAARYCDYLVALRGGEMI------AQGTPAEIMRGETLEMIYGIPMGILP
FhuC      GGKREKQRRIEIALWEH--VALDRRLAALWPHELSGGQQQRVALA-RALSQQPR------LMILDEFFSALDTGLRAATRKAVAELLTEAKVASILVTHDQSEALSFADQVAVMRSGRLA-------QVGAPQDLYLRPVDEPTASFLGETLV
SfuC      QERREKALDALRQVGL------ENYAHAYPDELSGGMRQRVGLA-RALAINPD------ILLMDEAFSALDPLIRTEMQDELVKLQAKHQRTIVFISHDLDEAMRIGDRIAIMQNGEVV-------QVGTPDEILNNPANDYVRTFFR---
ProV      QQIAERVKEAARILEI--------DGLLKRRPRELSGGQRQRVAMG-RAIVRDPA--------VFLFDEPLSNLDAKLRVQMRLELQQLHRRLKTTSLVTHDQVEAMTLAQRMVMNGGVAE--------QIGTPVEVYEKPASLFVASFIG---
UgpC      QERREKALDALRQVGL------ENYAHAYPDELSGGMRQRVGLA-RALAINPD------ILLMDEAFSALDPLIRTEMQDELVKLQAKHQRTIVFISHDLDEAMRIGDRIAIMQNGEVV-------QVGTPDEILNNPANDYVRTFFR---
PstB      ADMDERVQWALTKAAL--WNETKDKLHQSGYSLSGGQQRLCIA-RGIAIRPE--------VIMDEPTDALTDTETESLFRVIRELKSQ-GRVIVSHRMKEIFEICDDVTVFRDGQFIAE-REVASLTEDSLIEMMVGRKLEDQYPHLDK
RbsA      KTMYAEADKLLAKINL--RFKSDKLVG----DLSIGDQOMVEIA-KVLSFESK------IIAFDEPTSSLSAREIDNLFRVIRELRKE-GRVLLVSHRMEEIFALSDAITVFKDGRYVKTFTDMQQVHDLTVFKDGRYVKF
AraG      SLINYEAGLQLKHLGM--DIDPDTPLK---YLSIGQWOMVEIA-KALARNAK------LLIADEPTALDVTVQAQIMTLLNELKREFNTAIIMTHDLGVVAGICDKVLVMYAGRTM-----EYGKARDVFYQP----VHPY---S
OppD      MSKAEAFEESVRMLDAVKMPEARKRMKMYPHEFSGGMRQRVMIA-MALLCRPK------LLIADEPTTALDVTVQAQVNNLLQQLQREMGLSLIFIAHDLAVVKHISDRVLVMYLGHAV------ELGTYEHVHNP----LHPY---T
OppF      KLSRQDVDRVKAM-MLKVGLLPNLINRYPHEFSGGQCQRIGIA-RALITEPK------LIICDDAVSALDVSIQAQVNNLLQQLQEMGLSLIFIAHDLAVVKHISDRVLVMYLGHAV------ELGTYEHVHNP----LHPY---T
ChlD      KSMVDQFDKLVALLG-----IEPLLDRKPSGEKQKVAIG-RALLTAPE------LILLDEPLASLDIPRKRELLPYLQRITREINIPMLYVSHSLDEILHLADRMVLENGQVK------AFGALEEVWGSSV--MNFWLPKEQQ
MalK      EVINQRVNQVAEVIQL------AHLLDRKPKALSGGQORVAIG-RTLVAEPS------VFLLDEPLSNLDAALRVQMRIEISRLHKRLGRTMIYVTHDQVEAMTLADRIVLDAGRVAQVGKPLAV---PLSGRFCRRIYRFAKDELL
BtuD      KTRTELLNDVAGALAL------DDKLLGRSTNQLSGGEWQRVRLAAVVLQITPQANPAGQLLLLDEPMSLDVAQQSALDKIISALCQ-GLAIVMSSHDLNHTLRHAHRAWLLKGGRMLASGRREEVLTPNL-AQAYGMFRRLDIGHR
Cons #         160       170       180       190       200       210       220       230       240       250       260       270       280       290       300

Consensus  KSKREAVEDALRLLGL--KVELDPLLHRYPHELSGGQQQRVAIA-RALAIEPK------VLLLDEPTSALDLVALRAELLRLLQQLHRELGLTIIVVTHDLGEALR IADRVVMRGGRIV----EQVGTPDELFGNPVGRRLQDFLMEELK
Conserved  +++++++++++++++++  +++++++  ++++++  o  ooooo ooo  oo ooo oo  o   oooooO oooO         o   oo  oo   ooo       oo   o   o                o
Secondary  +++++++++++++++++  +++++++  o  ooooo ooo  o ooo o   ~~-~+++++++++++++    ^^^^^~-~+++++++++++++    ^^^^^~-+++++++++++++^^^^~  ^^^^~~-~+++++^^^^^
                                                H3        H4                 B3    T3     A3            B4  T4    A4            B5  T5   A5
```

FIG. 4 Sequence alignment, consensus sequence, secondary structure, and surface accessibility of conserved components. The methods for multiple sequence alignment and for all predictions are described in Mimura et al., 1991. Residues marked with o and O are >50% (at least 9 out of 17) and 75% (at least 13 out of 17) identical among all sequences, respectively. Due to differences in length and lack of sequence similarity at the extreme N and C termini, the alignment and the consensus sequence are presented only from the beginning to the end of HisP; the symbols < and > indicate that the sequence continues at the N and C terminus, respectively. Numbers refer to the consensus (cons # and HisP #) sequences. The consensus secondary structure prediction is shown below the consensus sequence. +, α Helix: ∧, β strand; ~, turn (indicated in the regions modeled as A1 to A5, B1 to B5, and T1 to T5, and H1 to H4 in the helical domain); all blank regions correspond to coils or turns. The top line indicates the sites where the 8-azido-ATP modifications have occurred (8). The second to fourth lines indicate residue replacements as derived either from in vivo mutagenesis, or by analysis of the homologous hisP gene in Escherichia coli (enclosed in a square). Changes resulting in suppression ability (see text) are underlined. Changes resulting in a decrease of 8-azido-ATP derivatization to less than 20% of wild type are underlined. The helical domain is underlined with a double line. Two black boxes on the bottom line mark the two strongly conserved aspects of the ATP binding motif (glycine-rich loop and conserved aspartate). (Modified from Ames et al., 1991, with permission.)

domain. Since this peptide may be involved in signal transduction (see Section III,A) it would not be surprising that it directed the MalK helical domain to act, but not the HisP domain. On the other hand, it is possible that the presence in this chimeric protein of the large portions of MalK 3' to the B5, T5, and A5 motifs is responsible for determining specificity in this case.

In conclusion, this model of the conserved component is a useful working hypothesis subject to modification and refinement by further experimentation, including site-directed mutagenesis, active site labeling, chemical cross-linking, and construction of deletion mutants and chimeric proteins.

III. Relevance to Eukaryotic Carriers

A. Sequence Homology

Much has been written on the MDR, CFTR, and STE-6 proteins and the reader is directed to some of these reviews for details (Kane *et al.*, 1990; Kuchler and Thorner, 1990; Juranka *et al.*, 1989; Ringe and Petsko, 1990; (Gros, this series, volume 137C). Here we attempt to draw a hypothetical unifying picture for traffic ATPases using the progress derived from studies on the structure and function of periplasmic permeases.

A consensus sequence was obtained for the sequences of the ATP binding domains of the eukaryotic transporters and it was compared with the consensus sequence for the 17 known prokaryotic conserved components (Ames *et al.*, 1992). A high level of conservation was obvious between the two consensus sequences. Due to this high degree of homology between the prokaryotic and eukaryotic sequences, especially in the nucleotide binding domain, it is justifiable to extrapolate the predicted secondary structure of the prokaryotic permeases to the eukaryotic proteins. In addition, the predicted tertiary structuree of the eukaryotic proteins can also be inferred from the sequence alignment by correlation with the framework given for the HisP protein (Mimura *et al.*, 1991). Figure 3 shows part of this alignment, including HisP and CFTR and the predicted separate consensus sequences for the prokaryotic and the eukaryotic traffic ATPases and the level of conservation.

From the full alignment (Ames *et al.*, 1992) it can be seen that the most highly conserved regions of both the eukaryotic and prokaryotic sequences correspond to the classical nucleotide binding motifs, around HisP positions 45 and 180. The most variable regions correspond to the large helical domain and the C terminus. In general, β strands and turns are more highly conserved than helices, and glycine and very hydrophobic amino acids are most highly conserved. The conservation of these residues is to be expected due to their importance in providing flexibility (glycine) and specific packing interactions (hydrophobic residues). Besides the ATP binding motifs, we have noted another well-conserved motif in the family of conserved components, which is particularly rich in glycine and glutamine residues, and additionally contains leucine and serine (glutamine-, glycine-rich motif). The prokaryotic consensus sequence for this motif is LSGGQQQRV, with the last Q 100% conserved (positions 154 to 162 of HisP; see Fig. 4). A study by Argos (1990) suggests that these residues are commonly found in "peptide linkers" that join together separate domains within proteins. In fact, the consensus sequence is similar to the domain-linking oligopeptide, SGAQQ in penicillinopepsin. In the predicted model this motif is located within the helical domain shortly before the sequence reenters the nucleotide binding pocket. Thus, the motif may be a linker joining the helical domain with the nucleotide binding domain and it is poised for action as a signal transducer between the postulated interaction with the hydrophobic domain of the transporter (or with the binding protein) and the energizing sector of the transporter. In contrast, in an alternative model (Hyde et al., 1990) the loop (loop 3) containing the glutamine-, glycine-rich motif is constrained to a location within the nucleotide binding pocket, which would be incompatible with our proposed function as a domain linker.

The homology between the prokaryotic conserved components and the hydrophilic domains of the eukaryotic traffic ATPases is therefore striking and undoubtedly must reflect a uniform mechanism of action. What about the differences between the other components of periplasmic permeases and the hydrophobic portions of the eukaryotic proteins? The establishment that the prokaryotic membrane bound complex has a stoichiometry of one each of the hydrophobic components of two of the conserved component suggests that the similarities may extend further than just the nucleotide binding domain. The secondary structure prediction and membrane topology of the hydrophobic portions of the eukaryotic transporters in remarkably similar to that of the two hydrophobic components of the histidine permease (except for the addition of an extra membrane spanner at the amino-terminal end) (Kerppola and Ames, 1992; see below). Since proteins that share a common function may be expected to maintain similar secondary and tertiary structures even though they may lose sequence similarity through evolution, it is possible that the hydrophobic domains of the eukaryotic transporters may have evolved from the same ancestor and serve the same function as the separate hydrophobic membrane proteins of periplasmic permeases. Thus, it has been proposed (Ames, 1986b) in the case of the MDR protein, and it is reasonable to include the other eukaryotic traffic ATPases as well, that a single protein molecule incorporates the features of the entire membrane bound complex of periplasmic permeases. It is not surprising to find that a fusion occurred during the evolution of eukaryotic transporters since it is a common finding that eukaryotes often incorporate into one protein what are separate proteins in prokaryotes. The structure of eukaryotic transporters probably reflects an initial fusion of a hydrophobic with a hydrophilic domain, which would have been followed by a duplication of the fused product, followed in turn by another fusion of the duplicated product [however, this view has been questioned (Chen *et al.*, 1990]. The duplication of the initial fusion product, which resulted in two hydrophobic domains and two ATP binding domains, is consistent with the basic structure of periplasmic permeases, which includes in the membrane bound complex two copies of the ATP binding component and one copy each of the hydrophobic proteins. A schematic representation of a possible evolutionary path taken by the genes encoding these transporters has been proposed (Ames *et al.*, 1991).

B. Is the Binding Protein an Essential Component of Transport?

One of the salient differences between periplasmic permeases and eukaryotic traffic ATPases is the absence of a substrate binding protein in the latter. Unless eukaryotic candidates that might be the equivalent of the substrate binding proteins (and they do not need to be extracellular) are uncovered in the future, a uniform hypothesis for a mechanism of action must include the possibility that the membrane bound complex has the inherent ability to function without the aid of a substrate binding protein in all members of the family. What evidence is available in this respect? Two periplasmic permeases, maltose and histidine, have been used to answer this question and mutants isolated in both indicate that transport in the absence of their respective binding proteins is indeed possible. In the case of the histidine permease, mutants have been identified in HisP that allow transport in the absence of HisJ (Speiser and Ames, 1991). Surprisingly, the mutations responsible for this phenotype were found to be located in similar sites and behaved similarly to mutations suppressing a defect in the interaction site of HisJ (see above). Vice versa, the suppressor mutations, when expressed at high level, allowed transport in the absence of binding protein. A possible explanation for all these mutations is derived from the observation that in these mutants ATP is hydrolyzed at a high rate and independently from the transport of histidine. Thus, these mutations have obviated the need for a signal from the binding protein in order to hydrolyze ATP, exposing spontaneously a substrate binding site located on the membrane bound complex, or opening a pathway through the membrane bound complex (Petronilli and Ames, 1991; Ames and Lecar, 1992). In the case of the maltose permease, multations in the hydrophobic components of the membrane bound complex allow transport of maltose in the absence of the maltose binding protein (Treptow and Shuman, 1985).

In conclusion, these results support the notion that the minimum traffic ATPase does not necessarily require a binding protein to carry on translocation. A possible explanation for the presence of substrate binding proteins in the prokaryotic systems is that, in contrast to eukaryotes, prokaryotes might have to survive in environments where nutrients are present in very low concentrations. Therefore, teleologically speaking, the binding, proteins might have evolved to provide the substrate in a concentrated form at the membrane surface and as a trapping mechanism to allow scavenging of precious nutrients (Ames, 1972). In this respect it is interesting that several gram-positive organisms, which do not have an outer membrane and therefore do not have a periplasm either, have transport systems entirely analogous to those of gram-negative bacteria, including the equivalent of a substrate binding protein. The latter is apparently anchored to the cell surface by a lipophilic posttranslational modification (Rudner et al., 1991; Perego et al., 1991; Alloing et al., 1990; Gilson et al., 1988). Conceivably, eukaryotic cells also have the equivalent of a substrate receptor anchored to the cell surface. On the other hand, the high concentration of substrates the eukaryotic ell is exposed to may have rendered the receptor obsolete and it may have been lost in evolution.

C. Directionality of Transport in Traffic ATPases

Are these transporters reversible? The fact that ATP hydrolysis is necessary for transport indicates that they are unlikely to be reversible. Using reconstituted proteoliposomes prepared with partially purified histidine permease components, it has been shown that histidine cannot exit once it has been transported to the interior (Petronilli and Ames, 1992). By using this system it was possible to determine that the factors responsible for the leveling off of transport (i.e., plateau conditions) are the accumulated histidine and the ADP produced by ATP hydrolysis, both of which inhibit the uptake process. This finding generated a hypothesis in which ATP hydrolysis results in a conformational change(s) that brings the substrate bound to the membrane complex from the outer to the inner surface of the membrane, thus allowing its release internally and the return of the empty site in the complex to the external surface (Fig. 5). Two important implications of this hypothesis come to mind. One is that transport can indeed occur in the absence of the binding protein, whose function is exclusively

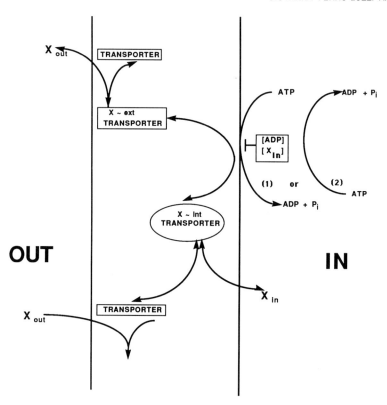

FIG. 5   Reversibility of a generic traffic ATPase. The transporter is represented as becoming liganded at either the outer or inner membrane surface. In the mechanism labeled (1) ATP hydrolysis causes a rearrangement that renders the transporter-bound substrate accessible at the inner surface, allowing its release. In the mechanism labeled (2) ATP hydrolysis causes the reverse rearrangement resulting in the extension of the substrate through an essentially identical transporter. The presence of a binding protein is not necessary for either mechanism. Accessibility of the transporter at the inner or outer membrane is hypothesized to be mediated through a series of conformational changes.

that of supplying this membrane-bound site with high concentrations of substrate and has no translocation function of its own. The other is that it could easily be imagined that in some carriers ATP hydrolysis effects conformational changes that are a reversal of those postulated for prokaryotic transporters. A loaded internal site would be exposed to the external surface on ATP hydrolysis, thus supplying the model with an explanation for the outward direction of transport. Therefore, this apparent difference between eukaryotic and prokaryotic traffic ATPases need not be an impediment to hypothesizing a common mechanism of action.

D. Structural and Functional Homologies

Even though the analogy and homology between the eukaryotic and pro-karyotic traffic ATPases is striking, is it justified to include all of them in the same family? On the basis of homology alone, it is clear that the conserved components of periplasmic permeases and the hydrophilic moieties of the eukaryotic transporters bear a relation beyond the nucleo-tide binding sequence motif. Indeed, the same nucleotide binding motif is present in several other proteins that are neither involved in transport nor otherwise related to each other (see, e.g., the three proteins used for modeling purposes); a comparison of the primary sequences of these other proteins with those of the transporters has not revealed any significant homology beyond the nucleotide binding motif, thus excluding them from the family of traffic ATPases. The high level of homology observed be-tween the eukaryotic and the prokaryotic members of this family must indicate that these proteins have additional functions in common and presumably reflect a case of divergent evolution from a common ancestral carrier. Among the other essential and common mechanistic characteris-tics a likely possibility is the signal transmission from the substrate-occupied site(s) to the energy-transducing triggering mechanism, or vice versa. A substrate binding site could be located either on the conserved component, in one of the nonconserved regions, or on the hydrophobic component(s). Similar signal transmission mechanisms might have re-sulted in conservation of secondary and topological structure, without necessarily maintaining primary sequence conservation. Comparisons of the hydrophobic domains of the eukaryotic transporters (two for each molecule) with the hydrophobic prokaryotic components do not reveal any significant primary sequence similarity, in agreement with the finding that there is relatively little sequence similarity in general among the prokaryo-tic hydrophobic components themselves (Kerppola and Ames, 1992). However, the similarity between the minimum prokaryotic topological structure and that of the eukaryotic transporters supports the contention that the hydrophobic moieties perform the same function in eukaryotes as the hydrophobic prokaryotic components.

In this respect, it is significant that in the eukaryotic hydrophobic moie-ties there is also a large cytoplasmic hydrophilic loop in a position that corresponds to the prokaryotic loop containing the conserved Glu-Ala-Ala sequence. Several of the eukaryotic loops also contain a conserved gluta-mate contiguous with residues similar to those near Glu-Ala-Ala in pro-karyotes. This loop may have the same function as the one postulated for the prokaryotic equivalents, i.e., interacting with the ATP binding moiety. It is possible that the eukaryotic hydrophobic moieties have not conserved an intact Glu-Ala-Ala motif because if the interaction capability of this sequence were necessary in prokaryotes for holding together separate subunits, the sequence would likely have evolved differently once the eukaryotic fusion event had occurred.

Thus, the hypothetical architecture of a typical eukaryotic traffic ATPase consists of two sets of membrane spanners (one in each half of the molecule), corresponding to the two hydrophobic subunits of a prokaryotic transporter; the two ATP binding moieties, by analogy to the prokaryotic conserved components, would be embedded partly within the hydrophobic domain with portions carrying the ATP binding sites well accessible to the cytoplasm. The entire structure may display a pseudo-twofold-symmetry. An artist's rendering of the hypothetical structure of a typical traffic ATPase is shown in Fig. 6.

The question as to whether eukaryotic proteins that are included in this superfamily are indeed involved in transport should be addressed. The evidence that MDR is involved in transport and that it uses ATP energy is very good. It has also been shown that the purified MDR protein has an ATPase activity, although very low (Hamada and Tsuruo, 1988). The *Saccharomyces cerevisiae* STE6 protein has been shown to be necessary for the secretion of **a** factor; its high similarity to mammalian MDR makes it very likely that it is indeed the transporter for **a** factor. The *white* gene product may be involved in the transfer of pigment, although no biochemical evidence concerning its direct involvement in transport is available. The function of the cystic fibrosis gene product, CFTR, has been controversial. Even though it has been postulated that CFTR is unlikely to function as a chloride channel (Hyde *et al.*, 1990), recent publications suggest that it is most likely a chloride channel (Anderson, *et al.*, 1991a; Kartner *et al.*, 1991). In this scenario it is unclear what function the hydrophilic conserved domains would have, since channels are not thought to be directly energized by ATP. However, it is possible that ATP energy is involved in opening the channel, once phosphorylation of the regulatory domain of CFTR (the R domain) has occurred. The R domain may act as a safety clasp that does not allow the ATP energy to be utilized to open the channel (Kartner *et al.*, 1991; Anderson *et al.*, 1991b). In addition, if our predicted model is correct, the large portion of this domain, which spans the membrane, would be intimately connected with the "channel" proper, and therefore may be directly involved in its function. It may be relevant that among the prokaryotic traffic APTases, several transport ions such as phosphate, molybdenum, sulfate, and potassium (Ames, 1986a), and none of them have been tested for channel-like properties. It is possible that there may be analogies in this respect also. An alternative suggestion proposes that the substrate of the CFTR transporter is a large molecule with hydrophobic properties (Ringe and Petsko, 1990), and among the arguments used was the large size of CFTR. However, if a

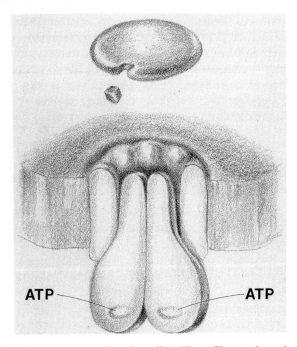

FIG. 6   Three-dimensional representation of a traffic ATPase. The membrane has been sliced
vertically to expose the ATP-binding domain (conserved component) as extending partly into
the membrane and encased by the hydrophobic domain (hydrophobic membrane components
for prokaryotic members). The ATP-binding sites are located in a cytoplasmic portion of the
ATP-binding domain. A substrate-binding protein is represented and is to be imagined as
present and necessary only for those traffic ATPases for which such a requirement has been
shown. (From Ames *et al.*, 1991, with permission.)

eukaryotic transporter is indeed the equivalent of an entire prokaryotic
permease, the size of the substrate transported cannot be predicted on the
basis of the size of the carrier. In fact, prokaryotic traffic ATPases, all of
which are essentially the same size as CFTR, translocate molecules of a
variety of sizes, from ions to large polypeptides. It should be mentioned
that the nomenclature implying that CFTR is a regulator rather than a
transporter may be misleading, even though it may be true that the R
domain located between the two homologous halves may have a regulatory function. The relationship between channels and transporters has
been discussed and the conclusion was reached that the analogies are
considerable, thus weakening objections to the notion that CFTR is a
channel (Ames and Lecar, 1992).

   Not much is known about the latest addition to the superfamily, the
product of a gene affecting class I major histocompatibility complex,

located in the MHC locus (summarized by Parham, 1990). This gene product, variously named depending on the organism, may be involved in transporting peptide fragments derived from cytoplasmic proteins from the cytoplasm to a compartment containing newly synthesized MHC molecules (Monaco et al., 1990). The fact that this gene codes for only one hydrophobic and one hydrophilic domain suggests that it must form a homodimer in the membrane, by analogy to other traffic ATPases.

IV. Structure—Function Analysis of Eukaryotic Transporters

It is not easy to perform a structure–function analysis of these eukaryotic transporters. However, it should be possible to draw relevant conclusions by examining the locations of naturally arising mutations and comparing them to those available in the easily manipulated prokaryotic systems. The mutations responsible for the cystic fibrosis phenotype are known for several classes of patients (Cutting et al., 1990, and references therein). The most common defect is the deletion of residue Phe-508, which accounts for 77% of Caucasian patients. This residue (corresponding to a gap between positions 114 and 115 of HisP in Fig. 3) would be located in the helical domain near the C terminus of helix H2 according to the structural model. A cluster of cystic fibrosis mutations (Cutting et al., 1990) has been placed in the highly conserved glycine-, glutamine-rich motif (around HisP position 160 in Fig. 3), which the model also places within the helical domain. Despite the fact that several defective CFTRs have been shown to be improperly located (Cheng et al., 1990), it is tempting to speculate that cystic fibrosis mutations alter the ability of CFTR to bind the substrate and/or the ability of the helical domain to interact with the ATP binding or the hydrophobic domains. Nevertheless, the consequences of these cystic fibrosis defects may still lie in the binding or hydrolyzing of ATP, since a HisP mutant (Trp-105 to Lys) located in the helical domain, not far from the residue corresponding to cystic fibrosis ΔPhe-508, is known to be unable to bind ATP (Shyamala et al., 1992b). In the case of the cluster of mutations in the glycine-, glutamine-rich motif, if this motif has the "domain linker" function we suggested in an earlier section, then this might result in inactivation of CFTR activity while maintaining ATP binding (and hydrolyzing?) activity. In support of this possibility is the existence of a hisP mutation within this motif, in which Ser-155 has been replaced with phenylalanine: this transport-negative HisP binds ATP essentially normally (Shyamala et al., 1992b).

Thus, it should be possible to design mutations in the prokaryotic sys-

tems that are specifically tailored to be related to the eukaryotic transporters.

V. Future Prospects

The realization that well-studied prokaryotic transporters have a mechanism of action that may be closely related to that of medically important
eukaryotic transporters is a gratifying example of the useful convergence
of basic research efforts. The study of the basic mechanism of action of
these transporters in both eukaryotes and prokaryotes will indeed benefit
from this convergence. What are the most obvious routes to follow at
present? For prokaryotic transporters, which being simpler are likely to be
very useful as model systems, these are obvious. The protein components
need to be purified, individually and/or as a complex, and characterized
biochemically. These studies are well under way in several laboratories
and promising in all respects. Overproduction of the proteins has been
achieved in several systems and antibodies against all the individual
proteins are available for the histidine and maltose permeases. The availability of the reconstituted proteoliposome system has provided an assay
for the purification and characterization of the complex. The crystallization of the complex is the most ambitious of these projects and would, of
course, be the most productive if successful. It is likely that the experience
recently gained in crystallizing the photosynthetic reaction center and
other membrane proteins will be useful for these systems. Analyses using
two-dimensional crystalline arrays are perhaps a more easily reached goal
in the near future and should provide important structural information.
Among the most exciting ways in which the prokaryotic systems can be
used is the easily performed genetic engineering approach to study
structure–function relationships. Specific changes can be inserted any-
where in any of the proteins and their effects studied both *in vivo* and *in
vitro,* in a multipronged approach to understand each of the numerous
aspects of their mechanism.

For eukaryotic transporters similar routes of analysis are also presumably to be followed, although the same aims may be harder to achieve.
However, with the availability of prokaryotic transporters as model systems, progress should be considerably accelerated in parallel. It is likely
that some of the techniques developed for the prokaryotic transporters
(e.g., reconstituted proteoliposomes) will be directly applicable and useful
for eukaryotic investigations. An important problem to be resolved for
most eukaryotic transporters is the identification of their physiological
transport substrate in most cases. This knowledge is obviously essential

for achieving progress in controlling the medical problems with which several of these transporters are associated.

Acknowledgments

The work done in the laboratory of GFLA was supported by the National Institutes of Health Grant DK12121. The author thanks all present and past members of her laboratory for their innumerable contributions to the work on the histidine permease.

References

Adams, M. D., and Oxender, D. L. (1989). *J. Biol. Chem.* **256,** 560–562.
Alloing, G., Trombe, M. C., and Claverys, J. P. (1990). *Mol. Microbiol.* **4,** 633–644.
Ames, G. F.-L. (1972). *In* "Biological Membranes. Proceedings of the 1972 ICN-UCLA Symposium in Molecular Biology" (C. F. Fox, ed.), pp. 409–426. Academic Press, New York.
Ames, G. F.-L. (1985). *Curr. Top. Membr. Transp.* **23,** 103–119.
Ames, G. F.-L. (1986a). *Annu. Rev. Biochem.* **55,** 397–425.
Ames, G. F.-L. (1986b). *Cell (Cambridge, Mass.)* **47,** 323–324.
Ames, G. F.-L. (1990a). *In* "Bacterial Energetics" (T. A. Krulwich, ed.), pp. 225–245. Academic Press, New York.
Ames, G. F.-L. (1990b). *Res. Microbiol.* **141,** 341–348.
Ames, G. F.-L., and Joshi, A. (1990). *J. Bacteriol.* **172,** 4133–4137.
Ames, G. F.-L., and Lecar, H. (1992). *FASEB J.* **66**
Ames, G. F.-L., and Nikaido, K. (1978). *Proc. Natl. Acad. Sci. U.S.A.* **75,** 5447–5451.
Ames, G. F.-L., and Nikaido, K. (1981). *Eur. J. Biochem.* **115,** 525–531.
Ames, G. F.-L., and Spudich, E. N. (1976). *Proc. Natl. Acad. Sci. U.S.A.* **73,** 1877–1881.
Ames, G. F.-L., Nikaido, K., Groarke, J., and Petithory, J. (1989). *J. Biol. Chem.* **264,** 3998–4002.
Ames, G. F-L., Mimura, C., and Shyamala, V. (1990). *FEMS Microbiol. Rev.* **75,** 429–446.
Ames, G. F.-L., Mimura, C., Holbrook, S., and Shyamala, V. (1992). *Adv. Enzymol.* **65,** 1–47.
Ames, G. F.-L., Nikaido, K., and Joshi, A. (1992). In preparation.
Anderson, M. P., Berger, H. A., Rich, D. P., Gregory, R. J., Smith, A. E., and Welsh, M. J. (1991a). *Cell (Cambridge, Mass.)* **67,** 775–786.
Anderson, M. P., Rich, D. P., Gregory, R. J., Smith, A. E., and Welsh, M. J. (1991b). *Science* **251,** 679–681.
Argos, P. (1990). *J. Mol. Biol.* **211,** 943–958.
Baichwal, V., and Ames, G. F.-L. (1992). Submitted.
Bell, A. W., Buckel, S. D., Groarke, J. M., Hope, J. N., Kingsley, D. H., and Hermodson, M. A. (1986). *J. Biol. Chem.* **261,** 7652–7658.
Bishop, L., Agbayani, R. J., Ambudkar, S. V., Maloney, P. C., and Ames, G. F.-L. (1989). *Proc. Natl. Acad. Sci. U.S.A.* **86,** 6953–6957.
Bordier, C. (1981). *J. Biol. Chem.* **256,** 1604–1607.

Boyd, D., and Beckwith, J. (1989). *Proc. Natl. Acad. Sci. U.S.A.* **86,** 9446–9450.

Brenner, M., and Ames, B. N. (1971). "Metabolic Regulation: Metabolic Pathways." Academic Press, New York.

Chen, C. J., Clark, D., Ueda, K., Pastan, I., Gottesman, M. M., and Roninson, I. B. (1990). *J. Biol. Chem.* **265,** 506–514.

Cutting, G. R., Kasch, L. M., Rosenstein, B. J., Zielenski, J., Tsui, L., Antonarakis, S. E., and Kazazian, H. J. J. (1990). *Nature (London)* **346,** 366–369.

Dalbey, R. E. (1990). *Trends Biochem. Sci.* **15,** 253–257.

Dassa, E., and Hofnung, M. (1985). *EMBO J.* **4,** 2287–2293.

Davidson, A. L., and Nikaido, H. (1990). *J. Biol. Chem.* **265,** 4265–4260.

Davidson, A. L., and Nikaido, H. (1991). *J. Biol. Chem.* **266,** 8946–8951.

Davies, K. (1990). *Nature (London)* **348,** 110–111.

Dean, D. A., Fikes, J. D., Gehring, K., Bassford, P. J. J., and Nikaido, H. (1989). *J. Bacteriol.* **171,** 503–510.

Deverson, E. V., Gow, I. R., Coadwell, W. J., Monaco, J. J., Butcher, G. W., and Howard, J. C. (1990). *Nature (London)* **348,** 738–741.

Dreusicke, D., Karplus, A., and Schultz, G. E. (1988). *Eur. J. Biochem.* **161,** 127–132.

Froshauer, S., Green, G., Boyd, D., McGovern, K., and Beckwith, J. (1988). *J. Mol. Biol.* **200,** 501–511.

Fry, D. C., Kuby, S. A., and Mildvan, A. S.. (1986). *Proc. Natl. Acad. Sci. U.S.A.* **83,** 907–911.

Furlong, C. E. (1987). *In* "*Escherichia coli* and *Salmonella typhimurium:* Cellular and Molecular Biology" (F. C. Neidhardt, J. L. Ingraham, K. B. Low, B. Magasanik, M. Schaechter, and H. E. Umbarger, eds.), pp. 768–796. Am. Soc. Microbiol., Washington, D.C.

Gallagher, M. P., Pearce, S. R., and Higgins, C. F. (1989). *Eur. J. Biochem.* **180,** 133–141.

Galloway, R. J., and Taylor, B. L. (1980). *J. Bacteriol.* **144,** 1068–1075.

Gilson, E., Alloing, G., Schmidt, T., Claverys, J.-P., Dudler, R., and Hofnung, M. (1988). *EMBO J.* **7,** 3971–3974.

Hamada, H., and Tsuruo, T. (1988). *J. Biol. Chem.* **263,** 1454–1458.

Higgins, C. F., Haag, P. D., Nikaido, K., Ardeshir, F., Garcia, G., and Ames, G. F.-L. (1982). *Nature (London)* **298,** 723–727.

Higgins, C. F., Hiles, I. D., Whalley, K., and Jamieson, D. K. (1985). *EMBO J.* **4,** 1033–1040.

Higgins, C. F., Hiles, I. D., Salmond, G. P. C., Gill, D. R., Downie, J. A., Evans, I. J., Holland, I. B., Gray, L., Buckel, S. D., Bell, A. W., and Hermodson, M. A. (1986). *Nature (London)* **323,** 448–450.

Higgins, C. F., Hyde, S. C., Mimmack, M. M., Gileadi, U., Gill, D. R., and Gallagher, M. P. (1990). *J. Bioenerg Biomembr* **22,** 571–592.

Hiles, I. D., Gallagher, M. P., Jamieson, D. J., and Higgins, C. F. (1987). *J. Mol. Biol.* **195,** 125–142.

Hobson, A., Weatherwax, R., and Ames, G. F.-L. (1984). *Proc. Natl. Acad. Sci. U.S.A.* **81,** 7333–7337.

Hyde, S. C., Emsley, P., Hartshorn, M. J., Mimmack, M. M., Gileadi, U., Pearce, S. R., Gallagher, M. P., Gill, D. R., Hubbard, R. E., and Higgins, C. (1990). *Nature (London)* **346,** 362–365.

Joshi, A., Ahmed, S., and Ames, G. F.-L. (1989). *J. Biol. Chem.* **264,** 2126–2133.

Juranka, P. F., Zastawny, R. L., and Ling, V. (1989). *FASEB J.* **3,** 2583–2592.

Kane, S. E., Pastan, I., and Gotteman, M. M. (1990). *J. Bioenerg. Biomembr.* **22,** 593–618.

Kang, C.-H., Kim, S.-H., Nikaido, K., Gokcen, S., and Ames, G. F.-L. (1989). *J. Mol. Biol.* **207,** 643–644.

Kang, C.-H., Shin, W.-C., Yamagata, Y., Gokcen, S., Ames, G. F.-L., and Kim, S.-H. (1991). *J. Biol. Chem.* **266**, 23893–23899.

Kartner, N., Hanrahan, J. W., Jensen, T. J., Naismith, A. L., Sun, S., Ackerly, C. A., Reyes, E. F., Tsui, L. C., Rommens, J. M., Bear, C. E., and Riordan, J. R. (1991). *Cell* (*Cambridge, Mass.*) **64**, 681–691.

Kerem, B., Rommens, J. M., Buchanan, J. A., Markiewicz, D., Cox, T. K., Chakravarti, A., Buchwald, M., and Tsui, L. C. (1989). *Science* **245**, 1073–1080.

Kerppola, R. E., and Ames, G. F.-L. (1992). *J. Biol. Chem.* **267**, 2329–2336.

Kerppola, R. E., Shyamala, V., Klebba, P., and Ames, G. F.-L. (1991). *J. Biol. Chem.* **266**, 9857–9865.

Klein, W. L., and Boyer, P. D. (1972). *J. Biol. Chem.* **247**, 7257–7265.

Kuchler, K., and Thorner, J. (1990). *Curr. Opin. Cell Biol.* **2**, (in press).

Kustu, S. G., and Ames, G. F.-L. (1973). *J. Bacteriol.* **116**, 107–113.

Kustu, S. G., and Ames, G. F.-L. (1974). *J. Biol. Chem.* **249**, 6976–6983.

Landick, R., and Oxender, D. L. (1985). *J. Biol. Chem.* **260**, 8257–8261.

Landick, R., Oxender, D. L., and Ames, G. F.-L. (1985). *In* "The Enzymes of Biological Membranes" (A. Martonosi, ed.), pp. 577–615. Plenum, New York.

Manoil, C., Mekalanos, J. J., and Beckwith, J. (1990). *J. Bacteriol.* **172**, 515–518.

Michel, H., Weyer, K. A., Gruenberg, H., Dunger, I., Oesterhelt, D., and Lottspeich, F. (1986). *EMBO J.* **5**, 1149–1158.

Milburn, M. V., Tong, L., de Vos, A. M., Brugner, A., Yamaizumi, Z., Nishimura, S., and Kim, S.-H. (1990). *Science* **247**, 939–945.

Mimmack, M. L., Gallagher, M. P., Pearch, S. R., Hyde, S. C., Booth, I. R., and Higgins, C. F. (1989). *Proc. Natl. Acad. Sci. U.S.A.* **86**, 8257–8260.

Mimura, C. S., Admon, A., Hurt, K. A., and Ames, G. F.-L. (1990). *J. Biol. Chem.* **265**, 19535–19542.

Mimura, C. S., Holbrook, S. R., and Ames, G. F.-L. (1991). *Proc. Natl. Acad. Sci. U.S.A.* **88**, 84–88.

Monaco, J. J., Cho, S., and Attaya, M. (1990). *Science* **250**, 1723–1726.

Oh, *et al.* (1992). In preparation.

Parham, P. (1990). *Nature* (*London*) **348**, 674–675.

Perego, M., Higgins, C. F., Pearce, S. R., Gallagher, M. P., and Hoch, J. A. (1991). *Mol. Microbiol.* **5**, 173–185.

Petronilli, V., and Ames, G. F.-L. (1991). *J. Biol. Chem.* **266**, 16293–16296.

Petronilli, V., and Ames, G. F.-L. (1992). Submitted.

Prossnitz, E., Nikaido, K., Ulbrich, S. J., and Ames, G. F.-L. (1988). *J. Biol. Chem.* **263**, 17917–17920.

Ringe, D., and Petsko, G. A. (1990). *Nature* (*London*) **346**, 312–313.

Riordan, J. R., Rommens, J. M., Kerem, B., Alon, N., Rozmahel, R., Grzelczak, Z., Zielenski, J., Lok, S., Plavsić, N., and Chou, J. L. (1989). *Science* **245**, 1066–1073.

Rommens, J. M., Iannuzzi, M. C., Kerem, B., Drumm, M. L., Melmer, G., Dean, M., Rozmahel, R., Cole, J. L., Kennedy, D., Hidaka, N., and Zsiga, M., Buchwald, M., Riordan, J. K., Tsui, L.-C., and Collins, F. S. (1989). *Science* **245**, 1059–1065.

Rudner, D. Z., LeDeaux, J. R., Ireton, K., and Grossman, A. D. (1991). *J. Bacteriol.* **173**, 1388–1398.

Schneider, E., and Walter, C. (1991). *Mol. Microbiol.* **5**, 1375–1383.

Shuman, H. A. (1987). *Annu. Rev. Genet.* **21**, 155–177.

Shyamala V., and Ames, G. F.-L. (1992a). Submitted.

Shyamala, V., Baichwal, V., Beall, E., and Ames, G. F.-L. (1992b). *J. Biol. Chem.* **266**, 18714–18719.

Speiser, D. M., and Ames, G. F. L. (1991). *J. Bacteriol.* **173,** 1444–1451.
Spies, T., Bresnahan, M., Bahram, S., Arnold, D., Blanck, G., Mellins, E., Pious, D., and DeMars, R. (1990). *Nature (London)* **348,** 744–747.
Treptow, N. A., and Shuman, H. A. (1985). *J. Bacteriol.* **163,** 654–660.
Trowsdale, J., Hanson, I., Mockridge, I., Beck, S., Townsend, A., and Kelly, A. (1990). *Nature (London)* **348,** 741–744.
von Heijne, G. (1986). *EMBO J.* **5,** 3021–3027.
Walker, J. E., Saraste, M., Runswick, M. J., and Gay, N. J. (1982). *EMBO J.* **1,** 945–951.
Woolley, P., and Clark, B. F. C. (1989). *Biol. Technology* **7,** 913–920.

Amino Acid Transport in Bacteria

Steven A. Haney[*][1] and Dale L. Oxender[*][†]

[*] Department of Biological Chemistry, University of Michigan Medical School, Ann Arbor, Michigan 48109
[†] Warner-Lambert/Parke-Davis, Ann Arbor, Michigan 48105

I. Introduction and Scope

Subjects to be covered in this article include studies that address the mechanism of transport as well as the genetic and regulatory interrelationships of multiple transport systems for an amino acid. Interrelationships of amino acid transport as a subset of the larger subject of bacterial metabolism and regulation will be discussed. We will focus on those features in amino acid transport that have been well characterized and therefore provide a comprehensive assessment of the system under study.

The discussion is divided into three main sections: First, a review of the high-affinity branched chain amino acid transport system in *Escherichia coli*, LIV-I; second, a discussion of studies that focus on the structure and mechanism of amino acid transport systems, including ATP-dependent transporters (except topics discussed by G. F.-L. Ames in this volume), and ion-coupled transport systems; third, some of the physiological and regulatory relationships of amino acid transporters will be presented. These include discussions of the roles different transporters for a given amino acid play in the cell, and how transport systems for a group of related amino acids are regulated. This latter section concludes with a discussion of the relationship of global control systems to amino acid transporters, including a newly recognized global regulator, the leucine-responsive regulatory protein, Lrp.

[1] Present address: Department of Molecular Biology, Princeton University, Princeton, New Jersey 08544.

II. LIV-I: The High-Affinity Branched Chain Amino Acid Transport System of *Escherichia coli*

A. Early Studies in Branched Chain Amino Acid Transport

The osmotic shock sensitivity of the LIV-I transport system had been shown to be the result of a periplasmic binding protein for leucine, isoleucine, and valine (Piperno and Oxender, 1966). Later, the leucine-specific binding protein was identified from osmotic shock fluids (Furlong and Weiner, 1970). The binding proteins were designated by the amino acids they bound: The LIV binding protein (LIV-BP) bound leucine, isoluecine, and valine and is a component of the LIV-I transport system, and the leucine-specific binding protein (LS-BP) binds D- and L-leucine as a component of the LS system.

Genetic characterization of branched chain amino acid transport in *E. coli* by Rahmanian *et al.* (1973) identified two systems, the LIV-I system for the high-affinity (0.2 μM), osmotic shock-sensitive system and LIV-II for the low affinity (2.0 μM), osmotic shock-insensitive system. A third system has been described, the high-affinity leucine-specific system, LS (Rahmanian *et al.*, 1973). This system was subsequently shown to be genetically and functionally related to the LIV-I system (Anderson and Oxender, 1977). In a study by Anderson and Oxender (1977), mutants in several genes involved in high-affinity transport were identified. The genes *livJ* and *livK* were identified in these studies and characterized as lacking the respective binding proteins; the *livJ* mutant does not express the LIV-BP and the *livK* mutant does not express the LS-BP. Therefore, the LIV-BP is also referred to as LivJ and the LS-BP is also referred to as LivK. The *livK* mutant resulted in the loss of both LIV-I and LS transport systems, despite the expression of the LIV-BP. This result suggested that additional components may be cotranscribed with *livK*, and that the mutation in *livK* (a Mu insertion) resulted in a loss of common components as a result of polarity. An additional mutant in a locus designated *livH* expressed both binding proteins but did not show high-affinity transport of either the branched chain amino acid or leucine-specific transport systems. Mutants in the *livP* locus, coding for the low-affinity LIV-II transport system, have been identified (Anderson and Oxender, 1978).

Oxender *et al.* (1980a,b) cloned the LIV-I locus containing the previously identified genes. Sequence analysis of the locus revealed seven genes in the LIV-I locus: the binding protein genes, *livJ* and *livK*, the membrane component genes, *livH, livM, livG,* and *livF,* and a gene of unknown function, *livL* (see below). The complete nucleotide sequence has been reported, as has the demonstration that all the above components, except *livL*, are required for branched chain amino acid transport (Adams *et al.*, 1990).

B. Studies in the Characterization of the Transport Components

With the cloning of the LIV-I regulon, a more detailed characterization became possible. In particular it has been shown that this system is a member of a superfamily of transport systems in prokaryotes and eukaryotes that share many conserved features. The conserved features are noted in Sections III,A,1 and III,A,3. and are discussed elsewhere extensively in this volume by G. F.-L. Ames. The components of the transport system have been characterized and their physical properties are listed in Table I.

Integrating these studies with those in related systems has allowed a model of branched chain amino acid transport to be developed. This is described in Fig. 1. This transport system has been characterized in several studies, described below.

1. Studies on the Periplasmic Binding Proteins LivJ and LivK

One study in the characterization of the transport components has examined the ligand specificity of the binding proteins. In the case of the leucine binding proteins, LivJ binds leucine, isoleucine, valine, and to a lesser extent, threonine. LivK binds leucine and some closely related analogs well, but not the other branched chain amino acids. The structures of both

TABLE I

Components of the LIV-I and LS Transport Systems

Protein	Residues	Mass[a]	Hydrophobic index[b]	Classification	pI
LivJ	367	39,006	−0.20	Binding protein	5.8
LivK	369	39,281	−0.19	Binding protein	5.7
LivH	308	32,928	0.92	Internal membrane protein	8.1
LivM	424	46,057	0.76	Internal membrane protein	9.7
LivG	255	28,502	−0.17	ATP binding comp.	7.6
LivF	237	26,166	−0.10	ATP binding comp.	5.7

[a] In kilodaltons.
[b] Kyte–Doolittle hydrophobicity.

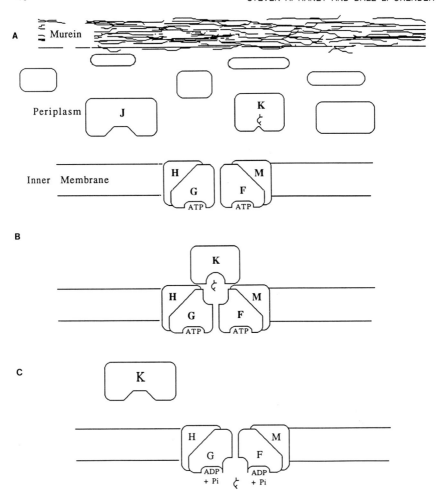

FIG. 1   Model for the transport of branched chain amino acids by the LIV-I transport system. (A) For Liv-I transport, branched chain amino acids are bound by the periplasmic binding protein LivJ, but only leucine is bound by LivK. When periplasmic proteins bind their substrate, they undergo a large conformational shift that almost completely envelopes a single amino acid ligand, such as leucine. Amino acids are believed to be able to diffuse freely through the porins of the outer membrane; no porin mutant affecting amino acid transport has been identified, such as has been for maltose and peptides. (B) These proteins then interact with the common membrane components. The membrane components consist of four proteins, two integral membrane components that interact with the membrane to form the outer structure of the channel, and two detergent-sensitive components that contain nucleotide binding sites that are believed to provide the energy required for transport through hydrolysis of ATP. (C) These nucleotide binding components may participate directly in the translocation of substrate.

proteins have been determined (Sack *et al.*, 1989a,b), and since the primary sequences are very similar, these proteins serve as good models for examining the residues that determine the ligand specificity for each binding protein. This has been studied by Adams *et al.* (1991). The putative leucine binding site of the LIV-BP (LivJ) was identified by Quiocho and co-workers by diffusing leucine into crystals of the LIV-BP and solving the structure of the leucine bound protein. Suitable crystals of the LIV-BP complexed to leucine have been difficult to prepare. Quiocho and co-workers identified six residues that appeared to bind the substrate from one cleft of the binding protein. Three of the six residues are conserved in the LS-BP (LivK), and three are different: Tyr-18, Ala-100, and Phe-276. The corresponding residues in the LS-BP are tryptophan, glycine, and tyrosine, respectively. It was reasoned that one or more of these residues may be involved in the specificity of the binding proteins for their substrates (Oxender *et al.*, 1990). Site-directed mutagenesis and the preparation of chimeric proteins produced several mutant binding proteins. These mutants were used to characterize the roles that the nonconserved residues play in the binding and discrimination of the branched chain amino acids and their analogs (Adams *et al.*, 1991). The results suggest that residue 18 plays a general role in binding, since a W18Y change in LivJ resulted in a 20-fold loss of binding of all three amino acids. Residue 100 may play a role in the discrimination between leucine and isoleucine. The binding of all three amino acids in the A100G mutant of LivJ was unchanged but the mutant protein could now bind the leucine analog trifluorolecuine, where the wild type cannot. In the converse mutant, a G100A change in LivK, the affinity for trifluoroleucine was slightly reduced. The Y276F mutant of LivJ showed reduced affinity for leucine and valine, but not isoleucine. The F276Y mutant of LivK showed a lowered affinity to leucine alone. These results provide a basis for the understanding of the determinants of specificity and recognition by the binding proteins. These studies on the interactions between the binding protein and their substrates would be greatly improved by a more complete characterization of the binding pocket, either through structural analysis of the leucine-liganded binding proteins, or through direct genetic selection of binding protein mutants with altered specificity.

2. Studies on the Hydrophobic Membrane Components LivH and LivM

The *livH* gene was first characterized as a mutant that did not show high-affinity branched chain amino acid transport, but still expressed the periplasmic binding protein (see above). DNA sequence analysis indicated that the *livH* gene coded for a large hydrophobic protein, affirming the presumption that it functions as a part of the inner membrane complex of the transport system (Nazos, 1984). Nazos (1984) further characterized the expression of *livH* as being significantly less than that of *livK*, despite the fact that the two are cotranscribed. Graddis (1990) has studied the expression of the cloned *livH* gene. The LivH protein has been difficult to study because it is highly hydrophobic, and overexpression tends to be toxic to cells. LivH is more stable when overexpressed together with the other membrane components of the LIV-I system. Coexpression of the *livH*, *livM*, *livG*, and *livF* genes allows better expression, presumably because the complex of these gene products resists degradation. The LivM component has not been isolated of characterized, although mutants created by deleting the *livM* gene show loss of transport function.

3. Studies on the Hydrophilic (ATP Binding) Membrane
Components LivG and LivF

In addition to *livH*, genetic studies on the LIV-I locus first described the existence of *livG*, one of two genes coding for the putative ATP binding components (Nazos *et al.*, 1985). Ethyl methane sulfonate (EMS)-mutagenized strains were selected for resistance to azaleucine and screened by rapid transport assay (Oxender and Quay, 1976). Complementation analysis identified a new gene, *livG*, that was located distally to *livM*, giving the organization of the operon as *livK livH livM livG*. DNA sequence analysis verified the location of *livG*, and identified another gene with a high sequence similarity to *livG*. This gene was named *livF* (Adams *et al.*, 1990). Deletion analysis and transport assay showed that all four of these membrane component genes (*livH*, *livM*, *livG*, and *livF*) were required for LIV-I or LS transport, in addition to the two binding protein genes, *livJ* and *livK*, that were required for the respective transport systems.

Currently, the ATP binding protein genes are under study for their relationship to the general class of membrane transport systems depicted in Fig. 2 and discussed in depth in this volume by G. F.-L. Ames. A. L. Gibson *et al.* (1991) (unpublished results) have compared the primary sequence of *livG* and *livF* to the cystic fibrosis transmembrane regulator (CFTR) gene. They have noted that the phenylalanine residue in the first ATP binding domain, Phe-507, when deleted, is responsible for 70% of the cases of cystic fibrosis in humans and is conserved in *livG*, the first ATP binding component of the LIV-I transport system. The general evolutionary relationship of members of this transport superfamily was shown to be conserved when the deletion of the conserved residue from *livG*, Phe-92, resulted in the loss of leucine transport in bacteria. Other homologous mutations that result in altered function of CFTR in humans also result in altered LIV transport in bacteria. The importance of these findings is that the power of bacterial genetics can be used to identify suppressors of these mutations, which should give valuable information about the mechanism of the ATP binding components in transport.

C. Regulation of LIV Transport

The high-affinity branched chain amino acid transport system has been shown to be regulated by both repression and transcription attenuation (Landick, 1984). Regulation of the activity of the low-affinity LIV-II system by leucine has been observed but it is not very significant (Anderson *et al.*, 1976). Repression control of the high-affinity systems was first identified by Rahmanian and Oxender (1972). In this study, they isolated an *E. coli* strain that was derepressed for the high-affinity transport system by selecting for the ability to utilize D-leucine. This locus was identified as *dlu*, for D-leucine utilizer. D-Leucine is presumed to alleviate a leucine auxotrophy by being converted to the other stereoisomer via a cytoplasmic racemase. The LS transport system can transport D-leucine, but not at a rate sufficient to alleviate the auxotrophy in a strain wild type for the regulation of transport. Growth on 200 mg/liter of D-leucine was used to select strains that were derepressed for transport. The gene was mapped to minute 20 on the *E. coli* chromosome and renamed *livR*, for leucine–isoleucine–valine transport repressor. A second allele, *lstR*, was also mapped to the minute 20 region of the *E. coli* chromosome (Anderson *et al.*, 1976). *lstR* is believed to be allelic to *livR*. The product of this locus has been identified as being involved in the control of many genes by leucine, and as such has been recognized as a global regulator. The gene coding for this regulator has been cloned in the laboratories of Short and Calvo (Austin *et al.*, 1989; Platko *et al.*, 1990), and has been renamed *lrp*, which codes for the leucine-responsive regulatory protein, Lrp, which is a more accurate acknowledgment of its pleiotropic nature (Platko *et al.*, 1990; Lin *et al.*, 1990). This regulatory factor is discussed in more detail in Section IV,E,4.

Transcription attenuation control of LIV-I expression was identified in our laboratory genetically, by observing that high-affinity transport activity was altered by mutations in several genes involved in formation of mature, charged leucyl-tRNA^{Leu}. These genes include the genes coding for pseudouridylate synthase, *hisT*, and a temperature-sensitive allele of the leucyl-tRNA synthetase gene, *leuS31* (Quay *et al.*, 1975a; Quay, 1976). These mutations affected both transport and leucine binding activity of the periplasmic osmotic shock fluid. Additionally, the relationship between a mutation in *hisT* and *rho* suggested that these gene products interact, suggesting that Rho factor is involved in the transcription attenuation control of transport genes (Quay *et al.*, 1978).

Cloning and sequence analysis of the LIV-I regulon provided a molecular framework that could account for the previous genetic results. As described by Landick (1984), the leader sequence of *livJ*, the LIV-BP gene, coded for two putative leader peptides, the first of which contains two leucine codons. Neither a factor-dependent terminator nor a series of overlapping stemloops, which are characteristic of the amino acid biosynthetic operons (reviewed by Landick and Yanofsky, 1987) could be identified. Instead, the leader transcript is low in any potential secondary structure, rich in cytosine residues, and low in guanosine. These features are characteristic of a Rho-dependent terminator (Alifano, 1991). A model that could account for the genetic data has been developed from the sequence of the regulatory region: The putative leader peptide in the mRNA can be translated by a ribosome, which in cases of leucine deficiency would be expected to pause at the leucine codons until a charged leucine tRNA becomes available for translation. This pausing by the ribosome prevents Rho factor from binding to the *livJ* message and terminating transcription of RNA polymerase, deattenuating transcription of the *livJ* gene.

Work in our laboratory has provided some support for this model. S. A. Haney and D. L. Oxender (unpublished results) have investigated the role of the leucine-coding leader peptide in transcription attenuation of *livJ*. We constructed mutant leader peptides by site-directed mutagenesis and fused the wild-type and mutant leader sequences to the β-galactosidase gene, generating *livJ'–lacZ* operon fusions for the wild-type and mutant *livJ* leader peptides. The constructs were crossed to phage λ and transduced to the chromosome such that each operon fusion was in single copy in a *lrp^c* strain (*livR*). Regulation by the addition of exogenous leucine was measured as changes in LacZ activity in exponentially growing cultures. The results show that addition of leucine to strains carrying operon fusions to *livJ* leader peptide sequences that code for leucine would attenuate expresssion of *lacZ* whereas fusions to leader peptides that either did not code for leucine or for the methionine initiation codon of the leader peptide did not attenuate expression when exogenous leucine was added to the medium. In addition, two of the fusions (the wild type and a mutant that coded for proline instead of leucine) were examined in a strain with a *lrp*::Tn*10 leuS31* background. This strain background did not express Lrp, and the tRNA synthetase for leucine was temperature sensitive. When the temperature of growing cultures was shifted from 30 to 39°C, the strain carrying the fusion to the wild-type *livJ* leader showed derepressed expression of *lacZ* relative to the fusion that did not code for leucine. In related work in our laboratory, the formation of the attenuated and full-length transcripts *in vivo* has been observed (R. M. Williamson and

D. L. Oxender, unpublished results). This work has shown that in strains carrying mutations in *rho* or *leuS*, the proportion of full-length message relative to the attenuated message is increased. These studies provide additional support for a transcription attenuation mechanism for the regulation of expression of *livJ*.

The physiological role of the attenuation mechanism is not fully understood. The regulation by Lrp is significant, and that of the *livJ* attenuator is relatively limited. This is made evident by the inefficient translation initiation region of the leader peptide. One possibility currently being examined is that attenuation may be functional only under specific circumstances. In particular, results have shown that attenuation and expression of the wild-type *livJ'–lacZ* operon fusion are stimulated by lower growth rates, whereas expression from mutant *livJ'–lacZ* fusions in which the leader peptide either does not code for leucine or for the methionine initiation codon is not stimulated by slower cell growth rates (S. A. Haney and D. L. Oxender, unpublished results). A *livK'–lacZ* operon fusion was also not stimulated by slower growth. This type of regulation has been observed before in the transcription attenuation control to the chromosomal β-lactamase gene, *ampC* (Jaurin *et al.*, 1981). In this latter system, expression of *ampC* by attenuation is a function of growth rate control. The authors conclude the *ampC* attenuator is dependent on growth rate control because the level of antitermination is dependent on the availability of ribosomes. In the case of *livJ*, it is difficult to determine with certainty that this is directly linked to growth rate control, since metabolic influences such as leucine pool size or other undescribed factors may also influence the regulation in ways that would not affect the *ampC* attenuator. However, it is clear that expression of *livJ* is affected by growth rate, and this may serve to increase the liganded *livJ* concentration. The role of such an increase may be either to maximize transport of branched chain amino acids in conditions where this is particularly beneficial for survival, or to increase the sensitivity of the cell (and therefore Lrp) to the presence of leucine, as described in Section IV,E,4.

III. Biochemistry and Mechanism of Amino Acid Transport

A. Structure and Organization of ATP-Dependent Transport Systems

1. Overview

The periplasmic binding protein, ATP hydrolysis-dependent, transporters are a well-characterized superfamily of substrate transport systems that have been identified in both prokaryotes and eukaryotes. The bacterial family members consist of high-affinity transporters for basic and branched chain amino acids, glutamine, peptides, some sugars, and small metabolites (Ames, 1987; Furlong, 1987; Higgins *et al.*, 1990). The eukaryotic members include the P-glycoprotein multidrug resistance factor, the cystic fibrosis transmembrane regulator CFTR, and a system involved in the transport of class 1 MHC antigens (for a more complete discussion see G. F.-L. Ames, this volume). Some representative transport systems are depicted in Fig. 2.

An important reason for this high level of characterization is that several of the components are soluble, making them easier to manipulate and analyze. The bacterial transporters have been used as models for related studies in protein secretion, folding, protein–protein interactions, and in studies on the physiology of the periplasm. The high degree of homology in

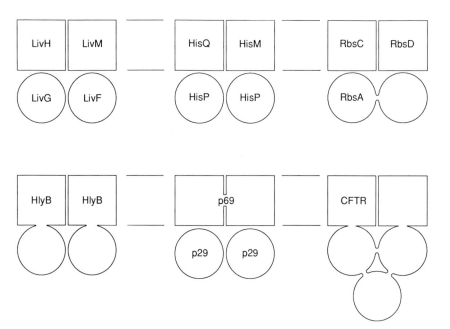

FIG. 2 Representative members of the ATP-dependent transporter superfamily. For systems that include binding proteins, only the membrane components are shown. *Top* (left to right): The LIV-I transport system: other members of this type of complex include the oligopeptide transport system; the histidine transport system: other members include the maltose system; The ribose transport system. *Bottom* (left to right): The hemolysin transport system: The White/Brown transporter from *Drosophila* is included in this family; The *Mycoplasma* p69 transporter: its substrate is uncharacterized; CFTR (the human cystic fibrosis transmembrane regulator): The additional domain depicted is the "R" domain, an acidic domain whose role is not yet established.

both prokaryotes and eukaryotes has allowed generalizations to be made from the characterization of the individual systems.

This section will cover many of the general characteristics of the bacterial transporters. However, since a detailed discussion of the mechanism of energy transduction, as well as the homologous relationship of these transport systems in prokaryotes and eukaryotes, is presented in another article of this volume by G. F.-L. Ames, these aspects will not be dealt with extensively here. This section will primarily discuss aspects of amino acid transport that are either unique to bacterial transport systems, or relevant to bacterial physiology.

2. The Periplasmic Binding Proteins

a. Structural Aspects of the Periplasmic Binding Proteins   The bacterial periplasmic binding proteins have been a very active area of research for protein structural analysis by Quiocho and co-workers (Quiocho, 1990). The binding proteins are easy to purify and crystalize, and at least six binding protein structures have been determined. One of the principle reasons for this sustained interest in the structure of the periplasmic binding proteins is that although the tertiary structures of the binding proteins from several different transport systems are quite similar, the primary sequences show little homology (Adams and Oxender, 1989). Thus, these proteins have turned out to be an attractive system for considering the segments of the primary amino acid sequence that are determinants of protein conformation. Three-dimensional structures have been obtained for the LIV, leucine, glutamine, maltose sulfate, and galactose binding proteins. The binding proteins have a bidomain structure, and the substrate is bound in the cleft between the domains. When substrate is bound, the binding protein undergoes a large conformational shift that closes around the substrate, virtually removing it from access by the solvent (Quiocho, 1990).

b. Specificity of the Binding Proteins and of the Transport System   The periplasmic binding proteins show very high specificity for their substrates, generally in the micromolar range. This is true even for the oligopeptide transport system, which, despite showing little specificity for the side chain residues of the peptides it transports, still shows an affinity for peptides in the range seen for the amino acid binding proteins. Residues critical to the substrate specificity for the leucine binding proteins have been investigated by Adams *et al.* (1991), as discussed in Section II,B,1. Studies on the specificity and mechanism of binding have benefited greatly from the structural studies on the binding proteins, as described above.

c. Regulation of Binding Protein Activity   Transcriptional regulation is the principal means of regulation of transport in bacteria, and many aspects of transcriptional regulation are presented in Section IV. However, regulation of transport by altering binding protein affinity through phosphorylation has also been described by Celis (1984). Samples of *E. coli* were incubated with $^{32}P_i$ and an osmotic shock fluid separated by gel chromotography. Binding of labeled arginine and ^{32}P incorporation were measured for the column fractions and were found to coelute in one fraction, shown to be the phosphorylated LAO-BP. Other arginine binding peaks were shown to be the Lys–Arg-BP and unphosphorylated LAO-BP. Other ^{32}P-labeled peaks were identified, but have not been investigated. The phosphorylated argeinine binding protein was shown to bind substrate 50-fold less tightly than the unphosphorylated protein. A mutant has been isolated that does not express the kinase that phosphorylates the BP (Celis, 1990). This kinase, which has been purified (Urban and Celis, 1990), serves to deactivate the binding protein and limit entry of basic amino acids into the cell. Study of the regulation of this kinase may contribute or our undersanding of a currently uncharacterized aspect of *E. coli* physiology.

d. Diffusion of Binding Proteins in the Periplasm   The environment of the periplasm has been difficult to characterize. A major component of the periplasmic space is the murein, or peptidoglycan, layer. The murein plays a significant role in determining the structure of the cell, and of protecting against damage to the cell membrane that could be caused by the high internal pressure. It is covalently attached to the outer membrane. Whether it fully extends across the periplasmic space and supports the inner membrane is unclear. Techniques used to characterize the membrane organization of gram-negative bacteria (such as electron microscopy) may alter the hydration of the murein, and therefore its structure (Beveridge, 1989). Even if it is located away from the inner membrane, it can have a significant impact on the movement of proteins in the periplasm, since it binds, and therefore orders, much of the water in the periplasm. Several investigations have studied the viscosity of the periplasm in order to better understand its role in transport and the physiology of the binding proteins. These studies suggest that the binding proteins are significantly impeded in their movement in the periplasm.

Hobot *et al.* (1984) studied different methods of fixing bacteria for electron microscopic studies, and concluded that methods of fixing bacteria that involved dehydration of the cell could result in shrinking of the murein, creating a specimen that has a partially collapsed or missing murein. The attachment of the murein to the outer membrane causes the dehydrated murein to appear to extend only partly across the periplasmic space. Using other methods, the murein can be characterized as spanning the periplasm, resulting in what the authors refer to as the periplasmic gel. The environment of the periplasm was studied *in vivo* by Brass *et al.* (1986). They showed that diffusion of binding proteins in the periplasm was 100-fold slower than in water, slower than the rate of diffusion of a protein in glycerol. The authors suggest several factors that may contribute to this high viscosity. These include the possibility of periplasmic structures that inhibit free movement, and that the concentration of murein and proteins in the periplasm causes the water of the periplasm to become highly ordered, and therefore unable to allow diffusion of the components of the periplasm. The net result of these studies is that the binding proteins do not float freely, but are suspended in a highly viscous solution.

3. The Membrane Components

a. General Organization Genetic analysis of several transport systems suggested that more than one gene was required for transport through the inner membrane. The first multicomponent transport system to be sequenced, the histidine/LAO system from *Salmonella typhimurium*, allowed this genetic data to be formalized into a physical model (Higgins *et al.*, 1982). In this model for the membrane components, two proteins formed the integral membrane complex and a third, less hydrophobic protein was peripherally associated with the integral membrane complex. The primary sequence of this peripherally associated component was shown to predict a nucleotide binding fold, suggesting this component might be involved in the energetic aspects of transport (Hobson *et al.*, 1984). The studies that have contributed to this area of the model for transport are numerous, and are discussed in detail in the chapter by Ames. Sequence analysis of a related system for maltose also showed that three gene products formed a membrane complex (Shuman *et al.*, 1980), again supporting the model of a three-component membrane complex.

When sequence information from other transport systems became available, this three-component model for the membrane complex changed somewhat. The *S. typhimurium opp* system revealed four membrane components (Hiles *et al.*, 1987), as did the *E. coli liv* system (Adams *et al.*, 1990). At the other extreme, the glutamate transport system of *E. coli* is coded for by a three-gene operon, *gltH gltP gltQ:* a binding protein gene, a hydrophobic membrane component gene, and an ATP binding component gene (Nohnó *et al.*, 1986). Analogies to related transport systems further complicate the picture. The *E. coli* ribose transport system showed a single peripheral membrane component, but this component was structurally similar to a dimer of the ATP binding components of the other systems (Bell *et al.*, 1986). It was becoming apparent that in many systems, these

sporulation, which is also controlled by this locus. These results indicate that SpoOKE plays a unique role in the competence signal transduction pathway, whereas the other members of the complex function in both the competence and sporulation signal transduction pathways (detailed in Section IV,E,3). Whether the two ATP binding components have identical roles in transport, or whether they serve different purposes, is not clear. Such biochemical analysis will not be possible until single turnover experiments can be developed.

4. Interactions between the Binding Proteins and the Membrane Components

a. Kinetic Studies on the Role of Binding Proteins in Transport Systems A clear understanding of the role of binding proteins in transport has been difficult to establish. Several factors account for this: First, our understanding of the dynamics of the periplasmic space is limited, due to few studies having been done in this area. Second, the studies that have been performed have indicated that the periplasm may be different than usually perceived, thus contradicting some of the conceptions of transport and of the periplasm (Section III,A,2,d). Third, rationales for the role and function of the binding proteins are easy to invoke, but do not necessarily agree with known kinetic parameters of binding and transport. Typically, the role of the periplasmic binding protein in transport is ascribed to that of increasing the concentration of substrate for transport. As was argued by Hennge and Boos (1983), this is a misstatement of the retention effect, first described by Silhavy *et al.* (1975). They point out that the effect of a high concentration of binding protein in the periplasm, with a high affinity for substrate, is to generate a high concentration of liganded binding protein, not of ligand itself, as is sometimes stated. While this is in essence a small point, its effect is to significantly obscure the role of the binding protein in transport. Estimates of the concentration of the binding protein in the periplasm vary from 0.5 to 10 mM (Silhavy *et al.*, 1975; Hennge and Boos, 1983; Ames, 1987). With dissociation constants for ligands in the range of 10^{-5} to 10^{-7} M, this means that even when low concentrations of ligand are present in the medium, the concentration of liganded binding protein in the periplasm can be in the millimolar range (Hennge and Boos, 1983). The concentration of the free ligand in the periplasm will be the same as it is outside the periplasm. This by itself effectively demonstrates that the membrane complex recognizes the liganded binding protein and not the free ligand, since strains not expressing binding proteins do not show appreciable transport from the high-affinity system, even when ligand is present in millimolar concentrations, whereas wild-type transport systems show transport affinities in the micromolar range (Hennge and Boos, 1983; Treptow and Shuman, 1985).

Rather than directly transporting substrate to the cytoplasm, as is done by the single subunit cotransporters, the binding proteins may function by providing a rapid increase in the concentration of liganded binding protein when substrate becomes available. Direct incorporation of low concentrations of substrate to the intracellular pool may not be as informative about changes in the environment to the cell, whereas a rapid increase in the liganded binding protein pool may serve to signal such a change. As discussed by Silhavy *et al.* (1975), the binding proteins may show increases in the extracellular concentration of substrate 10^3 to 10^5 times faster than corresponding decreases. The sudden availability of a substrate for a transport system will therefore result in a profound change in ratio of liganded to unliganded binding protein, even if the amount of substrate is low. This has already been shown to be important in the function of chemotactic systems mediated by binding proteins. Whether this functions in other transport-related processes is not understood at present. This can serve either in a sensory role in signal transduction or in scavenging very small amounts of amino acids. Scavenging has been presumed to be a principal function of the binding protein-dependent transport systems, but single subunit permeases have also been shown to scavenge small amounts of nutrients.

b. Genetic Analysis of Binding Protein–Membrane Complex Interactions

One of the fundamental steps of bacterial transport by the high-affinity, ATP-dependent transporters is the interaction between the binding proteins and the membrane components in the transfer of substrate. Studies of two binding protein-dependent systems have characterized these interactions genetically. Generally, these interactions have been approached through the identification of extragenic suppressors of transport mutants, many of which are subsequently located in gene products coded for by other genes for the transport system.

i. The Histidine Transport System The interaction of the histidien binding protein, HisJ, with the membrane complex was examined by Ames and Spudich (1976). In this study, they identified a suppressor of a mutation in *hisJ, hisJ5625,* that was previously characterized as coding for a mutant binding protein, HisJ*, able to bind histidine normally, but unable to transport it (Kustu and Ames, 1974). This strain was used to select for a second site mutation that would restore histidine transport. One mutant selected was identified as having a mutation in *hisP,* the gene for the ATP component of the membrane complex. This HisP mutant, HisP*, could transport histidine when either the mutant or the wild-type binding protein was expressed. Transport of histidine by the wild-type HisJ was more efficient than transport by the mutant HisJ* in the mutant HisP* system, but the opposite was true for the transport of arginine, where transport was improved by expression of the mutant HisJ* in the mutant HisP* system, indicating the mutant HisJ* competes less effectively than the wild-type histidine binding protein against the LAO binding protein. These results suggest that HisJ and HisP interact in transport, and that the mutant HisJ* is deficient in transport because it interacts less effectively with the transport complex.

Payne et al. (1985) characterized several previously identified histidine transport mutants that had an altered substrate specificity. The wild-type transport system cannot transport histidinol, whereas the mutants had now acquired this ability. These mutations were characterized as within the hisM gene, all of which code for proteins with small deletions of a repeated region of the hisM gene. The authors suggest this region, and therefore the HisM protein itself, plays a role in the substrate specificity of the transport mechanism.

Prossnitz et al. (1988) further characterized this interaction by using chemical cross-linking reagents to isolate subunits of the transport complex that were cross-linked to each other in vivo and in vitro. They identified cross-linked products by Western blot, using antibodies to HisJ and HisQ. The authors identified several cross-linked forms, including one that reacted with antibodies against both HisJ and HisQ. They further showed that the cross-linking of the mutant HisJ* (described above) to HisQ was significantly reduced relative to the wild-type HisJ, and that MalE could not be cross-linked to HisQ, supporting the suggestion that the interaction between HisJ and HisQ occurred during the transport event.

Speiser and Ames (1991) have characterized the specificity of histidine transporter by identifying mutants of the histidine transport complex that can transport D-histidine in the absence of either the histidine or the LAO binding proteins. Characterization of the mutations showed they were all located in hisP, as was the suppressor mutation characterized by Ames and Spudich (1976), described earlier. These mutations were isolated by expressing the membrane components on a plasmid from a promoter under the control of the temperature-sensitive phage λ repressor $cI.8_{57}$. Mutants were screened for those capable of transport at 37°C but not at 30°C. The cI_{857} repressor was functional at 30°C, and expression of the membrane components at this temperature was transcriptionally repressed. These mutants were mapped to hisP, sequenced, and shown to lie in a small area toward the carboxyl third of the hisP open reading frame. These mutants could transport histidine at very low levels when the binding protein was not expressed, and transport was significantly improved by the expression of either the wild-type HisJ or the mutant HisJ* binding proteins.

ii. The Maltose Transport System Although not an amino acid transporter, studies on the maltose transporter are considered here because of their general relevance to this class of transport systems. In particular, the studies on the maltose transport system provide clear genetic evidence of direct interactions between the binding protein and the hydrophobic membrane components.

Treptow and Shuman (1985) isolated mutants of a strain that, in spite of carrying a deletion of the maltose binding protein gene, *malE*, could use maltose as a carbon source. After screening to omit those that mapped outside of the maltose regulon, they determined the effects these mutations had on transport. Growth rates ranged from a $t_{1/2}$ of 75 min (the wild-type, *malE*$^+$ rate) to greater than 200 min. These mutations were mapped and shown to reside in either the *malF* or *malG* genes, which code for the hydrophobic membrane components, implicating MalF and MalG as the gene products that interact with the maltose binding protein. One of the most interesting findings was the effect of coexpressing the wild-type *malE* gene in these suppressor strains. In some cases, the presence of the maltose binding protein (MalE) increased the growth rate of the mutant, suggesting that MalE can function with the mutated transport system to increase the rate of transport. On the other hand, in most cases expressing *malE* served to inhibit transport and resulted in a Mal$^-$ phenotype. These data suggest that the maltose binding protein directly interacts with the mutant transport systems to interfere with transport.

A follow-up study explored the nature of this interference by the maltose binding protein on transport by the mutant membrane complexes (Treptow and Shuman, 1988). In this study, they isolated mutations in *malE* that restored transport when coexpressed in strains containing mutations in *malF* or *malG*. They isolated several mutant binding proteins that restored the loss of transport using strains carrying different mutations in *malF* or *malG*. What was remarkable about these mutant binding proteins was that many of these suppressors were allele specific. That is, a *malE* mutation selected in one mutant background would generally not be able to restore transport in other mutant backgrounds or in the wild type. These mutations have been sequenced and map widely over the *malE* gene, even though the authors identify a general pattern where mutants in the amino terminal of MalE tended to suppress *malF* alleles and mutants in the carboxyl terminal tended to suppress *malG* alleles. This study, in conjunction with the previously described study (Treptow and Shuman, 1985), provides strong genetic evidence that the binding proteins interact with the hydrophobic membrane components during transport. Other studies have identified a similar pattern of mutations and suppressor mutations in the MalE-dependent chemotaxis response mediated by the Tar protein (Kossmann *et al.*, 1988; Manson, 1991).

5. The "Periplasmic" Binding Protein-Dependent Transport Systems of Gram-Positive Bacteria

Experiments document the existence of binding proteins that are secreted from gram-positive bacteria across the outer membrane. In the gram-positive bacteria, extracellular binding proteins would be expected to diffuse away from the cell, or at least not be in high enough concentration to function in a manner analogous to those of gram-negative bacteria. This has led to the hypothesis that these extracellular binding proteins may be tethered to the outer membrane by lipylation through thioesterification to a fatty acid. Direct evidence for tethering has not yet been obtained. The cell wall of gram-positive bacteria is several times thicker than those of gram-negative bacteria such as *E. coli,* and therefore the retention of the binding proteins may result from this highly viscous outer layer (Beveridge, 1989).

The first report of extracellular binding proteins expressed by nonenteric bacteria was by Gilson *et al.* (1988). This report describes AmiA and MalX from *Streptococcus pneumoniae,* which are analogous to OppA, MalE, and p37 in *Mycoplasma,* whose substrate is unknown at present. These proteins are secreted and are believed to be lipoproteins because they contain consensus sequences in their amino-terminal regions for known cleavage and thioesterification sites of *E. coli* lipoproteins. For example, the deduced primary sequence of the OppA homolog in *S. pneumoniae,* AmiA, includes the amino acid sequence LAACSS early in the amino-terminal region, which corresponds to the consensus cleavage and thioesterification site of the lipoproteins of *E. coli.*

A detailed molecular and functional analysis of the *S. pneumoniae amiABCDEF* locus was described by Alloing *et al.* (1990). The components *amiACDEF* are homologous to *oppABCDF* from *S. typhimurium.* The *amiB* gene codes for a small, basic protein whose function is unclear. The evidence that this locus does indeed code for an oligopeptide transport system includes the observations that the *S. pneumoniae* auxotrophy for leucine or arginine could not be relieved by supplementation with peptides containing these amino acids in either *amiC* or *amiE* mutants, but could be relieved in a wild-type strain.

The *Bacillus subtilis spo0K* locus also codes for an oligopeptide transport system (Rudner *et al.,* 1991; Perego *et al.,* 1991). As discussed in Section IV,E,3, this locus was defined as one that is involved in induction of sporulation and competence. The discussion here concerns the physiology and function of the binding protein. This locus codes for a complete Opp-like system, and the binding protein gene in this species is the *spo0KA* gene product (referred to as OppA in Perego *et al.*). Perego and co-workers examined the association of the protein with the cell membrane and showed that it was found in the medium during growth, with 50% of the protein dispersed to the medium during rapid growth and 90% of the protein lost from the cells when the culture reached stationary phase. This gradual dissociation indicates that the binding protein has some mechanism for delaying dissociation, although examination of the binding protein isolated from both the medium and from cells, and OppA from *S. typhimurium*, showed them to be identical as determined by gel electrophoresis.

B. Ion-Coupled Amino Acid Transport Systems

1. Proton-Coupled Amino Acid Transport Systems

a. Introduction   A detailed exploration of the mechanism of proton-coupled transport is presneted by H. R. Kaback in this volume, and therefore will not be extensively reviewed here. This discussion will focus on the general characterization of proton-coupled amino acid transporters and some physiological aspects of these transporters. At one time, it was believed that all or most single subunit amino acid transporters were proton coupled. However, as discussed by Maloy (1990), careful work demonstrated that some of these were in fact sodium-coupled transporters. Studies to show the sodium independence can be difficult because the affinity for sodium can be very high, and efforts to sufficiently remove sodium from laboratory solutions can require extensive efforts.

b. Studies on Proton-Coupled Amino Acid Transport Using Membrane Vesicle Fusions   Work by Konings and co-workers has examined amino acid transport in a number of different prokaryotes through a variety of techniques based on vesicle fusions to proton motive force (pmf) generating systems. They have completed several studies on the proton-coupled amino acid transport systems of fermentive bacteria.

In one study (Driessen *et al.*, 1987), vesicle fusions were made between *Streptococcus cremoris* membranes and beef heart mitochondrial cytochrome oxidase. Addition of ascorbate and cytochrome *c* to the reaction mixture allows cytochrome oxidase to generate a proton gradient. The gradient is oriented positive on the outside even though the reconsituted cytochrome oxidase molecules face both sides of the membrane. Only the fraction of cytochrome oxidase that faced right side out would be expected to use the ascorbate/cytochrome substrate to generate a gradient. A schematic of this system is shown in Fig. 3. This system was used to examine the ratio of substrate to proton translocation for a number of amino acids. They found that for leucine transport, the apparent number of molecules of leucine translocated per number of protons (n_{app}) was 0.8. This value was

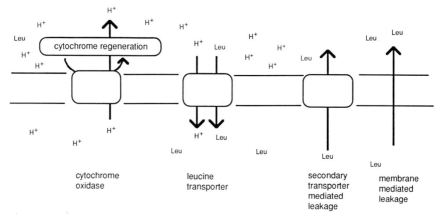

FIG. 3   Schematic representation of the bovine cytochrome oxidase-driven proton-coupled amino acid transporter. The factors affecting the measurement of the transport measurements are described in text.

fairly constant for an external pH range of 5.5 to 8.0. In contrast to this, the n_{app} for serine or alanine transport was 0.9–1.0 at pH 5.5, but fell to 0.0–0.2 at pH 8.0. They attribute the values less than unity as being due to the rate of amino acid exit from the vesicle being part of a steady state system with loss of the amino acid, resulting in the loss of accumulated amino acid. The rate of amino acid exit was shown to be independent of the external concentration of the solute, and was therefore attributed to either the membrane or another transporter catalyzing efflux. The dependence on external pH for this process in the case of serine and alanine transport suggested to the authors that at least a component of leakage was mediated by another transporter. The rate of amino acid exit from protein-free membrane liposomes showed that the rate of exit through the membrane was dependent on the hydrophobicity of the amino acid. The authors were able to calculate the rate of steady state transport as a function of transport, membrane-mediated exit, and carrier-mediated exit.

Another study used a system similar to that described above to examine amino acid transport by the obligate anaerobe *Clostridium acetobutylicum* (Driessen *et al.*, 1988). In this study, they examined leucine and lysine transport, which produce 40- and 100-fold accumulations, respectively. They further characterized this system by examining the role of the membrane lipids in transport. They found lipids from *E. coli* membranes gave high levels of transport, whereas lipids from soybeans or phosphatidylcholine diminished the function of the transporter.

Crielaard *et al.* (1988) developed a different assay to vary the pmf in a systematic way. They examined amino acid transport in vesicles of *S. cremoris* and *C. acetobutylicum* that were fused to vesicles of *Rhodopseudomonas palustris* photosynthetic reaction centers. In this system, the magnitude of the proton motive force could be varied as a function of light intensity. The initial rate of transport in these vesicles increased proportionally with the magnitude of the light intensity. The n_{app} for leucine of 0.8 was constant over the range of light intensities, and therefore the pmf range examined.

These and other studies have described the factors influencing the rate of uptake and the steady state accumulation of amino acids by proton-coupled transport systems. These include contributions to the understanding of the energetic factors that influence amino acid transport and the proton/solute stochiometry.

c. The Aromatic Amino Acid Transporters of Escherichia coli Aromatic amino acids are transported by several transport systems, which are discussed in more detail in Section IV,D. The tyrosine-specific permease, TyrP, has been identified as a proton-coupled transporter by Wookey *et al.* (1984). An analysis of the modeled membrane topology of this protein and the related tryptophan-specific transporters, Mtr and TnaB, has suggested these transporters may have a novel structure (Sarsero *et al.*, 1991a). Hydropathy analysis indicates that these transporters are clearly atypical in structure from other single subunit permeases such as the proline transporter PutP, and the general aromatic amino acid transporter AroP. Based on general criteria for previously characterized membrane protein structures, these proteins appear to have 11 membrane spanning helices, rather than the usual 12. The authors suggest that these proteins may be members of a new class of membrane proteins. Characteristics of membrane proteins used to model the tertiary structure of new membrane proteins also include the distribution of positive and negative charges. The regions of the polypeptide that link the membrane helices tend to be positively charged when they occur on the cytoplasmic side of the membrane and negatively charged when they occur on the periplasmic side of the membrane (von Heijne, 1986). Other factors include the locations of reverse turns (Levitt, 1977) and stretches of hydrophobic residues of sufficient length to form membrane spanning regions (White and Jacobs, 1989, 1990). An analysis of membrane protein structures by Maloney (1990) has revealed a general model of 12 membrane spanning helices as a basic structure, with some systems showing 10 or 14 helices (dimers of heptahelical subunits). It is possible that the unique topology of the Mtr, TyrP, and TnaB transporters may reflect unique characteristics of transport, such as the mechanism of proton/solute coupling or regulation.

2. Sodium-Coupled Transport Systems

a. Introduction Sodium-dependent transport has become an active area of research in amino acid transport for several reasons. These include the recognition that many of the putative proton-coupled transporters of enteric bacteria are actually sodium-coupled transporters, such as the proline, the serine-threonine, and the low-affinity branched chain amino acid transporters (Cairney *et al.*, 1984; Hama *et al.*, 1987; Hoshino *et al.*, 1990), and the study of sodium-coupled transporters in other bacteria such as marine bacteria, halophiles, and thermophiles. There is currently no structural model for sodium-coupled transport that has been examined to the extent of that of ATP-dependent transport systems (Section III,A and G. F.-L. Ames, this volume) and the H^+/lactose transporter (see H. R. Kaback, this volume). The detailed genetic studies of the proline transporter of *S. typhimurium* and *E. coli,* and the molecular biological studies of the *Atleromonas haloplanktis* D-alanine transporter and the *P. aeruginosa* PAO and PML branched chain amino acid transporter, are establishing these systems as models for the mechanism of sodium-coupled amino acid transport (see Sections III,B,2,b, III,B,2,d, and IV,A,3, respectively).

The sodium chemical potential used for these transporters can be generated in several ways. H^+/Na^+ antiporters can function to exchange the proton electrochemical potential for a sodium chemical potential. The proton chemical potential, $\Delta\bar{\mu}_{H^+}$, is a combination of the membrane potential ($\Delta\phi$) and the proton gradient (ΔpH). The sodium chemical potential, $\Delta\bar{\mu}_{Na^+}$, can be generated by exchanging the proton gradient for the sodium gradient. This can be done directly by exchanging the proton gradient for the sodium ion gradient through an antiporter, as depicted in Fig. 4 (Skulachev, 1985). Alternatively, bacteria possess several mechanisms capable of generating a sodium chemical gradient directly (Dimroth, 1987). One prominent class of such gradient-generating systems is referred to as the sodium ion transport decarboxylases. Sodium ions are pumped out of the cell as a consequence of reactions involved in the intermediary metabolism. As an example, in *Klebsiella pneumoniae*, two sodium ions are pumped out of the cell in the cell when oxaloacetate decarboxylase converts oxaloacetate to pyruvate (Dimroth and Thomer, 1986). Sodium transport decarboxylases are involved in many steps of the intermediary metabolism of the cell. A sodium gradient can also be generated by an ATPase-driven Na^+ pump. This system is structurally analogous to the F_0F_1 ATPases used in generating ATP from the membrane potential (Dimroth, 1990).

b. The Proline Porter I of Salmonella typhimurium and Escherichia coli

i. Proline Transport in Salmonella typhimurium Maloy and coworkers have used the proline transporter to examine the mechanism of its transport genetically. In a study to identify regions of PutP involved in solute transport, Dila and Maloy (1986) isolated mutants resistant to pro- line analogs and, after localizing the mutations to within *putP*, they used a sophisticated transduction scheme to map the location in the *putP* gene. Mutations giving rise to proline analog resistance could occur in either *putP* or the closely linked *putA*, coding for proline oxidase, which allows proline to be used as a sole carbon source. Missense mutations in *putP* were mapped to determine location in the *putP* gene. They genetically crossed each of the mutants by P22-mediated transduction to a set of strains carrying systematic deletions of the *putP* gene. Crosses to the *putP* deletion strains would localize the mutation to within an approximately 100-bp region on the 1.5-kb open reading frame. The 17 mutations mapped to essentially three regions of the gene. The regions were in internal segments of the open reading frame, at roughly 200, 400, and 800 bp inward from the 5' end.

In a related study, Myers and Maloy (1988) investigated the mechanism of sodium coupling to proline transport by identifying PutP mutants with an altered sensitivity to lithium ions. The proline transporter can co- transport both sodium and lithium ions, but lithium/proline cotransport can be moderately toxic to cells if too much Li^+ is accumulated. Thus, proline-driven lithium transport is deleterious to the cell, which can be used as a genetic screen for proline auxotrophs that could grow faster in the presence of proline and lithium ions. Mutagenized cells were screened for their ability to grow faster than control cells on plates containing proline as a sole carbon source and lithium as a cotransportable ion. They describe 24 lithium-resistant mutants with altered proline transport. Trans- duction analysis as described above revealed that all the mutations resided in the 5' and 3' ends of the *putP* open reading frame. The authors suggest from these results that the two ends of the PutP peptide may wrap around in a barrel so the amino- and carboxyl-terminal ends of the protein wrap around to form the sodium transporter. These studies show that distinct regions of PutP mediate transport of the substrates, sodium and proline.

The *putP* gene has recently been sequenced and analyzed. It is a very hydrophobic 54-kDa protein, and is predicted to form 12 membrane span- ning segments (Miller and Maloy, 1990).

ii. Proline Transport in Escherichia coli The mechanism of proline transport by the PP-I of *E. coli* has been studied by Anraku and co- workers. Mogi *et al.* (1986) isolated two mutants with altered proline transport and four other mutants were isolated by Motojima *et al.* (1978). Several of these mutants were subsequently cloned and characterized by Ohsawa *et al.* (1988). Two of these mutants showed binding of proline in the absence of sodium, and no change in affinity for proline on addition of sodium. These mutants appear to have become uncoupled in transport of the two ligands. One of these mutants was mapped by restriction fragment shuffling between the mutant and the wild-type alleles, and sequenced. The mutation was a missense transition mutation that resulted in an R257C mutant. This mutation maps to the middle region of the protein, correlating with the substrate specificity mutants isolated by Dila and Maloy (1986; and above). A structural model of the *E. coli* PP-I permease by Yamato *et al.* (1990) suggests the arginine residue may be exposed to the cytoplasmic side of the porter. Yamato and Anraku (1988) characterized two cysteine residues, at positions 281 and 344, as being sites affected by *N*-ethylmaleimide inhibition, whereas a third site, Cys-349, was still sensitive. All three sites, when individually changed to serine residues, could still function normally in transport.

Yamato *et al.* (1990) isolated mutants with altered sodium–proline transport. These mutants were selected as being resistant to the toxic analog 1-azetidine-2-carboxylic acid in the absence of sodium or lithium ions, and screened for sensitivity to the toxic compound as a function of sodium chloride concentration. These mutants were putatively mutant as a result of increased affinity for sodium. Transport was measured in these strains. Both mutants showed a strong dependence on sodium concentration for transport, and lithium was ineffective for transport. Sequencing of these alleles showed one change in each allele.

c. Transport in Thermophilic Bacteria Cytochrome oxidase from *Bacillis stearothermophilus* has been used to characterize energy-coupled transport *in vitro,* as depicted in Fig. 4. Speelmans *et al.* (1989) used this system to characterize the amino acid transport systems of *Clostridium fervidus.* Transport rates of many amino acids were measured at 40°C and pH 6.0. Serine was shown to have the fastest rate and was therefore chosen for further characterization. The serine transport rate was shown to be maximally stimulated by the presence of both a proton and a sodium gradient. A stochiometry of transport was shown to be approximately 1. The role of the sodium gradient as the driving force was shown by the use of the chemical Monesin, an Na^+/H^+ exchanger analogous to the antiporter described above. Raising the sodium chemical potential at the expense of the proton chemical potential through the addition of Monesin stimulated serine transport, demonstrating a role for sodium in the transport of serine.

Characterization of the role of the electrochemical potential of the cell membrane in *B. stearothermophilus* for amino acid transport was carried out by Heyne and co-workers (1991). Changes in amino acid transport as a consequence of the addition of different electrochemical inhibitors and decouplers were measured in membrane vesicles derived from *B. stearothermophilus.* The individual effects that ΔpH, $\Delta \phi$, and ΔpNa

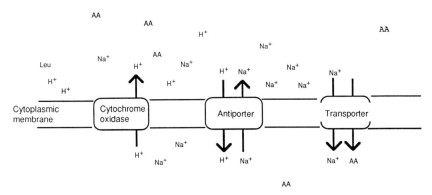

FIG. 4 Schematic representation of the heat-stable cytochrome oxidase-driven system for the study of sodium ion-coupled amino acid transport.

would have on amino acid transport were examined by comparing the effects of two ionophores on transport. The ionophore nonactin functions by dissipating both the membrane potential and the sodium gradient. In the presence of nonactin, transport of alanine was strongly inhibited. This suggests that one or both of these factors are important in alanine transport. The effect of monesin, which, as described above, increases the sodium gradient by exchanging it with the proton gradient, stimulated transport of alanine. This latter result indicates that the sodium gradient is responsible for energizing alanine transport in *B. stearothermophilus* vesicles. Through the use of artificially imposed gradients, transport of both alanine and leucine were shown to be additively stimulated by both the membrane potential and the sodium gradient, with the sodium gradient having a larger effect than the membrane potential when examined individually. All of these systems were shown to have a dependence on external pH, even though they do not use protons for transport. Raising the external pH from 5.5 to 9.0 was shown to dramatically decrease transport in a sigmoidal manner with a pK_a of 7.6 for alanine. The authors suggest that the deprotonation of the carrier on the inside of the vesicle results in an allosteric inhibition of transport.

d. Amino Acid Transport in Marine Bacteria Marine bacteria live in an environment that is about 600 mM sodium (MacLeod *et al.*, 1985). For many of these species, a sodium concentration of at least 15 to 50 mM is required for survival. MacLeod and co-workers have carried out studies on the physiology and molecular biology of amino acid transport in several marine bacteria. Droniuk *et al.* (1987) examined the role of sodium in the transport of nutrients in *A. haloplanktis* 214 and *Vibrio fischeri*. A factor in the motivation of these studies was to distinguish among the range of biological effects on the cell attributed to sodium. These include roles in osmoprotection, catalysis, and transport. Sodium-coupled transport was shown to be roughly maximal in media from 50 to 300 mM, a range significantly below the natural environment for these organisms. Concentrations of choline chloride equimolar to inhibiting concentrations of sodium chloride were generally not inhibitory, indicating that the sodium was directly inhibitory, even at concentrations found in the normal host range. Lithium ion could substitute for sodium to a large extent, but K^+ was largely ineffective. This might be expected since the K^+ concentration in the cell is 400 mM. Transport requires a membrane potential, as shown by the inhibition of transport by the decoupler 3,5,3′,4′-tetrachlorosalicylanilide (TCS).

Sodium-dependent D-alanine transport in *A. haloplanktis* has been further characterized by molecular biological methods. MacLeod *et al.* (1985) have prepared a genomic library from this strain and MacLeod and MacLeod (1986) were able to use the library to complement an *E. coli* strain unable to transport the substrate. Two independent transformants were shown to transport D-alanine in a sodium-dependent manner. Both clones were specific for *A. haloplanktis* DNA. The clones showed no homology to each other, despite their being indistinguishable kinetically.

e. Amino Acid Transport in a Facultative Alkalophile   The pH range of environments tolerated by *Bacillus* (YN-2000) is unusual. It can survive in environments that range in pH from 7.0 to 10.5. From pH 7 to pH 8.2, the proton chemical potential is outside negative, as in other bacteria. Above pH 8.2, however, the orientation of the proton gradient is reversed. Wakabayashi *et al.* (1988) examined the branched chain amino acid transport in this strain in an effort to characterize the driving force for transport. They were interested in determining what the effect of the reversal of the proton gradient would have on transport, and how the cell compensates for this change. For example, could two transport systems exist, where one would operate at moderate pH values and utilize the proton chemical potential and a second transporter operate at higher pH utilizing sodium, or the reverse proton gradient? The investigators used membrane vesicles to show that leucine transport was driven exclusively by the sodium chemical potential at all pH ranges. Lack of transport in the study of leucine transport by ΔpH supports this conclusion. They further characterized the system by showing that isoleucine and valine were the only effective competitive inhibitors, suggesting the system being studied is a general branched chain amino acid transporter.

IV. Physiology and Regulation of Amino Acid Transport in Bacteria

A. Branched Chain Amino Acid Transport in Enteric Bacteria

1. Introduction

The transport systems that mediate transport of the branched chain amino acids in *E. coli*, *S. typhimurium*, and *P. aeruginosa* PAO have been extensively characterized. Transport systems that have been characterized in more than one species have generally shown striking similarities, although each has unique features. The LIV transport systems from enteric bacteria that have been characterized to date are listed in Table II. The high-affinity branched chain amino acid transport systems from *E. coli*, LIV-I and LS, and the low-affinity LIV-II were discussed in Section II.

2. LIV Transport in *Salmonella typhimurium*

a. Transport Studies of transport of branched chain amino acids in *S. typhimurium* LT2 have characterized several branched chain amino acid transport systems biochemically and genetically. The K_m values of transport for the high-affinity system appear to be approximately 10-fold higher than that observed for *E. coli* (Kiritani and Ohnishi, 1977). As found for the

TABLE II

Leucine Transport Systems in Enterobacteria

		Genes[a]		
Transporter	Specificity	*E. coli*	*S. typhimurium*	*P. aeruginosa*
LIV-I	L, I, V	*livJ,* *livH,M,G,F*	*livB* ?	*braC* *braD,E,F,G*
LS	L	*livK,* *livH,M,G,F*	*livC* ?	(none?)
LIV-II	L, I, V	*livP*	*brnQ*	*braB*
LIV-III	I, V	?	?[b]	*braZ*

[a] For the LIV-I transporters, the top line is the binding protein gene, the second line denotes the membrane component genes.

[b] May transport leucine with equal affinity.

E. coli branched chain amino acid transport systems, transport was shown to be mediated by two binding proteins, which they designated LIVT-BP, which binds leucine, isoleucine, valine (and to a lesser extent, serine and threonine), and L-BP, which binds leucine (Ohnishi and Kiritani, 1983). A mutant in high-affinity transport, *livA*, was characterized and mapped to minute 76–77 in *S. typhimurium*, comparable to the LIV-I locus in *E. coli*. The binding protein genes have been cloned and sequenced (Ohnishi *et al.*, 1990). The *livB* gene codes for the LIVT-BP and the *livC* gene codes for the L-BP. The genes are highly homologous to the respective binding protein genes of *E. coli*. Interestingly, this region does not contain a homolog to the *E. coli* open reading frame between the two binding protein genes (which has been referred to as *livL*, and as ORF19), despite having a long intercistronic region between the binding protein genes.

The gene coding for the low-affinity LIV-II transporter, *brnQ*, has also been cloned and sequenced (Ohnishi *et al.*, 1988). The cloning of this gene was accomplished by the recognition that it maps near *gleR*, a regulatory mutant previously isolated as conferring glycyl-leucine resistance (discussed in the next section). The authors cloned the *gleR* locus, and the resulting clone was shown to code for *brnQ*. *gleR* was shown to act in cis with *brnQ*, and appears to be an operator mutation for *brnQ*.

b. Regulation   Repression of the high- and low-affinity transport systems has been studied in this system. Repression is mediated optimally by leucine and peptides containing leucine. They isolated two mutants, one named *gleR*, for glycl-leucine resistant, and one named *liv-231* (Ohnishi *et al.*, 1980). In this study, the authors suggest the *gleR* allele regulates the low-affinity system, since it regulates low-affinity transport, but the high-affinity binding protein activity of osmotic shock fluids were still regulated normally. The cloning of *brnQ*, described above, confirmed the conclusion that it was regulated separately from LIV-I. *liv-231* resulted in the loss of regulation of the high-affinity transport and binding protein activity. A *livR3*::Tn*10* allele was isolated and shown to constitutively increase the level of binding protein over the level found in a wild-type derepressed strain. This mutation mapped to minute 76 on the *S. typhimurium* chromosome, near the locus of the structural components (Ohnishi *et al.*, 1983), and may represent a mutation in the operator of the LIV-I locus. Further characterization of the *liv-231* allele showed that it mapped to minute 19, the position corresponding to that of the *livR*/*lrp* gene in *E. coli* (Murata-Matsubara *et al.*, 1985). The *liv-231* allele was named *livS*. One observation made by this group suggested that this may also be a pleiotropic gene: The authors noted that the structure of the ribosomes is altered in a *livS* strain. This argument is based on three results: One, extracts of *livS* strains were markedly reduced in their ability to function in an *in vitro* translation system, as measured by the incorporation of leucine into acid-precipitable material. This was shown to not be an effect of the leucyltRNA synthetase. Two, RNase I, a periplasmic enzyme that copurifies with ribosomes during purification, is more loosely associated in a *livS* strain. Three, there is a marked increase in the concentration of proteins in the supernatant of a ribosome preparation of a *livS* strain, particularly for proteins with a molecular weight of less than 20,000. While none of these results definitively shows that the formation of the ribosomes is affected, they clearly indicate that many factors are altered in the *livS* (*lrp?*) strain.

Investigations into the possible role of attenuation in the control of *livB* have not been reported. The leader region of the *livB* gene is, however, highly homologous to the corresponding LIV-BP gene from *E. coli*, *livJ*. The leucine-coding leader peptide is completely conserved, and the leader region is high in cytosine and low in guanosine, with a low potential to form secondary structures, as found for *E. coli*. Interestingly, the second putative, nonleucine coding, leader peptide in the *livJ* gene is not conserved in *livB*.

3. Transport in *Pseudomonas aeruginosa* PAO

Branched chain amino acid transport in *P. aeruginosa* PAO has also been extensively characterized. The LIV-I system has been cloned and sequenced (Hoshino and Kose, 1989, 1990a). The deduced protein sequences from the *braC*, *braD*, *braE*, *braF*, and *braG* genes are highly homologous to the *E. coli livJ*, *livH*, *livM*, *livG*, and *livF* genes, respectively. There are no homologs for the *livK* or *livL* genes in this system. It is unlikely that a *livK* homolog exists, since a *braC* mutant does not exhibit high-affinity, leucine-specific transport. Mutants in all of the above alleles have been identified and all the genes have been shown to be required for transport (Hoshino and Kose, 1990b). Biochemical analysis of the LIV-II transport system has shown that it transports with either sodium or lithium ions, but not protons, as a cosubstrate (Uratani *et al.*, 1989). The LIV-I locus has been cloned and sequenced (Hoshino *et al.*, 1990). The *braB* gene codes for a protein that is predicted to form 12 membrane spanning regions according to hydropathy analysis. BraB is 76% similar to BrnQ of *S. typhimurium* at the protein level but the genes share little homology at the nucleotide level.

One interesting study performed by this group was to compare the transport from this LIV-II system with that of the related *p. aeruginosa* PML (Uratani and Hoshino, 1989). They characterized the transport system in both strains and found that for *P. aeruginosa* PAO the affinity for sodium is 3 mM, and for *P. aeruginosa* PML the affinity for sodium is 95 mM. The affinity for leucine, however, was comparable in both strains. They cloned the *P. aeruginosa* PLM LIV-II locus, the *braB* gene, and sequenced it. The *braB* genes were identical except for one amino acid, at position 292 of the 437-residue protein, as deduced from the nucleic acid sequence. Position 292 in LIV-II from *P. aeruginosa* PAO is a threonine, and in the gene from *P. aeruginosa* PML position 292 is an alanine.

The LIV-III locus of *P. aeruginosa* PAO has been cloned and characterized. Its properties are somewhat different from the LIV-III locus of *S. typhimurium* (Hoshino *et al.*, 1991). The *braZ* gene codes for a protein that appears to be structurally similar to *Pseudomonas* LIV-II transporter, BraB. The presence of the LIV-III transporter was not expected, since kinetic analysis on *braC braB* mutants did not show appreciable leucine transport. The specificity for leucine is, however, markedly less than for isoleucine and valine (150 vs. 12 μM), in contrast to the LIV-III locus from *S. typhimurium*, which has an equivalent affinity for all three branched chain amino acids.

B. Genetic Analysis of Peptide Transport in Enteric Bacteria

1. Oligopeptide Transport in *Salmonella typhimurium*

The Opp system transports peptides from two to five amino acid residues with little residue specificity (Higgins, 1984). Opp$^-$ strains are easily selectable on the basis of resistance to the toxic analog triornithine. In the characterization of this system, mutants mapped over a 6-kb region, giving the first indication that the Opp system may be a multicomponent transport system (Higgins *et al.*, 1983). Genetic analysis revealed four genes, *oppA*, *oppB*, *oppC*, and *oppD* (Hogarth and Higgins, 1983). The locus mapped at minute 34, between *trp* and *galU*. A periplasmic protein was identified as the product of the *oppA* gene (Hiles and Higgins, 1986). This protein, M_r 52,000, is considerably larger than the previously identified binding proteins (for example, LivJ is M_r 37,000). The binding protein showed little substrate specificity, which is similar to the transport system, and distinct from the characteristic specificity of many other binding proteins. Despite the lack of specificity for the side chains of the peptide, the binding affinity for peptides in general is quite high, at 1 μM. Sequence and biochemical analysis of the cloned genes revealed five genes were required for transport and were transcribed as a single operon: *oppABCDF* (Hiles *et al.*, 1987). The fifth gene coded for a second ATP binding component, in addition to the *oppD* gene. The transport system is constitutively expressed, suggesting it has a "housekeeping" function in the cell. Higgins and co-workers investigated this possibility and showed that the Opp system is uniquely capable of transporting and recycling the cell wall peptides. In contrast to this system, the Opp system from *E. coli* is regulated by leucine (see below). The physiological implications of this difference are not yet clear.

Since the presence of the fifth gene of the *opp* operon was not anticipated, Gallagher *et al.* (1989) further characterized the gene product, OppF. They identified the *oppF* gene product as a peripheral inner membrane protein that was exposed to the cytoplasmic side, as would be expected for its presumed role as the ATP binding component of the Opp transport complex. One unexpected result was the finding of an intracellular location of the OppF protein in a strain that was deleted for the other transport genes. In a strain that was missing the chromosomal *oppABCDF* locus, a plasmid-borne *oppF* gene expressed a gene product that was still associated with the inner membrane, in contrast to that found for the *malK* gene product in the absence of *malE* (Shuman and Silhavy, 1981). Whether this indicates that OppF can associate with another transport complex or if it has an intrinsic affinity for the inner membrane is not clear.

2. The Oligopeptide Transport System from *Escherichia coli*

Short and co-workers have extensively characterized the Opp transport systems in *E. coli*. The *oppABCDF* system is linked to *trp*, as in *S. typhimurium*, but in *E. coli* this locus is at minute 27 (Andrews and Short, 1985). As a part of the characterization of peptide transport, Short showed that the outer membrane porins OmpC and OmpF are the primary means of entry of oligopeptides into the periplasmic space. The authors showed that an *ompC ompF* strain is unable to transport oligopeptides, making the strains phenotypically Opp⁻, despite the expression the *oppABCDF* and *oppE* systems. Hydrophobic peptides are predominantly transported through the outer membrane by OppF. Short was able to isolate mutants at two loci, the *trp*-linked *OppABCEF* system and the *OppE* system, which maps at minute 98.5. OppE differs biochemically in many ways from the other *E. coli* Opp system, and is thought to be a separate system, as opposed to a component of the *trp*-linked system. *oppE* is not involved in the transcriptional regulation of *oppABCDF* (Andrews and Short, 1986). It is not known if this is related to the Tpp system of *S. typhimurium* (see below). *oppABCDF* in *E. coli* is regulated by leucine and by anaerobiosis (Andrews *et al.*, 1986). Leucine control of *oppABCDF* has been shown to be dependent on the Lrp global control system, described in Section IV,E,4.

3. Tripeptide Transport in *Salmonella typhimurium*

The tripeptide transport permease (TppB) was identified by resistance to alafosfalin, a peptide antibiotic preferentially taken up by this system (Gibson *et al.*, 1984). The *tppB* locus was cloned, as was a gene coding for a positive regulator, *tppA*. Subsequent analysis revealed that *tppA* is *ompR*, the response regulator that is activated by the osmolarity sensor/kinase EnvZ (Gibson *et al.*, 1987). The OmpR/EnvZ system regulates expression of the outer membrane porins OmpC and OmpF as a function of medium osmolarity. Expression of *tppB* is dependent on *ompR*, but is apparently not affected by changes in the osmolarity of the medium or by certain *envZ* mutants. These results suggest that regulation of *tppB* by *ompR* is not dependent on OmpR being phosphorylated, or that a basal level of phosphorylation is sufficient for maximal expression of *tppB*.

The study of *lac* fusions to *tppB* showed that expression of *tppB* is also regulated by leucine and by anaerobiosis (Jamieson and Higgins, 1984). The anaerobic control is independent of the Fnr anaerobic-sensitive global regulator. It is not known if the anaerobic regulation of *tppB* is related to the anaerobic control of the *E. coli oppABCDF* operon. Jamieson and Higgins (1986) constructed a *tppB–lacZ* gene fusion by random insertion of a Mu phage derivative carrying the *lacZ* gene. This strain was then mutagenized by random mutagenesis with the transposon Tn5. Identification of anaerobic gene expression was made through examination of colony behavior. Colonies were incubated on MacConkey indicator plates aerobically, but due to the poor oxygen diffusion of a bacterial colony the middle of each colony is anaerobic. Anaerobic induction of the *tppB–lacZ* fusion in the middle of a colony turns the center of the colony red, giving the colony what the authors refer to as a "fish eye" appearance. On the basis of this screen, they identified two genes involved in the anaerobic induction of the tripeptide transport gene, *tppR* and *oxrC*. *tppR*::Tn5 results in one or more auxotrophies in addition to inactivating the anaerobic induction of *tppB*. *oxrC*::Tn5 is a pleiotropic mutation, causing the loss of anaerobic induction of several enzymes involved in anaerobic metabolism, such as formate hydrogen lyase, hydrogenases 1 and 3, in addition to *tppB* and the tripeptide peptidase gene, *pepT*. This pleiotropic regulator was shown to be independent of the *oxrA/fnr* regulatory system.

4. Dipeptide Transport in *Escherichia coli*

Chemotaxis in *E. coli* is mediated by four membrane proteins, Tar, Tsr, Trg, and Tap. The Tap protein was identified as the signal transducer toward dipeptides. Chemotaxis was shown to be signaled by Tap when it interacts with the dipeptide binding protein, DppA (Manson *et al.*, 1986).

In the initial study, chemotaxis was shown to be dependent on the expression of this binding protein, but a Dpp⁻ mutant that still expressed the DppA binding protein still showed chemotaxis toward dipeptides, indicating that the membrane complex for dipeptide transport is independent from Tap and not necessary for chemotaxis. The chemotactic response was shown to be strongest with peptides that contain leucine. The *dppA* gene was cloned and characterized by Olson and co-workers (1991). Among the binding protein genes, the deduced amino acid sequence of *dppA* shows the greatest similarity to the oligopeptide binding protein. The gene is not expressed in all strains of *E. coli*, but in strains where it is expressed it is a major protein. The gene is expressed strongly when cells are grown in minimal medium, but is repressed by casamino acids. The repressing agent in casamino acids has not yet been determined, although it does not appear to be a single amino acid.

Manson, Higgins, and co-workers have further characterized the *dpp* locus and the Dpp system in peptide transport. Abouhamad *et al.* (1991) have extensively characterized the genetics of the *dpp* system. They have identified a number of dipeptide substrates that can be used as selective agents. Selections have been performed in strains lacking the Opp and Tpp systems since dipeptides are transported by these systems to a significant extent. The authors were able to select Dpp⁻ strains by using bacilycin (a dipeptide antibiotic secreted by some species of *Bacillus*), by using valine-containing dipeptides, and by using bialaphos, a tripeptide antibiotic from *Streptomyces hygroscopicus*. Selections for bacilysin and bialaphos resistance were quite effective. Valine-containing dipepetides are toxic in *E. coli* and in *ilvG* strains of *S. typhimurium*. The authors noted that valine-dipeptide sensitivity did not select for *dpp* mutants unless a high concentration of leucine was also present, to prevent the transport of free valine that is released through branched chain amino acid exchange (Quay *et al.*, 1977). Without the addition of leucine, selection for valine-dipeptide resistance results in valine resistance due to mutations in genes other than for the dipeptide transport system. The *dpp* loci of both *E. coli* and *S. typhimurium* were mapped to mintue 79.2 and near minute 78, respectively. The system was found to be unregulated by casamino acids in strains from both species, in contrast to the regulation seen by Olson *et al.* (1991). It appears that along with expression of *dppA,* regulation also shows some variability among strains.

The Manson laboratory is continuing the characterization of the *dpp* locus (M. Manson, personal communication). The next significant open reading frame, *dppB,* is highly homologous to the corresponding gene from the oligopeptide transport system, *oppB*. One unusual aspect of the gene structure of the *dpp* operon is the long intercistronic region between *dppA* and *dppB,* 313 bases. No other significant open reading frames can be determined in the intervening region in either direction, and no other promoter has been determined from sequence analysis.

C. Proline Transport in Nutrition and Osmoprotection

The complexity of the proline transport systems in *E. coli* and *S. typhimurium* parallels the many roles of proline in the cell. Transported proline can be used in protein biosynthesis, as a carbon and nitrogen source, and as an osmoprotectant in situations of osmotic stress (Wood, 1988; Czonka, 1989). These transport needs are met by several different transport systems. The main proline transporter, PP-I, is coded for by the *putP* gene, described in Section II,B,2. This system is essential for the cell to use proline as a carbon or nitrogen source, which has been attributed to the direct coupling of PutP to PutA, the proline/glutamate decarboxylase (Maloy, 1987). A futile cycle of proline synthesis and catabolism is also avoided by different sensitivities of the regulator of the proline biosysnthesis genes and PutA activity (Ekena and Maloy, 1990).

The ProP transport system, or PP-II, is less well characterized. It is induced by osmotic pressure, and has a moderate affinity for proline, 0.3 mM, and a higher affinity for glycine betaine, 0.044 mM. A trans-acting factor has been identified genetically in *E. coli, proQ,* which maps at minute 40.4. Insertions at this locus result in the loss of expression of the proline porter II gene, *proP* (Milner and Wood, 1989).

The ProU transport system of *S. typhimurium* and *E. coli* is an ATP-dependent, multicomponent transport system referred to as the proline porter III system (Higgins *et al.,* 1987; Stirling *et al.,* 1989). This system is also induced by increases in osmotic pressure. Induction by osmotic pressure has been demonstrated *in vitro,* through the use of S-30 extracts with added potassium and from extracts prepared from strains grown in high osmolarity medium (Jovanovich *et al.,* 1989). The system can transport proline to give a high concentration gradient in the cell, but the affinity for proline is rather low, at 0.2 mM. The affinity of the ProU system for another osmoprotectant, N,N,N-trimethylglycine or glycine betaine, is significantly higher, 1 μM.

A factor involved in the regulation of *proU* expression has been identified, *osmZ* (Higgins *et al.,* 1988). The authors identified *osmZ* as a pleiotropic locus affecting the expression of a number of genes. Subsequent analysis has shown this locus is allelic to several other mutations, all of which are located in the gene coding for the bacterial histone H1 or H-NS (Hulton *et al.,* 1990). These results show that changes in osmotic pressure resulting in induction of the *proU* locus are mediated, at least in part, by changes in the topology of the chromosome.

D. Aromatic Amino Acid Transport

Five systems have been well characterized for the transport of aromatic amino acids in *E. coli:* a general aromatic amino acid permease coded for by the *aroP* gene, the tyrosine permease coded for by the *tyrP* gene, the phenylalanine permease, *pheP,* and the high-affinity tryptophan permease, coded for by the *mtr* gene, and a fifth locus, *tnaB,* codes for the transporter of tryptophan when tryptophan is catabolized to make indole, pyruvate, and ammonia. This fifth permease is cocistronic with the catabolic enzyme tryptophanase, coded for by *tnaA.* Other permeases have been identified, but are not well characterized at present: the *aroT* locus transports aromatic amino acids, glycine, and alanine, and the *azaT* locus transports phenylalanine (Pittard, 1987).

The aromatic amino acid permeases form a unique class of transporters because of their extensively interrelated regulation, both within the transport systems and between the transport and biosynthetic genes. In other systems where regulation has been investigated, the regulation of transport function is generally separate from that of biosynthesis, such as has been seen for the branched chain amino acids and for arginine (Quay *et al.,* 1975b; Celis, 1977). The regulatory interrelationships between the biosynthetic and transport systems are described in Table III.

1. AroP: The General Aromatic Acid Permease of *Escherichia coli*

AroP transports tyrosine, phenylalanine, and tryptophan (Ames, 1964; Brown, 1970), and its biosynthesis is repressed by the presence of any of the three amino acids (Pittard, 1987). The *aroP* gene has been cloned and characterized (Chye *et al.,* 1986; Honoré and Cole, 1990). Its deduced amino acid sequence is highly hydrophobic, which is consistent with its altered mobility by sodium dodecyl sulfate-polyacrylamide gel electrophoresis (SDS-PAGE) analysis, where it runs with a molecular mass 12 kDa smaller than the predicted 49.6 kDa. It is predicted to form an integral membrane protein with 12 membrane spanning regions. Curiously, the most significant homology it shares with other permeases is with the arginine and histidine transporters from *Saccharomyces cerevisiae* (Honoré and Cole, 1990).

Chye and Pittard (1987) have characterized the regulation of transcription of *aroP.* They characterized *aroP* as a member of the TyrR regulon, a family of genes regulated by the TyrR regulator, which includes genes for the biosynthesis and transport of several aromatic amino acids (see Table IV). Repression by each of the aromatic amino acids was shown to be mediated by TyrR. Regulation was further characterized by the isolation

TABLE III

Regulation of Aromatic Amino Acid Biosynthesis and Transport

	Regulation					
	TyrR			TrpR	Attenuation control by	
Genes	Tyr[a]	Phe	Trp	Trp	Trp	Phe
Tyrosine biosynthesis						
aroF tyrA	Rep[b]					
TyrB	Rep					
Phenylalanine biosynthesis						
aroG	Rep	Rep				
pheA						Negative
Tryptophan biosynthesis						
aroH				Rep		
trpEDCBA				Rep	Negative	
Transporters						
aroP	Rep	Rep	Rep			
tyrP	Rep	Ind				
pheP	----------------------------None-------------					
mtr	Ind	Ind	(Ind)[c]	Rep		
tnaB					Positive	

[a] Tyr, Phe, and Trp are ligands to TyrR.
[b] Rep, Repression; Ind, induction.
[c] Only in the absence of TrpR.

of mutants derepressed for transcription from the *aroP* promoter. These regulatory mutations mapped within sequence elements homologous to previously identified Tyr R boxes from *aroF* and *tyrP* (see below). Any of the aromatic amino acids can act as a corepressor of TyrR, inhibiting transcription. Repression is mediated by TyrR because the Tyr R boxes are located downstream from the promoter, presumably acting by blocking transcription.

2. TyrP: The Tyrosine-Specific Permease

The TyrP tyrosine-specific transporter and the gene that codes for it have been extensively characterized. Wookey *et al.* (1984) cloned the *tyrP* gene and characterized TyrP-mediated transport. They further characterized the previous findings that the transporter is sensitive to uncouplers of the pmf, suggesting that this is the energy source for the TyrP transporter (Whipp, 1977). Sequence analysis of the structural gene for *tyrP* indicates TyrP is a membrane protein (Wookey and Pittard, 1988). As with other membrane proteins, its apparent molecular weight is underestimated by SDS-PAGE analysis, running at just over one-half of its deduced molecular weight of 43,000.

The regulation of *tyrP* was shown to be dependent on TyrR. Repression occurred when tyrosine was present, but when phenylalanine was present, expression of *tyrP* was increased (Whipp and Pittard, 1977). This regulation was further characterized by Kasian and Pittard (1984). They studied the regulation of a *tyrP–lacZ* fusion and characterized several regulatory mutants. Several classes of mutants were isolated: mutants that were no longer repressed by tyrosine but were still induced by phenylalanine, mutants unaffected by either phenylalanine or tyrosine, and those in which tyrosine did not repress *tyrP*, but did antagonize the stimulation by phenylalanine when supplemented together. Sequence analysis of the *tyrP* promoter was done by Kasian *et al.* (1986). They found that aromatic amino acid control of tyrP was mediated at the major, downstream promoter, p_1. Two Tyr R boxes were identified by sequence analysis, one of which overlaps the -35 region of the promoter. The Tyr R boxes of *tyrP* are organized in a pair, separated by a single adenine residue. This arrangement is observed in many (but not all) other TyrR-regulated promoters (*aroF, aroP, aroLM,* and others; see Pittard, 1987, for review). Two mutants that were no longer repressible by tyrosine, but could still be induced by phenylalanine, were examined further. The mutations were sequenced and shown to be in the downstream Tyr R box that overlaps the -35 region of p_1.

Andrews and Pittard (1991) have shown that the activation by phenylalanine is the result of the binding of TyrR to a single Tyr R box, whereas tyrosine-mediated repression results from the cooperative binding of two TyrR proteins to both Tyr R boxes. In this and other TyrR-regulated promoters, activation and repression is mediated by different cooperative arrangements of TyrR. Activation could be enhanced by altering the spatial arrangement of the activating Tyr R box relative to the promoter, but this would mean a loss of repression control by TyrR. As discussed by Sarsero *et al.* (1991b) and Andrews and Pittard (1991), the arrangement of the Tyr R boxes represents a compromise between the requirements of TyrR to function in both induction and repression.

3. PheP: The Phenylalanine Permease

Phenylalanine can be transported by the PheP permease, which is a single subunit transporter that is highly homologous to the AroP permease (Pi *et al.,* 1991). PheP shares 59.6% amino acid sequence identity with AroP, and the hydropathy plots of the two proteins are largely superimposable. Both proteins have distinct, well-defined hydrophobic regions that are indicative of multiple membrane spanning regions. The authors note that the other members of the aromatic transporter family do not share this similarity, except for a stretch of aromatic amino acids along one side of one of the putative membrane spanning reigons, which is conserved in TyrP. The affinity of PheP for phenylalanine is somewhat lower than the affinity of the general aromatic acid permease. The K_m for phenylalanine is 2 μM for PheP and 0.5 μM for AroP.

The *pheP* gene is a unique member of the aromatic amino acid transport family with respect to regulation. Expression of *pheP* is unaffected by mutations in *tyrR* or by the addition of any aromatic amino acid to the growth medium. PheP may represent an unregulated member of this transport family.

4. Mtr: The High-Affinity Tryptophan Permease

Recently, there has been much work on the characterization of tryptophan-specific transport systems. The *mtr* locus was identified originally as conferring methyl-tryptophan resistance, resulting from the loss of the tryptophan-specific transport system (C. Yanofsky, in Oxender, 1975). The sequence and structural analysis of Mtr, as well as TyrP and TnaB, are discussed by Sarsero *et al.* (1991a) and in Section III,B,1,c of this article.

Expression of the *mtr* gene is highly regulated. Heatwole and Somerville (1991) identified two Tyr R boxes and a TrpR (tryptophan repressor) consensus sequence. The arrangement of the Tyr R boxes in *mtr* identified by these authors is different from that of other systems regulated by a pair of Tyr R boxes. In *mtr* the Tyr R boxes are separated by eight nucleotides, whereas many other Tyr R box pairs are separated by a single base, usually an adenosine residue. Whether this affects regulation, the cooperative interactions of TyrR, or if the upstream site is even functional, is not yet clear.

Regulation of *mtr* has been further characterized by Sarsero *et al.* (1991b). They have shown that *mtr* is moderately repressed by TrpR in the presence of tryptophan and is induced by TyrR in the presence of phenylalanine or tyrosine. Induction by addition of tyrosine or phenylalanine was not present in a *tyrR* mutant. Induction was also not found in a strain that expressed TyrR but the (downstream) putative Tyr R box was altered by site-directed mutagenesis. Induction was increased in a strain that carried the *tyrR* gene on a plasmid. Regulation by tryptophan was also characterized. A strain carrying a mutation in *trpR* was no longer repressed by the addition of tryptophan. In fact, in a strain that carried a functional *tyrR* gene and an inactivating mutation in *trpR*, addition of tryptophan induced expression of *mtr*. A *trpR* binding site, described as either a pair of imperfect palindromes or a single imperfect palindrome (Klig *et al.*, 1988; Staacke *et al.*, 1990), was altered by site-directed mutagenesis. The mutations altered residues critical to either model. Expression of this mutated *mtr* promoter was not repressed by tryptophan in a strain that expressed TrpR, implicating the role of TrpR in the regulation of the *mtr* gene.

5. TnaB: The Catabolic Tryptophan Permease

The nucleotide sequence of the *tnaB* gene was reported recently (Sarsero *et al.*, 1991b). In this report the deduced amino acid sequence was shown to be atypical from other transporters. As discussed in Section III,B,1, the hydrophobicity character of TnaB is shared by Mtr and TyrP, suggesting this may form a different class of transporters.

Regulation of the tryptophanase/low-affinity tryptophan transporter operon is unique. The *tnaAB* operon is subject to catabolite repression and is induced by tryptophan (Bilezikian *et al.*, 1967; Botsford and DeMosse, 1971). Induction by tryptophan is mediated by Rho-dependent transcription attenuation. The attenuator requires translation of a putative 24-residue leader peptide that codes for a single tryptophan in the twelfth codon (Steward and Yanofsky, 1985). Similar to the *livJ* attenuator (Landick, 1984, and Section II,A,2), its leader sequence does not code for a factor-independent terminator. Instead, it codes for a long sequence with little secondary structure potential, and therefore may function as a Rho factor binding site. Mutants isolated that constitutively expressed tryptophanase were isolated by Steward and Yanofsky (1985). In general, these mutants were confined to a small region corresponding to the carboxyl-terminal region of the leader peptide that the authors identified as being similar to BoxA, a sequence previously identified as important to the mechanism of N-mediated antitermination of phage λ (Friedman and Olson, 1983). It is therefore conceivable that changes in the translation of the single Trp codon result in changes in the interaction of factors responsible for the BoxA-mediated antitermination.

The role of Rho factor was investigated by Stewart *et al.* (1986). In this study, the authors demonstrate that Rho factor can terminate transcription *in vitro* and *in vivo*. Evidence that the leader peptide is in fact critical to this regulation by tryptophan was described by Stewart and Yanofsky (1986). They showed that altering the initiating methionine codon (AUG) to an isoleucine codon (AUC) reduced the uninduced expression from the *tna* promoter and significantly reduced the induction by tryptophan.

Gollnick and Yanofsky (1990) carried out studies designed to address aspects related to the mechanism of the *tna* attenuator. The role of the single Trp codon of the leader peptide was investigated by altering it to a

UAG stop codon, a UGA stop codon, or a CGG arginine codon. The leader region was fused to the *lacZ* gene and transferred to the chromosome in single copy. Basal level (uninduced) expression from the constructs was reduced from 5- to 10-fold, and none of the constructs was induced by the addition of tryptophan to the medium. Suppressor tRNAs that code for amino acids other than tryptophan allow the restoration of the basal level of expression, to varying extents. A tryptophan-coding suppressor restored the basal level of expression and partially restored the induction by tryptophan. Other experiments by Gollnick and Yanofsky further demonstrated the role of tRNA^{Trp} in the induction of *tnaAB*. The tryptophanase attenuator provides a contrast to the transcription attenuation of the tryptophan amino acid biosynthetic operon both in its mechanism and in its physiological role.

6. Physiological Studies on Tryptophan Transport

Characterization of the roles played by each of the tryptophan permeases in contributing to the cellular pool of tryptophan has been studied by C. Yanofsky *et al.* (1991). Utilizing a *lacZ* gene fusion to the *tnaA* promoter and leader region (see above), whose induction is proportional to increasing tryptophan concentration, they developed an *in vivo* assay for measuring the tryptophan pool in the cell. This gene fusion was used to evaluate the contribution AroP, TnaB, and Mtr made to the tryptophan pool under various circumstances. They found that full induction of the *tnaA'–'lacZ* fusion required high levels of tryptophan, even in rich media. When they deleted the *tnaA* gene, coding for tryptophanase, lower levels of tryptophan were required for induction. Addition of 1-methyl tryptophan, a substrate that induces tryptophan-regulated systems, but is poorly catabolized by tryptophanase, was substantially more effective than tryptophan in stimulating β-galactosidase activity. These two results suggest that the high level of tryptophan required for induction is a result of its rapid catabolism by tryptophanase. High levels of induction of *tnaAB* requires the *tnaB* gene, suggesting that TnaB tranports with a high V_{max}, relative to Mtr. TnaB is essential for growth on tryptophan as a sole carbon source, again suggesting this transporter is unique in its capacity to transport sufficient levels of tryptophan. When 1-methyl tryptophan was used as an inducer, Mtr was shown to be an effective inducer in addition of *tnaA'–lacZ*. AroP was effective for transport of tryptophan only in minimal medium, and then only in the absenece of tryosine and phenylalanine. As discussed above, it plays a role in the cell in overcoming severe shortages of aromatic amino acids, due to its high affinity and its repressibility by any of the three amino acids.

E. Global Regulatory Systems and Amino Acid Transport

1. Nitrogen Control of Amino Acid Transport Systems

Nitrogen control is one of the fundamental systems of transcription control in the cell. Low levels of nitrogen induce a number of genes that can incorporate ammonia into glutamate more effectively. This regulation is mediated by the Ntr system, a two-component system consisting of Nr_I, the response regulator, and Nr_{II}, the histidine kinase. Aspects of this regulatory system, and of many other members of the two-component signal transduction systems that respond to various stimuli, have been extensively reviewed by Stock et al. (1989). The Ntr system has been shown to regulate two transport systems, the S. typhimurium histidine/ LAO and the E. coli glutamate high-affinity transport systems.

For the histidine system, Higgins and Ames (1982) studied the two promoters for the histidine/LAO system: argTr, the promoter for the LAO binding protein gene, and dhuA, the promoter for the HisJ binding protein gene and the common membrane components. They showed that in a mutation in ntrB, the gene for the Nr_{II} histidine kinase, expression of hisJ was somewhat derepressed, suggesting that Nr_I may act as a repressor of dhuA. Stern et al. (1984) isolated fusions of these promoters to lacZ, and studied their regulation. The regulation by nitrogen of argTr was strongly influenced by carbon source. Expression from argTr was significantly repressed by a nitrogen-rich glucose medium, but was somewhat dere-pressed in a nitrogen-rich succinate medium. A minor trend in the opposite direction seen in nitrogen-poor medium gave an induction ratio by nitrogen of 16 in glucose and 1.3 in succinate. The site of action by Nr_I on the argTr promoter was described by Schmitz et al. (1988). In this work, the authors describe the effect of the Nr_I/Nr_{II} system on the argTr promoter and the role of σ factor specific for genes under the Ntr-mediated nitrogen control of NtrA, or $σ^{54}$. They also show that NtrA can be replaced by NifA, another σ factor that controls a subset of nitrogen-regulated promoters, and can substitute for NtrA in a number of cases. In summary, the pro-moter for the LAO binding protein is recognized by the $σ^{54}$ factor that is activated directly by the Nri response regulator. Regulation of dhuA, the histidine binding protein promoter, appears to be regulated by Nr_I as a classical repressor.

The glutamine transport system has been used as a model for transcrip-tion activation by Nr_I/Nr_{II}. The transport system was cloned and charac-terized by Nohnó et al. (1986), and was introduced in Section III of this article. Overexpression of this system is apparently lethal. Claverie-Martin and Magasanik (1991) have used this system to investigate the role of integration host factor (IHF) in the mechanism of transcriptional activation by Nr_I. They showed that IHF is critical for the activation of *glnHPQ* by Nr_I. Integration host factor could stimulate transcription when the IHF was bound to a binding site that bisected the distance between the Nr_I and the RNA polymerase binding sites. If the IHF binding site is changed such that it no longer bisects this segment, IHF interferes with transcription initiation, presumably by misdirecting the transcriptional activator away from the promoter.

2. Regulation of Amino Acid Transport Systems by the Cpx Signal Transduction System

Studies on the formation of the membrane potential identified several mutants that appeared to have altered several proton-mediated transport systems. One of these mutants was in the *eup* locus, giving rise to the energy-uncoupled phenotype (Plate and Suit, 1981). In this mutant, transport of several amino acids, as well as that of thio-methyl-galactoside (a melibiose analog), was impaired. It was suggested that the *eup* locus affected formation of the membrane potential. Subsequent analysis showed that the membrane potential was unaltered in an *eup* mutant (Hitchens *et al.*, 1982), which in turn led to the theory that the product of the *eup* locus was required for the utilization of the membrane potential in transport, perhaps being a common component that complexes with many tranporters. Several lines of evidence ruled that possibility out, including the observation that the *eup* gene product is insufficiently expressed to function stochiometrically in transport (Albin *et al.*, 1984).

The *eup* locus has since been shown to be an allele of *cpxA*, a sensor/kinase factor of a two-component signal transduction system, which has also been identified as *ecfB* and *ssd* (Rainwater and Silverman, 1990). *cpxA* has been characterized as a gene involved in expression of the plasmid-borne F pili genes (McEwen and Silverman, 1980). It was also identified as required for expression of the *ilvBN* operon, coding for AHAS1 (Sutton *et al.*, 1982). The *ssd* phenotype was characterized by derepression of the L-serine deaminase (Newman and Walker, 1982). CpxA is the sensor for the ArcA response regulator, which is also affected by the ArcB sensor/kinase-regulator (Iuchi *et al.*, 1989a). CpxA and ArcB act through ArcA, but stimulation of the respective kinase activities involves other factors, since mutations in these genes can be distinguished phenotypically (Iuchi *et al.*, 1989b).

It was first recognized that mutations in *cpxA* affect regulation of many transport systems, and it was recognized subsequently that many of these systems are sodium cotransporters rather than proton cotransporters. As such, it is possible that *cpxA* mutations have an effect on the sodium chemical potential instead of the proton chemical potential. The mechnaism of other response regulators, however, suggests that the effect of CpxA on transport is a result of reduced transcription of the transport genes. The nature of the signal sensed by CpxA is unclear. It is thought to be stimulated by anaerobiosis.

3. Peptide Transport and Signal Transduction: Regulation of Sporulation and Competence in *Bacillus subtilis*

When *B. subtilis* reaches steady state in glucose minimal medium, several gene cascades are initiated that result in changes in cell morphology, competence, and antibotic production, ending ultimately in formation of the endospore. Mutations in this process have generally been grouped by the developmental stage at which they are blocked, identified as stage 0 (initiation), stage I, stage II, etc. Within these stages, mutations have been subdivided as well. For the stage 0 mutants, several complementation groups have been characterized (*spo0A, 0B, 0E, 0F, 0H, 0J, 0K*). The *spo0A* gene is homologous to the response regulator components such as OmpR, CheY, and NtrC (Stock *et al.*, 1989). Spo0A is phosphorylated by the SpoIIJ transmembrane kinase. Spo0F is similar to Spo0A, except that it consists of only the conserved domain, and may function by competing with Spo0A for the kinase. Missense mutations in *spo0A* can suppress mutations in *spo0B, 0E, 0F, 0J*, and *0K*, suggesting that it functions downstream of these genes. Spo0H is a σ factor, participating directly in transcription initiation.

Competence, or the ability of *B. subtilis* to take up DNA, is mediated by the *com* genes, some of which are regulated by the *spo0A* gene. Competence is also regulated by an additional two-component system. The ComA response regulator is activated by the ComP and ComB transmembrane histidine kinases. A deletion of either ComP or ComA leads to a loss of competence. In addition to *spo0A*, several other *spo0* genes are involved. These include *spo0H* and *spo0K*.

Both sporulation and competence are regulated by the *spo0K* locus. This locus was recently cloned, sequenced, and found to encode an oligopeptide transport system (Rudner *et al.*, 1991; Perego *et al.*, 1991). The signal for sporulation is at least partly pheromone dependent, and this pheromone is protease sensitive. Therefore, the role of the *spo0K* locus in signal transduction may involve the transport of a peptide that is a ligand for an intracellular receptor, such as one of the other *spo0* genes.

Sporulation and peptide transport are not affected by a mutation in the *spo0KE/oppF* component. This may be the first example of a transport complex that still functions in transport when it is missing one of its membrane components. Such mutations have been shown to inactivate transport in the Opp and Liv systems. These results raise several possibilities. The first possibility is that another ATP binding component, possibly even Spo0KD/OppD, can complement the *spo0KE/oppF* deletion by taking the place of Spo0KE/OppF in the transport complex. Another possibility is that a trimeric complex, without Spo0KE, is still a functional transporter.

In contrast to the induction of sporulation, compentence is directly affected by the Spo0KE/OppF protein. A *spo0KE* mutant is Com⁻, suggesting that this component is specifically involved in the signal for compentence. Rudner *et al.* (1991) suggest that Spo0KE is a part of the signal transduction pathway, perhaps involved directly in signaling ComB to phosphorylate ComA. The possible mechanisms of action by the Spo0K transport complex are depicted schematically in Fig. 5, and have been discussed by Grossman *et al.* (1991).

These results have implications on our current understanding of the mechanism of transport. While it has seemed that solute transport required hydrolysis of two molecules of ATP per molecule of substrate, presumably one by each ATP binding component, this work raises the possibility that the two components may serve different functions. In the case of Spo0KE, this component would mediate the transduction of the signal for compentence. It is possible that the peptide transport function may not be the signal for sporulation, but that this sporulation signal could be transmitted directly by Spo0KD.

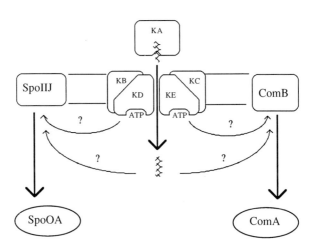

FIG. 5 Schematic representation of the possible modes of signal transduction of peptide transport to the sporulation and competence responses. Peptide transport can be signaled either by the transporter directly (upper pathway), or through the action of the peptide(s) such as in the lower pathway. Either pathway may be mediated by other factors.

4. Lrp: The Leucine-Responsive Regulatory Protein of *Escherichia coli*

a. The Central Role of Leucine in Metabolic Regulation   A central role of leucine in the cell has been observed for many years. One of the very early reports of a regulatory role for leucine was by Pardee and Prestidge (1955). They examined the role of several energy sources and metabolites as inducers in a search of enzyme systems that could be contrasted with the contemporary studies on the regulation of β-galactosidase. They identified leucine as an inducer of threonine and serine deaminase(s). This activity was not yet separable into distinct enzymes. Isenberg and New man (1974) characterized the induction of L-serine deaminase by leucine and glycine and indicated its enzyme activity was distinct from that of threonine deaminase, showing that these were two enzymes. Fraser and Newman (1975) described the role of leucine in the conversion of threonine to glycine by threonine deaminase. They observed that labeled carbon atoms of threonine will appear in purines and in glycine when leucine is added, suggesting that leucine stimulates this conversion. Newman *et al.* (1976) indentified L-threonine dehydrogenase, Tdh, as the catalyst for this conversion, and further demonstrated that this enzyme activity was specifically induced by leucine. The authors suggest the induction of enzymes involved in the degradation of serine and threonine by leucine may mean that leucine serves as a signal for the catabolism of amino acids.

Another suggestion that leucine may be a central regulatory molecule arose from studies of the regulation of branched chain amino acid transport. As opposed to the multivalent regulation of branched chain amino acid biosynthesis, LIV transport was regulated solely by leucine (and independently by leucyl-tRNA^{Leu}, through an attenuation mechanism described in Section II,C). Quay and Oxender (1976) discussed the implications of univalent regulation of transport. The regulation of LIV-I contrasted with what was a generally accepted tenet of prokaryotic regulation, that a system that affected more than one amino acid would be regulated by each amino acid with which the system interacted. The absence of isoleucine and valine regulation of LIV-I transport suggested another regulatory model. Apparently, leucine was critical to the regulation of the metabolism of the cell, and as such, leucine regulation is of special importance. The careful regulation of branched chain amino acid transport is consistent with this global role of leucine in regulation. Quay (1976) further described the role of leucine in the regulation of many other systems.

These studies are summarized by Oxender *et al.* (1980c). In 1980, the list included 10–12 enzymes or families that were sensitive to the leucine level in the cell. These included activities involved in the transport of branched chain amino acids, glycine, and D-alanine, the metabolism of proline, serine, threonine, S-adenylmethionine, and several tRNAs. The regulation by leucine was clearly not limited to genes directly concerned with leucine metabolism.

The role of leucine in regulation has been observed in related enterobacteria as well. Leucine has been identified as a specific inducer of the 56-kDa extracellular protease in *Serratia marcescens*. *Serratia marcescens* is an enterobacterial species related to the more commonly studied species, *E. coli* and *S. typhimurium*. Regulation of the extracellular protease has been studied for two reasons. One, as a secreted enzyme it has received interest from the biotechnology community as a model of protein secretion (Nakahama *et al.*, 1986). Second, *Serratia* is a pathogen involved in some eye diseases, and the severity of these infections has been shown to be a direct result of the level of the extracellular protease secreted (Kamata *et al.*, 1985a,b). For *Serratia*, extracellular leucine appears to signal the availability of an extracellular protein source that can be degraded into peptides for transport and intracellular catabolism. Loriya and co-workers (1977) were interested in investigating the induction of the protease. It had been previously shown that it was induced by casamino acids and some protein nutrients such as gelatin and albumin. Further analysis showed that this induction was specifically mediated by leucine. There is at present no molecular characterization of this regulation.

b. Lrp: A Newly Recognized Global Control System in Escherichia coli

A study of the regulation of several systems sensitive to leucine revealed some surprising similarities. Genetic studies of the *E. coli oppABCDF* oligopeptide transport system (discussed in Section IV,B,2) resulted in the cloning and sequencing of the leucine-dependent inducer of this system, which was named *oppI* (Austin *et al.*, 1989). The genetic locus was mapped to minute 19.5, the same general region identified by Anderson *et al.* (1976) as the regulator for the LIV-I system, *livR*.

Biochemical studies of the *ilvIH* promoter showed that a protein, IHB, bound the promoter in the absence of leucine, but could not bind when leucine was present (Ricca *et al.*, 1989). The amino acid sequence of the IHB protein were compared with *oppI* open reading frame as deduced by Short and co-workers, and was shown to be identical. The gene coding for this protein was cloned and further characterized (Platko *et al.*, 1990). The gene was shown to map near minute 19, as determined by genetic transduction and identification of the gene on the physical map of the *E. coli* chromosome (Kohara *et al.*, 1987). This gene was identical to *ihb*, as was indicated by the primary sequence of IHB.

Coincident with the cloning of *oppI* and *ihb*, the locus conferring leucine regulation of serine deaminase was identified by Lin *et al.* (1990). This locus, named by the authors as *rbl* (regulated by leucine), was identified by a Tn*10*-derived vector for random insertional mutagenesis. They characterized this locus as being responsible for the leucine-dependent regulation of serine deaminase and threonine dehydrogenase, thus addressing the molecular basis of the leucine regulation seen in these enzymes by Pardee and Prestidge in 1955. The authors also showed this locus repressed serine biosynthetic activity and demonstrated that this locus was allelic to the *ihb* locus, in collaboration with Calvo and co-workers. Aronson *et al.* (1989) also demonstrated that the effect of leucine on threonine dehydrogenase was at the level of transcriptional control on the *tdh–kbl* operon. It was clear at this point that the regulatory protein common to all of these systems should be renamed to reflect better its pleiotropic nature. It was collectively agreed that the gene should be renamed *lrp*, the gene that codes for the leucine-responsive regulatory protein, Lrp (Platko *et al.*, 1990; Lin *et al.*, 1990).

Work described in Section II,C concerns two mutations affecting the trans-acting regulation of branched chain amino acid transport, and that mapped to minute 19.5 on the *E. coli* chromosome: *livR* and *lstR*. The colocalization of *livR* with *lrp* was examined (S. A. Haney *et al.* (1992). The results show that a *livR* strain contains mutations in the *lrp* gene, and that these mutations in *lrp* account for all the observed phenotypes seen in the regulation of *livJ*, *livK*, and *ilvIH*. Mutants selected as having altered regulation of *ilvIH* were shown to have altered regulation of *liv* promoters as well. Thus, *livR* is allelic to *lrp* and has been renamed to reflect this fact.

These studies have established that many genes regulated by leucine (the leucine stimulon) are regulated by a common regulator, Lrp, and are therefore defined as members of the Lrp regulon (see Neidhardt, 1987, for a discussion of these terms). The complete identification of the genes that are included in this Lrp-regulated subset of the leucine stimulon is still in progress. At present, it is clear that the regulon includes 10–15 members, and this number is likely to increase. The Lrp regulon also includes genes not regulated by leucine. These genes are affected by changes in the expression of *lrp* independent of the effect of leucine (see below). This makes the leucine stimulon and the Lrp regulon overlapping regulatory systems, rather than one being a subset of the other.

c. Mehanism of Lrp   The best characterized Lrp-mediated leucine-regulated promoter at present is that for the Aceto-hydroxyacid synthase (AHAS) III gene, *ilvIH* (Ricca *et al.*, 1989; Platko *et al.*, 1990). Studies by Calvo and co-workers have shown that *in vitro* Lrp binds to the promoter in the absence of leucine as determined by gel retardation assays. The presence of leucine causes Lrp to lose its affinity for specific binding sites on the promoter. The loss of binding of Lrp to the *ilvIH* promoter *in vitro* has been correlated with the loss of activation of *ilvIH* expression *in vivo*. This work has helped establish a model for the action of Lrp at other promoters.

Other promoters have been characterized genetically as regulated by leucine through Lrp, but in an apparently different manner. In the case of the genes for the high-affinity branched chain amino acid transport systems, LIV-I and LS, an *lrp*::Tn*10* insertion results in constitutive expression, which is the *opposite* effect seen for the addition of leucine to a wild-type strain, indicating that for the promoters of the *liv* operons, Lrp is involved in the repression by leucine rather than in the loss of activation by leucine, as is the case for the *ilvIH* operon. It is clear that concomitant with the pleiotropic role of Lrp in the cell, there are also several modes for enacting this regulation.

d. Regulation of Two Transport Systems by Lrp   What is known about the regulation of the two transport systems currently characterized as part of the Lrp regulon, the high-affinity branched chain amino acid transporter, LIV-I, and the oligopeptide transport system, Opp, is compatible with what is known about the effects of leucine and peptide transport in *E. coli*. Leucine represses the LIV-I and LS transport systems and stimulates expression of the oligopeptide transport system. The repression of LIV/ LS transport and derepression of Opp transport by leucine in *E. coli* can be explained by examining the effects each has on the cell physiology in different circumstances.

Leucine is rare in the free-living environment (Savageau, 1983), and would require a high-affinity transport system to scavenge what leucine is available, and to do this the cell should be capable of derepressing the LIV/LS transport systems. However, leucine can also be quite toxic to the cell, and unregulated high-affinity transport can cause the cell to accumulate growth-inhibitory levels of leucine (Quay *et al.*, 1977). Bacteria in the host intestine would benefit from limiting the entry of free amino acids when peptides are available. These would be better able to balance leucine transport with the other amino acids. Therefore, repression of the branched chain amino acid transport systems, when high concentrations of amino acids are present, functions to prevent toxic complications of high levels of branched chain amino acids.

Since much of the amino acid transport by the host is at the level of small peptides (Gardner and Wood, 1989), *E. coli* would compete for nutrition best at this level, and as such derepression of peptide transport would benefit the cell in the host. Thus, induction of the oligopeptide transport system in a nutritionally rich environment would best benefit the bacterium. However, for a bacterium in the free-living environment, it would be detrimental to transport all peptides since many of them are toxic, including peptide antibiotics from other microorganisms such as members of the bacillus family (Katz and Demain, 1977; Kleinkauf and von Dohren, 1987). In this case, it seems to be of greatest benefit to restrict peptide transport in favor of the high specificity of the amino acid transport systems, which can precisely determine the structure of the liganded substrate.

e. Further Roles of Lrp in Regulation The role of Lrp in the regulation of cellular systems in *E. coli* is still in the early stages of characterization. Much attention is focused on the role of Lrp in the regulation of the intermediary metabolism of amino acids. Genes regulated by the addition of leucine shift much of the intermediary metabolism to catabolism of amino acids obtained from transport of oligopeptides, as shown in Table IV.

Some genes have been identified that are regulated by Lrp, but are not regulated by leucine. In a study using two-dimensional gel electrophoresis to observe changes in gene expression in the cell by added leucine in the

TABLE IV
Genes Regulated by Exogenous Leucine

Genes	Gene products/functions	Regulation on addition of leucine	Effect of leucine regulation on cell
livJ, livKHMGF	LIV-I, LS transport systems	Repression	Prevent imbalances of branched chain amino acids
oppABCDF	Oligopeptide transport system	Activation	Mediate primary source of amino acids in the host
ilvIH	AHAS III, isoleucine, valine biosynthesis	Repression	Amino acids provided by transport
serA	Phosphoglycerate synthase	Repression	Serine provided by transport
sdaA	Serine deaminase	Activation	Convert excess serine to pyruvate, energy stores
tdh-kbl	Threonine catabolism	Activation	Convert excess threonine to glycine for use in energy and one-carbon units
avtA[a]	Transaminase C	Repression	Block shunt of pyruvate to alanine (provided by host); prevent synthesis of α-amino butyrate (toxic) from α-keto-butyrate

[a] A member of the leucine stimulon, although not yet characterized as a member of the Lrp regulon (Whalen and Berg, 1982, 1984).

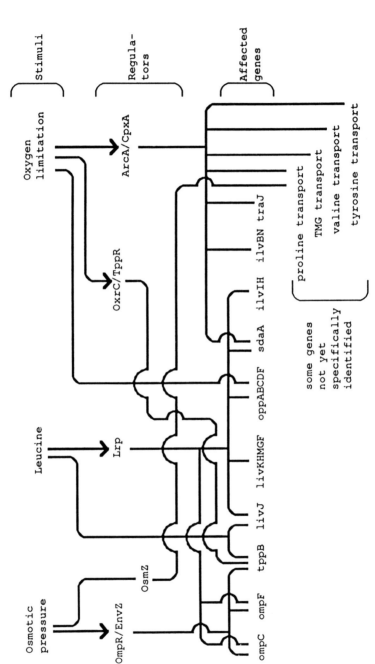

FIG. 6 Interrelationships between global regulatory systems. The regulatory system mediating the indicated response is indicated if known, except for the leucine regulation of *livJ* by attenuation, which is not indicated. Modes of regulation are discussed in text. All examples are in *E. coli* except *tppB*, which was characterized in *S. typhimurium*.

presence and absence of the *lrp* gene product, B. Ernsting and R. G. Matthews (personal communication) have identified many genes that are regulated by Lrp independent of leucine. These include *ompF*, which is positively regulated by Lrp, and *ompC*, which is negatively regulated by Lrp. Leucine-independent regulation of genes by Lrp represents an additional type of control by Lrp. An additional question of regulation of Lrp concerns the role of alanine. Several systems have identified regulation by alanine in addition to that of leucine. This has been seen in the regulation of oligopeptide transport, AHAS III synthesis, and other systems.

f. Interactions between Lrp and Other Global Regulatory Systems   A complete understanding of the leucine regulon will clearly require a description of the interrelationships between it and other global regulatory systems in the cell. As can be seen in Fig. 6, a number of genes are affected by several regulatory systems; a description of the regulation of even a few genes by different global regulatory systems becomes quite complex. These relationships suggest several questions about the role of these genes, and of their regulators. For example, why are there two distinct control systems for isoleucine–valine biosynthesis, one regulated by leucine and one by anaerobiosis? Why do both of these stimuli alter the expression of serine deaminase and oligopeptide transport? Is the leucine-independent regulation of the porin gene expression by Lrp related to the preferences of hydrophobic peptides transported by OmpF? These questions ultimately lead to questions about the roles of the regulatory systems themselves. As an example, one question concerns whether there are circumstances in which either leucine or anaerobiosis is found, since two regulatory systems exist for an overlapping set of genes. The interacting and combinatorial regulatory networks that were once thought to be prevalent only in eukaryotes are clearly important systems of regulation in bacteria as well. Addressing these and other aspects of the global regulatory networks that control amino acid metabolism in the cell will provide new insights into the coordination of metabolic pathways of *E. coli.*

Acknowledgments

Work done in this laboratory has been supported by Public Health Service Grant GM11024 to D.L.O.

We would like to thank the many people who sent reprints and preprints for consideration in this article. In addition, we would like to thank those who were willing to discuss their work in progress and who were willing to share their insights into the study of amino acid transport. In particular, we would like to thank Brian Ernsting, Alan Grossman, Mike Manson, Rowena Matthews, and Elaine Newman for their contributions.

References

Abouhamad, W. N., Manson, M., Gibson, M. M., and Higgins, C. F. (1991). *Mol. Microbiol.* **5**, 1035–1047.

Adams, M. D., and Oxender, D. L. (1989). *J. Biol. Chem.* **264**, 15739–15742.

Adams, M. D., Wagner, L. M., Graddis, T. J., Landick, R., Antonucci, T. K., Gibson, A. L., and Oxender, D. L. (1990). *J. Biol. Chem.* **265**, 11436–11443.

Adams, M. D., MaGuire, D. J., and Oxender, D. L. (1991). *J. Biol. Chem.* **266**, 6209–6214.

Albin, R., Weber, R., and Silverman, P. (1984). *J. Biol. Chem.* **261**, 4698–4705.

Alifano, P., Rivallini, F., Linauro, D., Bruni, C. B., and Carlomango, M. S. (1991). *Cell* **64**: 553–563.

Alloing, G., Trombe, M.-C., and Claverys, J. P. (1990). *Mol. Microbiol.* **4**, 633–644.

Ames, G. F.-L. (1964). *Arch. Biochem. Biophys.* **104**, 1–18.

Ames, G. F.-L. (1987). *Annu. Rev. Biochem.* **55**, 397–425.

Ames, G. F.-L., and Spudich, E. N. (1976). *Proc. Natl. Acad. Sci. U.S.A.* **73**, 1877–1881.

Anderson, J. J., and Oxender, D. L. (1977). *J. Bacteriol.* **130**, 384–392.

Anderson, J. J., and Oxender, D. L. (1978). *J. Bacteriol.* **136**, 168–174.

Anderson, J. J, Quay, S. C., and Oxender, D. L. (1976). *J. Bacteriol.* **126**, 80–90.

Andrews, A. E., Lauley, B. and Pittard, A. J. (1991). *J. Bacteriol.* **173**: 5068–5078.

Andrews, J. C., and Short, S. A. (1985). *J. Bacteriol.* **161**, 484–492.

Andrews, J. C., and Short, S. A. (1986). *J. Bacteriol.* **165**, 434–442.

Andrews, J. C., Blevins, T. C., and Short, S. A. (1986). *J. Bacteriol.* **165**, 28–433.

Aronson, B. D., Levinthal, M., and Somerville, R. L. (1989). *J. Bacteriol.* **171**, 5503–5511.

Austin, E. A., Andrews, J. C., and Short, S. A. (1989). *Mol. Genet. Bacteria, Phages* Abstr., p. 153.

Bell, A. W., Buckel, S. D., Groarke, J., Hope, J. N., Kingsley, D. H., and Hermodson, M. A. (1986). *J. Biol. Chem.* **261**, 7652–7658.

Beveridge, T. J. (1989). *In* "Bacteria in Nature" (J. S. Poindexter and E. R. Leadbetter, eds.), vol. 3, pp. 1–27. Plenum, New York.

Bilezikian, J., Kaempfer, R., and Magasanik, B. (1967). *J. Mol. Biol.* **27**, 495–506.

Botsford, J. L., and DeMoss, R. D. (1971). *J. Bacteriol.* **105**, 303–312.

Brass, J. M., Higgins, C. P., Foley, M., Rugman, P. A., Birmingham, J., and Garland, P. B. (1986). *J. Bacteriol.* **165**, 787–794.

Brown, K. D. (1970). *J. Bacteriol.* **104**, 177–188.

Cairney, J., Higgins, C. F., and Booth, I. R. (1984). *J. Bacteriol.* **160**, 22–27.

Celis, R. T. F. (1977). *J. Bacteriol.* **130**, 1244–1252.

Celis, R. T. F. (1984). *Eur. J. Biochem.* **145**, 403–422.

Celis, R. T. F. (1990). *J. Biol. Chem.* **265**, 1787–1783.

Chye, M. L., and Pittard, J. (1987). *J. Bacteriol.* **169**, 386–393.

Chye, M. L., Guest, J. R., and Pittard, J. (1986). *J. Bacteriol.* **167**, 749–753.

Claverie-Martin, F., and Magasanik, B. (1991). *Proc. Natl. Acad. Sci. U.S.A.* **88**, 1631–1635.

Crielaard, W., Dreissen, A. J. M., Molenaar, D., Hellingwerf, K. J., and Konings, W. N. (1988). *J. Bacteriol.* **170**, 1820–1824.

Czonka, L. (1989). *Microbiol. Rev.* **53**, 121–147.

Davidson, A. L., and Nikaido, H. (1991). *J. Biol. Chem.* **266**, 8946–8951.

Dila, D. K., and Maloy, S. R. (1986). *J. Bacteriol.* **168**, 590–594.

Dimroth, P. (1987). *Microbiol. Rev.* **51**, 320–340.

Dimroth, P. (1990). *Res. Microbiol.* **141**, 332–336.

Dimroth, P., and Thomer, A. C. (1986). *Biol. Chem. Hoppe-Seyler* **367**, 813–823.

Driessen, A. J. M., Hellingwerf, K. J., and Konings, W. N. (1987). *J. Biol. Chem.* **262**, 12438–12443.

Driessen, A. J. M., Ubbink-Kok, T., and Konings, W. N. (1988). *J. Bacteriol.* **170,** 817–820.

Droniuk, R., Wong, P. T. S., Wisse, G., and MacLeod, R. A. (1987). *Appl. Environ. Microbiol.* **53,** 1487–1495.

Ekena, K., and Maloy, S. (1990). *Mol. Gen. Genet.* **220,** 492–494.

Fraser, J., and Newman, E. B. (1975). *J. Bacteriol.* **122,** 810–817.

Friedman, D. I., and Olson, E. R. (1983). *Cell (Cambridge, Mass.)* **34,** 143–149.

Furlong, C. E. (1987). *In* "*Escherichia coli* and *Salmonella typhimurium:* Cellular and Molecular Biology" (F. C. Neidhardt, J. L. Ingraham, K. B. Low, B. Magasanik, M. Schaechter, and H. E. Umbarger, eds.), pp. 768–796. Am. Soc. Microbiol., Washington, D.C.

Furlong, C. E., and Weiner, J. H. (1970). *Biochem. Biophys. Res. Commun.* **38,** 1076–1083.

Gallagher, M. P., Pearce, S. R., and Higgins, C. F. (1989). *Eur. J. Biochem.* **180,** 133–141.

Gardner, M. L., and Wood, D. (1989). *Biochem. Soc. Trans.* **17,** 934–937.

Gibson, A. L., Wagner, L. M., Collins, F. S., and Oxender, D. L. (1991). *Science* **254,** 109–112.

Gibson, M. M., Price, M., and Higgins, C. F. (1984). *J. Bacteriol.* **160,** 122–130.

Gibson, M. M., Ellis, E. M., Graeme-Cook, K. A., and Higgins, C. F. (1987). *Mol. Gen. Genet.* **207,** 120–129.

Gilson, E., Alloing, G., Schmidt, T., Claverys, J.-P., Dudler, D., and Hofnung, M. (1988). *EMBO J.* **7,** 3971–3974.

Gollnick, P., and Yanofsky, C. (1990). *J. Bacteriol.* **172,** 3100–3107.

Graddis, T. J. (1990). Ph.D. Thesis, University of Michigan, Ann Arbor.

Grossman, A. D., Ireton, K., Hoff, E. F., LeDeaux, J. R., Rudner, D. Z., Magnuson, R., and Hicks, K. A. (1991). *Semin. Dev. Biol.* **2:** 31–36.

Haney, S. A., Platko, J. V., Oxender, D. L., and Calvo, J. M. (1992). *J. Bacteriol.* **174,** 108–115.

Hama, H., Shimamoto, T., Tsuda, M., and Tsuchiya, T. (1987). *Biochim. Biophys. Acta* **905,** 231–239.

Heatwole, V. M., and Somerville, R. L. (1991). *J. Bacteriol.* **173,** 108–115.

Hennge, R., and Boos, W. (1983). *Biochim. Biophys. Acta* **737,** 443–478.

Heyne, R. I. R., De Vrij, W., Crieland, W., and Konings, W. N. (1991). *J. Bacteriol.* **173,** 791–800.

Higgins, C. F. (1984). *In* "Microbiology—1984" (D. Schlessinger, ed.), pp. 17–20. Am. Soc. Microbiol., Washington, D.C.

Higgins, C. F., and Ames, G. F.-L. (1982). *Proc. Natl. Acad. Sci. U.S.A.* **79,** 1083–1087.

Higgins, C. F., Haag, P. D., Nikaido, K., Ardeshir, F., Garcia, G., and Ames, G. F.-L. (1982). *Nature (London)* **298,** 723–727.

Higgins, C. F., Hardie, M. M., Jamieson, D., and Powell, L. M. (1983). *J. Bacteriol.* **153,** 830–836.

Higgins, C. F., Sutherland, L., Cairney, J., and Booth, I. R. (1987). *J. Gen. Microbiol.* **133,** 305–310.

Higgins, C. F., Dorman, C. J., Stirling, D. A., Waddell, L., Booth, I. R., May, G., and Bremer, E. (1988). *Cell (Cambridge, Mass.)* **52,** 569–584.

Higgins, C. F., Hyde, S. C., Mimmack, M. M., Gileadi, U., Gill, D. R., and Gallagher, M. P. (1990). *J. Bioenerg. Biomembr.* **22,** 571–591.

Hiles, I. D., and Higgins, C. F. (1986). *Eur. J. Biochem.* **158,** 561–567.

Hiles, I. D., Gallagher, M. P., Jamieson, D., and Higgins, C. F. (1987). *J. Mol. Biol.* **195,** 125–142.

Hitchens, G. D., Kell, D. B., and Morris, J. G. (1982). *J. Gen. Microbiol.* **128,** 2207–2209.

Hobot, J. A., Carlemalm, E., Villiger, W., and Kellenberger, E. (1984). *J. Bacteriol.* **160,** 143–152.

Hobson, A., Weatherwax, R., and Ames, G. F.-L. (1984). *Proc. Natl. Acad. Sci. U.S.A.* **81**, 7333–7337.

Hogarth, B. G., and Higgins, C. F. (1983). *J. Bacteriol.* **153**, 1548–1551.

Honoré, N., and Cole, S. T. (1990). *Nucleic Acids Res.* **18**, 653.

Hoshino, T., and Kose, K. (1989). *J. Bacteriol.* **171**, 6300–6306.

Hoshino, T., and Kose, K. (1990a). *J. Bacteriol.* **172**, 5531–5539.

Hoshino, T., and Kose, K. (1990b). *J. Bacteriol.* **172**, 5540–5543.

Hoshino, T., Kose, K., and Uratani, Y. (1990). *Mol. Gen. Genet.* **220**, 461–467.

Hoshino, T., Kose-Terai, K., and Uratani, Y. (1991). *J. Bacteriol.* **173**, 1855–1861.

Hulton, C. S. J., Seirafi, A., Hinton, J. C. D., Sidebotham, J. M., Waddell, L., Pravitt, G. D., Owen-Hughes, T., Spassky, A., Buc, H., and Higgins, C. F. (1990). *Cell (Cambridge, Mass.)* **63**, 631–642.

Isenberg, S., and Newman, E. B. (1974). *J. Bacteriol.* **118**, 53–58.

Iuchi, S., Cameron, D. C., and Lin, E. C. C. (1989a). *J. Bacteriol.* **171**, 868–873.

Iuchi, S., Furlong, D., and Lin, E. C. C. (1989b). *J. Bacteriol.* **171**, 2889–2832.

Jameison, D. J., and Higgins, C. F. (1984). *J. Bacteriol.* **160**, 131–136.

Jameison, D. J., and Higgins, C. F. (1986). *J. Bacteriol.* **168**, 389–397.

Jaurin, B., Grunstrom, T., Edlund, T., and Normark, S. (1981). *Nature (London)* **290**, 221–225.

Jovanovich, S. B., Record, M. T., and Burgess, R. R. (1989). *J. Biol. Chem.* **264**, 7821–7825.

Kamata, R., Yamamoto, T., Matsumoto, K., and Maeda, H. (1985a). *Infect. Immun.* **48**(3), 747–753.

Kamata, R., Matsumoto, K., Okamura, R., Yamamoto, T., and Maeda, H. (1985b). *Ophthalmology (Rochester, Minn.)* **92**, 1452–1459.

Kerppola, R. E., Shymala, V. K., Klebba, P., and Ames, G. F.-L. (1991) *J. Biol. Chem.* **266**, 9857–9865.

Kasian, P. A., and Pittard, J. (1984). *J. Bacteriol.* **160**, 175–183.

Kasian, P. A., Davidson, B. E., and Pittard, J. (1986). *J. Bacteriol.* **167**, 556–561.

Katz, E., and Demain, A. L. (1977). *Bacteriol. Rev.* **41**, 449–474.

Kiritani, K., and Ohnishi, K. (1977). *J. Bacteriol.* **129**, 589–598.

Kleinkauf, H., and von Dohren, H. (1987). *Annu. Rev. Microbiol.* **41**, 259–289.

Klig, L. S., Carey, J., and Yanofsky, C. (1988). *J. Mol. Biol.* **202**, 769–777.

Kohara, Y., Akiyama, K., and Isono, K. (1987). *Cell (Cambridge, Mass.)* **50**, 495–508.

Kossmann, M., Wolff, C., and Manson, M. (1988). *J. Bacteriol.* **170**, 4516–4521.

Kustu, S. G., and Ames, G. F.-L. (1974). *J. Biol. Chem.* **249**, 6976–6983.

Landick, R. C. (1984). *In* "Microbiology—1984" (D. Schlessinger, ed.), pp. 71–74. Am. Soc. Microbiol., Washington, D.C.

Landick, R. C., and Yanofsky, C. (1987). *In* "*Escherichia coli* and *Salmonella typhimurium:* Cellular and Molecular Biology." (F. C. Neidhardt, J. L. Ingraham, K. B. Low, B. Magasanik, M. Schaechter, and H. E. Umbarger, eds.), pp. 1276–1301. Am. Soc. Microbiol., Washington, D.C.

Levitt, M. (1977). *Biochemistry* **17**, 4277–4285.

Lin, R.-T., D'Ari, R., and Newman, E. B. (1990). *J. Bacteriol.* **172**, 4529–4535.

Loriya, Z. K., Bryukner, B., and Egorov, N. S. (1977). *Mikrobiologiya* **46**, 440–446.

MacLeod, P. A., and MacLeod, R. A. (1986). *J. Bacteriol.* **165**, 825–830.

MacLeod, R. A., Hadley, R. G., Szalay, A. A., Vink, B., and MacLeod, P. R. (1985). *Arch. Microbiol.* **142**, 248–252.

Maloney, P. C. (1990). *Res. Microbiol.* **141**, 374–383.

Maloy, S. R. (1987). *In* "*Escherichia coli* and *Salmonella typhimurium:* Cellular and Molecular Biology" (F. C. Neidhardt, J. L. Ingraham, K. B. Low, B. Magasanik, M. Schaechter, and H. E. Umbarger, eds.), pp. 1513–1519. Am. Soc. Microbiol., Washington, D.C.

Maloy, S. R. (1990). *In* "The Bacteria" (T. A. Krulwich, ed.), Vol. 12, pp. 203–224. Academic Press, San Diego, California.

Manson, M. D. (1991). *Adv. Microbiol. Physiol.* **33** (in press).

Manson, M. D., Blank, V., Brade, G., and Higgins, C. F. (1986). *Nature (London)* **321**, 253–256.

McEwen, J., and Silverman, P. (1980). *J. Bacteriol.* **144**, 60–67.

Miller, K., and Maloy, S. (1990). *Nucleic Acids Res.* **18**, 3057.

Milner, J. L., and Wood, J. M. (1989). *J. Bacteriol.* **171**, 947–951.

Mogi, T., Yamamoto, H., Nakao, T., Yamato, I., and Anraku, Y. (1986). *Mol. Gen. Genet.* **202**, 35–41.

Motojima, K., Yamato, I., and Anraku, Y. (1978). *J. Bacteriol.* **136**, 5–9.

Murata-Matsubara, K., Ohnishi, K., and Kiritani, K. (1985). *Jpn. J. Genet.* **60**, 11–25.

Myers, R. S., and Maloy, S. R. (1988). *Mol. Microbiol.* **2**, 749–755.

Nakahama, K., Yoshimura, K., Marumoto, R., Kikuchi, M., Lee, I. S., Hase, T., and Matsubara, H. (1986). *Nucleic Acids Res.* **14**, 5843–5855.

Nazos, P. M. (1984). Ph.D. Thesis, University of Michigan, Ann Arbor.

Nazos, P. M., Mayo, M. M., Su, T.-Z., Anderson, J. J., and Oxender, D. L. (1985). *J. Bacteriol.* **163**, 1196–1202.

Neidhardt, F. C. (1987). *In* "*Escherichia coli* and *Salmonella typhimurium:* Cellular and Molecular Biology" (F. C. Neidhardt, J. L. Ingraham, K. B. Low, B. Magasanik, M. Schaechter, and H. E. Umbarger, eds.), pp. 1313–1317. Am. Soc. Microbiol., Washington, D.C.

Newman, E. B., and Walker, C. (1982). *J. Bacteriol.* **151**, 777–782.

Newman, E. B., Kapoor, V., and Potter, R. (1976). *J. Bacteriol.* **126**, 1245–1249.

Nohnó, T., Saito, T., and Hong, J.-S. (1986). *Mol. Gen. Genet.* **205**, 260–269.

Ohnishi, K., and Kiritani, K. (1983). *J. Biochem. (Tokyo)* **49**, 433–441.

Ohnishi, K., Murata, K., and Kiritani, K. (1980). *Jpn. J. Genet.* **55**, 349–359.

Ohnishi, K., Murata-Matsubara, K., and Kiritani, K. (1983). *Jpn. J. Genet.* **58**, 107–119.

Ohnishi, K., Hasegawa, A., Matsubara, K., Date, T., Okada, T., and Kiritani, K. (1988). *Jpn. J. Genet.* **63**, 343–357.

Ohnishi, K., Nakazima, A., Matsubara, K., and Kiritani, K. (1990). *J. Biochem. (Tokyo)* **107**, 202–208.

Ohsawa, M., Mogi, T., Yamamoto, H., Yamato, I., and Anraku, Y. (1988). *J. Bacteriol.* **170**, 5185–5191.

Olson, E.R., Dunyak, D. S., Jurss, L. M., and Poorman, R. A. (1991). *J. Bacteriol.* **173**, 234–244.

Oxender, D. L. (1975). *In* "Biological Transport" (H. N. Christiansen, ed.), 2nd ed., pp. 214–231. Benjamin, Reading, Massachusetts.

Oxender, D. L., and Quay, S. C. (1976). *Methods Membr. Biol.* **6**, 183–242.

Oxender, D. L., Anderson, J. J., Daniels, C. J., Landick, R., Gunsalus, R. P., Zurawski, G., Selker, E., and Yanofsky, C. (1980a). *Proc. Natl. Acad. Sci. U.S.A.* **77**, 1412–1416.

Oxender, D. L., Anderson, J. J., Daniels, C. J., Landick, R., Gunsalus, R. P., Zurawski, G., and Yanofsky, C. (1980b). *Proc. Natl. Acad. Sci. U.S.A.* **77**, 2005–2009.

Oxender, D. L., Quay, S. C., and Anderson, J. J. (1980c). *In* "Microrganisms and Nitrogen Sources" (J. W. Payne, ed.), pp. 153–169. Wiley, New York.

Oxender, D. L., MaGuire, D. X., and Adams, M. A. (1990). *In* "Protein and Pharmaceutical Engineering," (Craik, C. S., ed.) pp. 145–158. Wiley-Liss, Inc., New York.

Pardee, A. B., and Prestidge, L. S. (1955). *J. Bacteriol.* **70**, 667–674.

Payne, G. M., Spudich, E. N., and Ames, G. F.-L. (1985). *Mol. Gen. Genet.* **200**, 493–496.

Perego, M., Higgins, C. F., Pearce, S. R., Gallager, M. P., and Hoch, J. A. (1991). *Mol. Microbiol.* **5**, 173–185.

Pi, J., Wookey, P. J., and Pittard, A. J. (1991). *J. Bacteriol.* **173,** 3622–3629.
Piperno, J. R., and Oxender, D. L. (1966). *J. Biol. Chem.* **241,** 5732–5734.
Pittard, A. J. (1987). In "*Escherichia coli* and *Salmonella typhimurium:* Cellular and Molecular Biology" (F. C. Neidhardt, J. L. Ingraham, K. B. Low, B. Magasanik, M. Schaechter, and H. E. Umbarger, eds.), pp. 368–394. Am. Soc. Microbiol., Washington, D.C.
Plate, C. A., and Suit, J. L. (1981). *J. Biol. Chem.* **256,** 12974–12980.
Platko, J. V., Willins, D. A., and Calvo, J. M. (1990). *J. Bacteriol.* **172,** 4563–4570.
Prossnitz, E., Nikaido, K., Ulbrich, S. J., and Ames, G. F.-L. (1988). *J. Biol Chem.* **263,** 17917–17920.
Quay, S. C. (1976). Ph.D. Thesis, University of Michigan, Ann Arbor.
Quay, S. C., and Oxender, D. L. (1976). *J. Bacteriol.* **127,** 1225–1238.
Quay, S. C., Kline, E. L., and Oxender, D. L. (1975a). *Proc. Natl. Acad. Sci. U.S.A.* **72,** 3921–3924.
Quay, S. C., Oxender, D. L., Tsuyumu, S., and Umbarger, H. E. (1975b). *J. Bacteriol.* **122,** 994–1000.
Quay, S. C., Dick, T. E., and Oxender, D. L. (1977). *J. Bacteriol.* **129,** 1257–1265.
Quay, S. C., Lawther, R. P., Hatfield, G. W., and Oxender, D. L. (1978). *J. Bacteriol.* **134,** 683–686.
Quiocho, F. (1990). *Philos. Trans. R. Soc. London, Ser. B* **326,** 341–351.
Rahmanian, M., and Oxender, D. L. (1972). *J. Supramol. Struct.* **1,** 55–59.
Rahmanian, M., Claus, D. R., and Oxender, D. L. (1973). *J. Bacteriol.* **116,** 1258–1266.
Rainwater, S., and Silverman, P. M. (1990). *J. Bacteriol.* **172,** 2456–2461.
Ricca, E., Aker, D. A., and Calvo, J. M. (1989). *J. Bacteriol.* **171,** 1658–1664.
Rudner, D. Z., LeDeaux, J. R., Ireton, K., and Grossman, A. D. (1991). *J. Bacteriol.* **173,** 1388–1398.
Sack, J. S., Saper, M. A., and Quiocho, F. A. (1989a). *J. Mol. Biol.* **206,** 171–191.
Sack, J. S., Trakhanov, S. D., Tsigannik, I. H., and Quiocho, F. A. (1989b). *J. Mol. Biol.* **206,** 193–207.
Sarsero, J. P., Wookey, P. J., Gollnick, P., Yanofsky, C., and Pittard, A. J. (1991a). *J. Bacteriol.* **173:** 3231–3234.
Sarsero, J. P., Wookey, P. J., and Pittard, A. J. (1991b). *J. Bacteriol.* **73:** 4133–4143.
Savageau, M. A. (1983). *Am. Nat.* **122,** 732–744.
Schmitz, G., Nikaido, K., and Ames, G. F.-L. (1988). *Mol. Gen. Genet.* **215,** 107–117.
Shuman, H. A., and Silhavy, T. J. (1981). *J. Biol. Chem.* **256,** 560–562.
Shuman, H. A., Silhavy, T. J., and Bexkwith, J. R. (1980). *J. Biol. Chem.* **255,** 168–174.
Silhavy, T. J., Szmelcman, S., Boos, W., and Schwartz, M. (1975). *Proc. Natl. Acad. Sci. U.S.A.* **72,** 2120–2124.
Skulachev, V. P. (1985). *Eur. J. Biochem.* **151,** 199–208.
Speelmans, G., De Vrij, W., and Konings, W. N. (1989). *J. Bacteriol.* **171,** 3788–3795.
Speiser, D. M., and Ames, G. F.-L. (1991). *J. Bacteriol.* **173:** 1444–1451.
Staacke, E., Walter, B., Kisters-Woike, B., von Wilken-Bergmann, B., and Muller-Hill, B. (1990). *EMBO J.* **9,** 1963–1967.
Stern, M. J., Higgins, C. F., and Ames, G. F.-L. (1984). *Mol. Gen. Genet.* **195,** 219–227.
Stewart, V., and Yanofsky, C. (1985). *J. Bacteriol.* **164,** 731–740.
Stewart, V., and Yanofsky, C. (1986). *J. Bacteriol.* **167,** 383–386.
Stewart, V., Landick, R. C., and Yanofsky, C. (1986). *J. Bacteriol.* **166,** 217–223.
Stirling, D. A., Hulton, C. S. J., Waddell, L., Park, S. F., Stewart, G. S. A. B., Booth, I. R., and Higgins, C. F. (1989). *Mol. Microbiol.* **3,** 1025–1038.
Stock, A., Ninfa, A., and Stock, J. (1989). *Microbiol. Rev.* **53,** 450–490.
Sutton, A., Newman, T., McEwen, J., Silverman, P. M., and Freundlich, M. (1982). *J. Bacteriol.* **151,** 976–982.

Treptow, N. A., and Shuman, H. A. (1985). *J. Bacteriol.* **163,** 654–660.

Treptow, N. A., and Shuman, H. A. (1988). *J. Mol. Biol.* **202,** 809–922.

Uratani, Y., and Hoshino, T. (1989). *J. Biol. Chem.* **264,** 18944–18950.

Uratani, Y., Tsuchiya, T., Akamatsu, Y., and Hoshino, T. (1989). *J. Membr. Biol.* **107,** 57–62.

Urban, C., and Celis, R. T. F. (1990). *J. Biol. Chem.* **265,** 1783–1786.

von Heijne, G. (1986). *EMBO J.* **5,** 3021–3027.

Wakabayshi, K., Koyama, N., and Nosoh, Y. (1988). *Arch. Biochem. Biophys.* **262,** 19–26.

Wang, R., Seror, S. J., Blight, M., Pratt, J. M., Broome-Smith, J. K., and Holland, B. I. (1991). **212:** 441–454.

Whalen, W. A., and Berg, C. M. (1982). *J. Bacteriol.* **150,** 739–746.

Whalen, W. A., and Berg, C. M. (1984). *J. Bacteriol.* **158,** 571–574.

Whipp, M. J. (1977). Ph.D. Thesis, Univ. of Melbourne, Parkville.

Whipp, M. J., and Pittard, A. J. (1977). *J. Bacteriol.* **132,** 453–461.

White, S. H., and Jacobs, R. E. (1989). *Biochem.* **28,** 3421–3437.

White, S. H., and Jacobs, R. E. (1990). *J. Membr. Biol.* **115:** 145–158.

Wood, J. M. (1988). *J. Membr. Biol.* **106,** 183–202.

Wookey, P. J., and Pittard, A. J. (1988). *J. Bacteriol.* **170,** 4946–4949.

Wookey, P. J., Pittard, J., Forrest, S. M., and Davidson, B. E. (1984). *J. Bacteriol.* **160,** 169–174.

Yamato, I., and Anraku, Y. (1988). *J. Biol. Chem.* **263,** 16055–16057.

Yamato, I., Ohsawa, M., and Anraku, Y. (1990). *J. Biol. Chem.* **265,** 2450–2455.

Yanofsky, C., Horn, V., and Gollnick, P. (1991). *J. Bacteriol.* **173,** 6009–6017.

In and Out and Up and Down with Lac Permease

H. Ronald Kaback
Howard Hughes Medical Institute, Departments of Physiology and
Microbiology and Molecular Genetics, Molecular Biology Institute, University
of California, Los Angeles, Los Angeles, California 90024

An unsolved biochemical problem of critical importance is the mechanism of energy transduction in biological membranes at the molecular level. Over the past 20 years, it has become apparent that the immediate driving force for a wide range of seemingly unrelated phenomena such as solute accumulation against a concentration gradient (i.e., secondary active transport), oxidative phosphorylation, and rotation of the bacterial flagellar motor consists of bulk-phase, transmembrane electrochemical H^+ or Na^+ gradients. However, the molecular mechanism by which free energy stored in such gradients is transduced into work or into other forms of chemical energy (i.e., ATP) remains enigmatic. In order to gain insight into this general problem, studies in this laboratory have focused on the lactose (lac)[1] permease of *Escherichia coli* as a paradigm for membrane transport proteins.

Accumulation of β-galactosides against a concentration gradient in *E. coli* is carried out by the lac permease, a hydrophobic polytopic cytoplasmic membrane protein that catalyzes the coupled translocation of β-galactosides and H^+ with a stoichiometry of unity (i.e., β-galactoside/H^+ symport or cotransport) (cf. Kaback, 1983, 1986, 1990; Kaback *et al.*, 1990, for reviews). Under physiological conditions, where the H^+ electrochemical gradient across the cytoplasmic membrane ($\Delta\bar{\mu}_{H^+}$) is interior negative and/or alkaline, lac permease utilizes free energy released from downhill translocation of H^+ to drive accumulation of β-galactosides against a concentration gradient. In the absence of $\Delta\bar{\mu}_{H^+}$, the permease catalyzes the converse reaction, utilizing free energy from downhill translocation of

---

[1]Abbreviations: lac, lactose; $\Delta\bar{\mu}_{H^+}$, the proton electrochemical gradient; TDG, β, D-galactopyranosyl 1-thio-β, D-galactopyranoside; NPG, *p*-nitrophenyl-α, D-galactopyranoside; mel, melibiose.

97

β-galactosides to drive uphill translocation of H^+ and generating Δ_{H^+}, the polarity of which depends on the direction of the substrate concentration gradient.

Lac permease is encoded by the *lacY* gene, the second structural gene in the *lac* operon, and it has been cloned into a recombinant plasmid (Teather *et al.*, 1978) and sequenced (Büchel *et al.*, 1980). By combining overexpression of *lacY* with the use of a highly specific photoaffinity probe for the permease (Kaczorowski *et al.*, 1980) and reconstitution of transport activity in artificial phospholipid vesicles (I.e., proteoliposomes) (Newman and Wilson, 1980), the permease has been solubilized from the membrane, purified to homogeneity (Newman *et al.*, 1981; Foster *et al.*, 1982; Viitanen *et al.*, 1986; cf. Wright *et al.*, 1986, in addition), and shown to catalyze all the translocation reactions typical of the β-galactoside transport system *in vivo* with comparable turnover numbers (Matsushita *et al.*, 1983; Viitanen *et al.*, 1984). Therefore, a single gene product—the product of *lacY*—is solely responsible for all of the translocation reactions catalyzed by the β-galactoside transport system.

This article discusses selected observations with a specific membrane transport protein at the molecular level, but it should be emphasized that there are a huge number of proteins that catalyze similar transport reactions in virtually all biological membranes from archaebacteria to the mammalian central nervous system. Furthermore, it should be stated at the outset that structural information at the atomic level is particularly difficult to obtain with hydrophobic membrane proteins (Deisenhofer and Michel, 1989). Since the great majority of membrane proteins, lac permease in particular, have yet to be crystallized and it is becoming increasingly apparent that high-resolution structure is a prerequisite for mechanistic considerations, conclusions derived from some of these studies must be regarded as speculative.

I. Permease Structure

Circular dichroic measurements on purified lac permeas indicate that the protein is over 80% helical in conformation, an estimate consistent with the hydropathy profile of the permease, which suggests that approximately 70% of its 417 amino acid residues is found in hydrophobic domains with a mean length of 24 ± 4 residues (Foster *et al.*, 1983). Based on these findings, a secondary structure was proposed in which the permease is composed of a hydrophilic N terminus followed by 12 hydrophobic segments in α-helical conformation that traverse the membrane in zig-zag fashion connected by hydrophilic domains (loops) with a 17-residue C-terminal hydrophilic tail (Fig. 1). Support for the general features of the

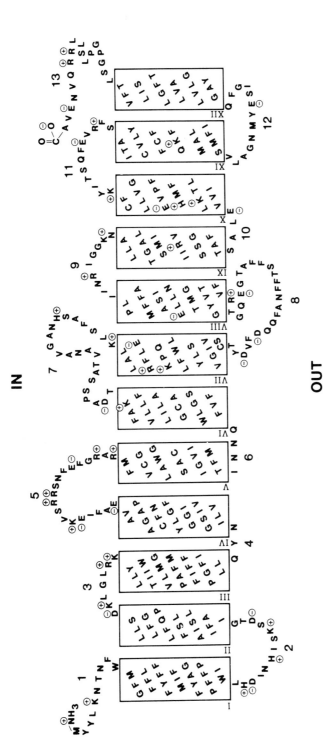

FIG. 1 Secondary structure model of lac permease based on the hydropathy profile of the protein. The single-letter amino acid code is used, and hydrophobic segments are shown in boxes as transmembrane α-helical domains connected by hydrophilic loops.

model and evidence that both the N and C termini of the permease are exposed to the cytoplasmic face of the membrane were obtained subsequently from laser Raman (Vogel et al., 1985) and Fourier transform infrared (P. D. Roepe, H. R. Kaback, and K. J. Rothschild, unpublished information) spectroscopy, immunological studies (Carrasco et al., 1982; 1984a,b; Seckler et al., 1983, 1986; Seckler and Wright, 1984; Herzlinger et al., 1984; Danho et al., 1985), limited proteolysis (Goldkorn et al., 1983; Stochaj et al., 1986), and chemical modification (Page and Rosenbusch, 1988). However, none of these approaches differentiates between the 12-helix structure and other models containing 10 (Vogel et al., 1985) or 13 (Bieseler et al., 1985) putative transmembrane helices.

Calamia and Manoil (1990) have provided elegant and exclusive support for the topological predictions of the 12-helix model by analyzing an extensive series of lac permease–alkaline phosphatase (lacY–phoA) chimeras. Alkaline phosphatase is synthesized as an inactive precursor in the cytoplasm of E. coli with a short signal sequence that directs its secretion into the periplasmic space, where it dimerizes to form active enzyme. If the signal sequence is deleted, the enzyme remains in the cytoplasm in an inactive form. When alkaline phosphatase devoid of the signal sequence is fused to the C termini of fragments of a cytoplasmic membrane protein, enzyme activity reflects the ability of the N-terminal portions of the chimeric polypeptides to translocate alkaline phosphatase to the outer surface of the membrane (Manoil and Beckwith, 1986). Alkaline phosphatase activity in cells independently expressing each of 36 lacY–phoA fusions exclusively favors the model of lac permease with 12 transmembrane domains.

In addition, it was demonstrated (Calamia and Manoil, 1990) that approximately half of a transmembrane domain is needed to translocate alkaline phosphatase to the external surface of the membrane. Thus, the alkaline phosphatase activity of fusions engineered at every third amino acid residue in putative helices III and V (Fig. 1) increases as a step function as the fusion junction procedes from the eighth to the eleventh residue of each of these transmembrane domains. Furthermore, when fusions are constructed at each amino acid residue in putative helix X of the permease, a sharp discontinuity in alkaline phosphatase activity is observed at His-322–Met-323, thereby implying that these residues are located in the middle of the membrane (M. L. Ujwal, E. Bibi, C. Manoil, and H. R. Kaback, unpublished information).

Purified lac permease reconstituted into proteoliposomes exhibits a notch or cleft (Costello et al., 1984, 1987), an observation independently documented by Li and Tooth (1987) using completely different techniques. The presence of a solvent-filled cleft in the permease may have important implications with regard to the mechanism of β-galactoside/H^+ symport, as the barrier within the permease may be thinner than the full thickness of the membrane. Therefore, the number of amino acid residues in the protein directly involved in translocation may be fewer than required for lactose and H^+ to traverse the entire thickness of the membrane.

II. Membrane Insertion and Stability of Lac Permease

Stochaj *et al.* (Stochaj and Ehring, 1987; Stochaj *et al.*, 1988) demonstrated that sequences within the N-terminal 170 amino acid residues of lac permease may be important for insertion. Moreover, a truncated permease containing only the N-terminal 50 amino acid residues is inserted into the membrane, and it was proposed that this region contains an internal "start transfer" sequence resulting in the insertion of the N terminus as a "helical hairpin" (von Heijne and Blomberg, 1979; Engleman and Steitz, 1981).

With respect to the C terminus, the 17-amino acid C-terminal hydrophilic tail is not involved in insertion of the permease into the membrane, its stability, or its ability to catalyze transport. On the other hand, a three-amino acid sequence at the end of the last putative transmembrane helix (. . . VFT . . . ; see Fig. 1) is critical for stability and hence activity once the protein is inserted into the membrane (Roepe *et al.*, 1989; McKenna *et al.*, 1991). When stop codons (TAA) are placed sequentially at amino acid codons 396–401, permease truncated at residue 396 or 397 is completely defective with respect to lactose transport, while molecules truncated at residues 398, 399, 400, and 401, respectively, exhibit 15–25, 30–40, 40–45, and 70–100% of wild-type activity. As judged by pulse-chase experiments with [35S]methionine, wild-type permease or permease truncated at residue 401 is stable, while permease molecules truncated at residue 400, 399, 398, 397, or 396 are degraded at increasingly rapid rates. Finally, recent experiments (McKenna *et al.*, 1992) demonstrate that replacement of residues 397–399 with three leucine residues yields a stable, fully functional permease, while replacement with Gly-Pro-Gly yields an unstable molecule with minimal transport activity. The results indicate that the last turn of putative helix XII is important for proper folding and protection against proteolytic degradation.

The overall topology of polytopic membrane proteins like lac permease is thought to result from either the oriented insertion of the N-terminal α-helical domain followed by passive, serpentine insertion of subsequent helices (Wickner and Lodish, 1985; Rapoport, 1986; Singer *et al.*, 1987) or from the function of independent topogenic determinants dispersed throughout the molecules (Blobel, 1980; Friedlander and Blobel, 1985; Popot and Engleman, 1990; Popot and de Vitry, 1990). In order to test these alternatives, even or odd numbers of putative transmembrane domains in lac permease were deleted, and the effect of the deletions on insertion, stability, and the orientation of the C terminus with respect to the plane of the membrane was examined (Fig. 2) (Bibi *et al.*, 1991). The strategy is that deletion of odd numbers of transmembrane domains might be expected to alter the position of the C terminus relative to the plane of the membrane, while deletion of even numbers of transmembrane domains would not be expected to do so. As demonstrated, so long as the first N terminal and the last four C-terminal putative α-helical domains are retained, stable polypeptides are inserted into the membrane, even when an odd number of helical domains is deleted. Moreover, even when an odd number of helices is deleted, the C terminus remains on the cytoplasmic surface of the membrane, as judged by the activity of C-terminal lac permease–alkaline phosphatase fusions. Interestingly, although none of the deletions catalyzes lactose accumulation against a concentration gradient, permease molecules devoid of even or odd numbers of putative transmembrane helices retain a specific pathway for "downhill" lactose translocation. One construct, in particular, which is devoid of putative helices II–V, exhibits about 80% of the downhill transport activity of intact permease, suggesting that the pathway for lactose translocation may be largely contained within the last six transmembrane domains. The results indicate that relatively short C-terminal domains in the permease contain topological information sufficient for insertion in the native orientation regardless of the orientation of the N terminus. This conclusion has been confirmed and extended by Calamia and Manoil (1992), who demonstrated recently that many individual membrane-spanning domains of lac permease act as independent export signals for attached alkaline phosphatase.

III. Oligomeric State of Lac Permease

Although early evidence (see Kaback, 1989) suggested that oligomerization might be important for activity, lac permease is probably fully functional as a monomer, as demonstrated by rotational diffusion measurements with eosinylmaleimide-labeled permease or by direct observation. Using the former approach, Dornmair *et al.* (1985) utilized fluorescence anisotropy to demonstrate that purified, reconstituted lac permease exhibits the rotational diffusion constant of a 46.5-kDa particle and that the diffusion constant is not altered in the presence of $\Delta\bar{\mu}_H^+$. Regarding the latter approach, Costello *et al.* (1987) purified lac permease and cytochrome *o*, reconstituted the proteins into proteoliposomes under conditions in which both are fully functional, and examined the preparations by

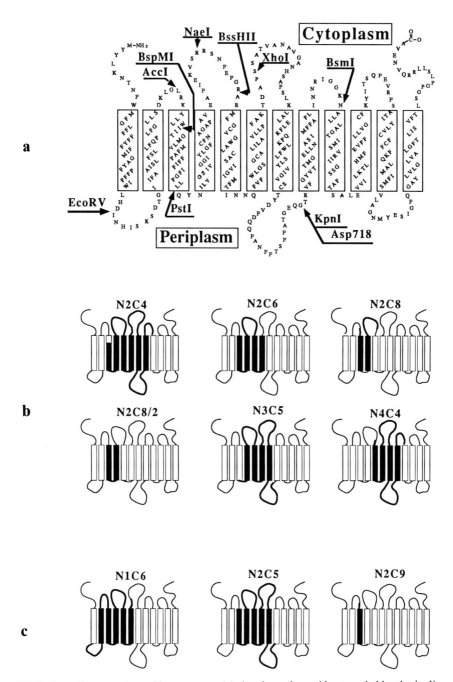

FIG. 2 Secondary structure of lac permease (a) showing polypeptides encoded by the *lacY* deletion constructs described in (b) and (c). The single-letter amino acid code is used, hydrophobic transmembrane helices are shown in boxes, and deleted regions are solid boxes and thick lines. (a) Wild-type lac permease with restriction sites in the cassette *lacY* DNA indicated. (b) Constructs in which even numbers of putative transmembrane helices were deleted. (c) Constructs in which odd numbers of putative transmembrane helices were deleted. (From Bibi *et al.*, 1991.)

freeze-fracture electron microscopy. In nonenergized proteoliposomes, each protein appears to reconstitute as a monomer based on (1) the variation of intramembrane particle density with protein concentration, (2) the ratio of particles corresponding to each protein in proteoliposomes reconstituted with a known ratio of permease to oxidase, and (3) the dimensions of the particles observed in tantalum replicas. None of the parameters is altered in the presence of $\Delta\bar{\mu}_{H^+}$. Importantly, moreover, the initial rate of $\Delta\bar{\mu}_{H^+}$-driven lactose transport in proteoliposomes varies linearly with the ratio of lac permease to phospholipid, particularly over the range at which there is statistically between zero and three molecules of permease per proteoliposome. If more than a single molecule of lac permease is required for active lactose transport, an exponential relationship would be expected, particularly at low protein to phospholipid ratios.

IV. In Vivo Expression of the LacY Gene in Two Fragments Leads to Functional Lactose Permease

Bibi and Kaback (1990) restricted the lacY gene into two approximately equal-sized fragments that were subcloned individually or together under separate lac operator/promoters. Under these conditions, lac permease is expressed in two portions: (1) the N terminus, the first six putative transmembrane helices, and most of putative loop 7, and (2) the last six putative transmembrane helices and the C terminus (Fig. 3). Cells expressing both fragments transport lactose at about 30% the rate of cells expressing intact permease to a comparable steady state level of accumulation. In contrast, cells expressing either half of the permease independently do not transport lactose. [35S]Methionine labeling and immunoblotting experiments demonstrate that intact permease is completely absent from the membrane of cells expressing lacY fragments either individually or together. Thus, transport activity must result from an association between independently synthesized pieces of lac permease. When the gene fragments are expressed individually, the N-terminal portion of the permease is observed sporadically and the C-terminal portion is not observed. When the gene fragments are expressed together, polypeptides identified as the N- and C-terminal moieties of the permease are found in the membrane. The results are consistent with the conclusion that the N- or C-terminal halves of lac permease are proteolyzed when synthesized independently and that association between the two complementing polypeptides leads to a more stable, catalytically active complex. More recent experiments demonstrate that coexpression of independently cloned fragments of the lacY gene encoding N_2 and C_8 (Wrubel et al., 1990), N_1 and C_{11}, or $N_{6.5}$ and $C_{5.5}$

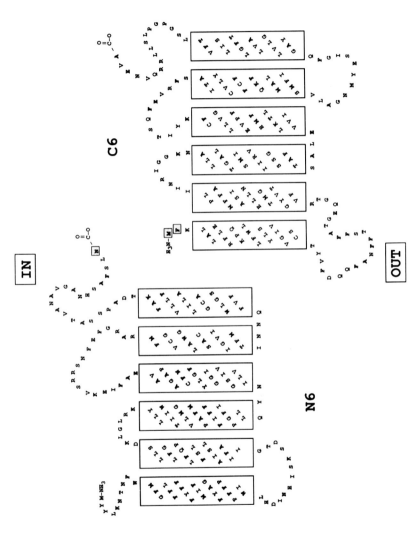

FIG. 3 Secondary structure of lac permease showing portions encoded by the *lacY* gene fragments described in the text. The single-letter amino acid code is used, hydrophobic transmembrane helices are shown in boxes, and new amino acid residues inserted are shown in shaded boxes. (From Bibi and Kaback, 1990.)

(E. McKenna, D. Hardy, E. Bibi, and H. R. Kaback, unpublished information) results in the formation of stable molecules in the membrane that interact to form functional permease, while expression of the fragments by themselves yields polypeptides that are relatively unstable and exhibit no transport activity. In addition, the demonstration that polypeptides corresponding to N_1 and C_{11} form a relative stable, functional complex argues against the hypothesis that the N terminus of the permease inserts into the membrane as a helical hairpin.

V. Functional Complementation of Deletion Mutants

As discussed above, convincing evidence has been presented indicating that lac permease is functional as a monomer. Nonetheless, recent experiments (Bibi and Kaback, 1991) demonstrate that certain paired in-frame deletion mutants expressed from separate but compatible plasmids exhibit functional reconstitution when transformed into an appropriate host cell. The nomenclature of the constructs (N_xC_y) describes the number of putative transmembrane helices retained in the N-terminal (N_x) and C-terminal (C_y) portions of the permease before and after the deletion (Fig. 4). Although cells expressing each deletion individually are unable to catalyze active lactose accumulation, cells simultaneously expressing N_2C_8/N_8C_2 catalyze lactose transport 60% as well as cells expressing wild-type permease, while cells expressing N_2C_6/N_8C_2 or N_4C_6/N_8C_2 exhibit diminished but significant transport activity. On the other hand, cells expressing N_4C_6/N_6C_4 or N_2C_6/N_6C_2 may exhibit only marginal activity, and the combinations N_4C_4/N_8C_2, N_2C_4/N_8C_2, and N_6C_4/N_8C_2 exhibit no activity whatsoever. Moreover, the following pairs of missense mutations or single amino acid deletions also exhibit no activity: P28S/H322K, E325C/H322K, E325C/K319L, E325C/R302L, or ΔW38/ΔH322. Importantly, it has been shown that complementation between N_2C_8/N_8C_2 does not occur at the DNA level, but probably at the protein level. Therefore, the ability to complement functionally is apparently a specific property of pairs of permease molecules containing relatively large deletions separated by at least two transmembrane hydrophobic domains, and it is not observed with pairs of missense mutations or point deletions.

FIG. 4 Secondary structure of lac permease (a) showing polypeptides encoded by the *lacY* deletion mutants described in (b). The single-letter amino acid code is used, hydrophobic transmembrane helices are shown in boxes, and deleted regions are shown in bold. (a) Wild-type lac permease with restriction sites in the cassette *lacY* DNA indicated; (b) constructs in which putative transmembrane helices were deleted. A cassette *lacY* gene

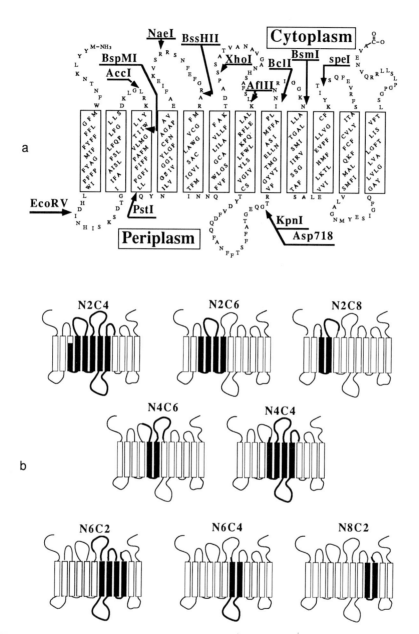

(EMBL-X56095) containing the *lac* promoter/operator was cloned into plasmid pT7-5 and used for all *lacY* gene manipulations. The cassette gene contains unique restriction sites in each segment of the gene encoding a putative loop (i.e., approximately every 100 bp). In most of the constructs, the plasmid was digested with appropriate restriction enzymes to remove the desired segment of the gene, treated with DNA polymerase (Klenow fragment), and ligated to itself. In one construct (N_2C_6), a linker was synthesized and inserted, and the cohesive ends were ligated. The fusion junctions of each construct were sequenced using the dideoxyoligonucleotide method. (From Bibi and Kaback, 1991.)

One possible interpretation of the results is that specific interactions occur between transmembrane helices in wild-type permease and that disruption of these interactions by deletion leaves a "potential gap" in the structure that can be filled by interaction with another molecule containing the deleted segment. For instance, perhaps putative helix VIII has a high affinity for helix IX (see Fig. 4; note that the loop between putative transmembrane helices VIII and IX is relatively short) and poor affinity for helix XI. By this means, a permease molecule lacking helices IX and X (eg., N_8C_2) might "accept" these helices from a "donor" molecule lacking helices III and IV (eg., N_2C_8) and/or vice versa. However, *E. coli* T184 transformed with plasmids encoding P28S and N_8C_2 or H322K and N_2C_8 as potential donor/acceptor pairs do not complement functionally. Similarly, ΔH322 does not exhibit functional complementation with N_2C_8 nor does ΔW78 complement functionally with N_8C_2. Thus, the simplistic explanation does not appear to be the case.

As discussed above, certain independently cloned fragments of the *lacY* gene yield functionally active complexes when expressed together. Since these observations may be related to the functional complementation phenomenon, it is suggested that permease mutants containing missense mutations or point deletions, like the intact wild-type molecule, form relatively compact structures that are unable to form intermolecular complexes. On the other hand, molecules containing deletions in certain hydrophobic transmembrane domains (e.g., N_2C_8 and N_8C_2) may be in a more "relaxed" state and therefore able to interact to form functional dimers. In any case, the ability of a given set of deletion mutants to complement functionally is clearly dependent on the specific nature of the deleted domains (i.e., N_2C_8/N_8C_2, N_2C_6/N_8C_2, and N_4C_6/N_8C_2 exhibit significant transport activity, N_4C_6/N_6C_4 and N_2C_6/N_6C_2 exhibit marginal activity, and N_4C_4/N_8C_2, N_2C_4/N_8C_2, and N_6C_4/N_8C_2 exhibit no activity).

VI. Insertional Mutagenesis

Recently, by using a cassette version of the *lacY* gene containing a unique restriction site in the DNA encoding each loop of the permease (EMBL-X56095), two or six histidinyl residues have been inserted individually into each hydrophilic domain of the molecule (E. McKenna, D. Hardy, and H. R. Kaback, unpublished information). Remarkably, with the exception of domains 8, 9, and 10, permease with of two or six histidinyl residues inserted into each of the other hydrophilic domains catalyzes significant lactose accumulation (Fig. 5). When two or six histidinyl residues are inserted into domains 8, 9, or 10, lactose transport is dramatically decreased, although

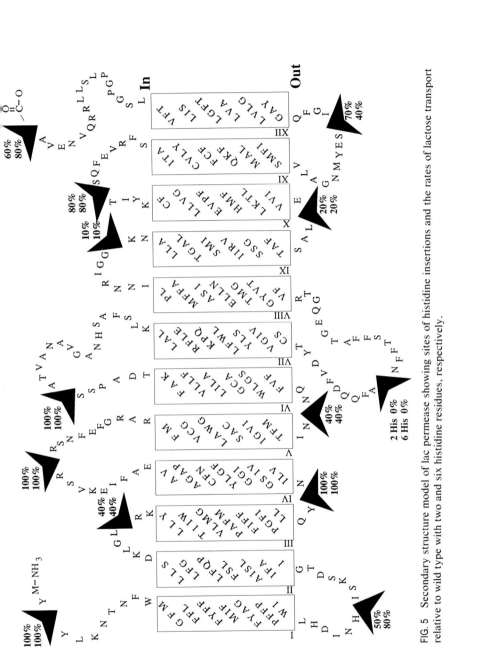

FIG. 5   Secondary structure model of lac permease showing sites of histidine insertions and the rates of lactose transport relative to wild type with two and six histidine residues, respectively.

the polypeptides are stably inserted into the membrane, as demonstrated by either ^{35}S labeling or immunoblotting. Furthermore, similar lactose transport activity is observed with each of the insertion mutant at pH 5.5 (i.e., below the pK of histidine) or at pH 7.5 (i.e., above the pK). Thus, it appears that out of all of the hydrophilic domains in the permease, only domains 8, 9, and 10 play an essential role in the activity of the permease.

VII. Use of Site-Directed Mutagenesis to Probe the Mechanism of Active Transport

Notwithstanding the importance of high-resolution structure, it has become apparent that oligonucleotide-directed site-specific mutagenesis can be used to delineate amino acid residues that play an important role in active transport (see Kaback, 1990; Roepe et al., 1990; Roepe and Kaback, 1992, for reviews). Out of the more than 150 site-directed mutants in lac permease, approximately 90% exhibit significant activity. Therefore, it is unlikely that individual amino acid replacements indiscriminately cause conformational changes in the protein.

A. Cysteinyl Residues

Fox and Kennedy (1965; Kennedy et al., 1974) demonstrated initially that lac permease is irreversibly inactivated by N-ethylmaleimide (NEM) and that protection is afforded by substrates such as β,D-galactopyranosyl 1-thio-β,D-galactopyranoside (TDG). On the basis of these findings, it was postulated (Fox and Kennedy, 1965) that a cysteinyl residue is at or near the substrate-binding site of lac permease, and Beyreuther et al. (1981) later showed that the substrate-protectable residue is Cys-148 (Fig. 6). In addition, the permease is reversibly inactivated by other sulfhydryl reagents, like p-chloromercuribenzenesulfonate, or by sulfhydryl oxidants such as diamide (Kaback and Patel, 1978) or plumbagin (Konings and Robillard, 1982), and TDG blocks inactivation by these reagents as well.

In view of the importance attributed to sulfhydryl groups in lac permease (see Kaback and Barnes, 1971; Konings and Robillard, 1982; Robillard and Konings, 1982, in addition), particularly Cys-148, site-directed mutagenesis was used initially to replace Cys-148 with glycine (Trumble et al., 1984; Viitanen et al., 1985) or serine (Neuhaus et al., 1985; Sarkar et al., 1986). Surprisingly, although Cys-148 is required for substrate protection against alkylation by NEM, it is not important for lactose/H+ symport.

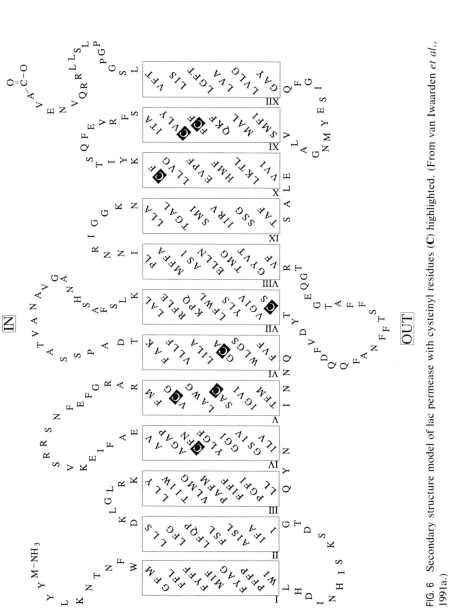

FIG. 6   Secondary structure model of lac permease with cysteinyl residues (C) highlighted. (From van Iwaarden *et al.*, 1991a.)

Subsequently, it was shown (Menick *et al.*, 1985; Kaback, 1989) that replacement of Cys-154 with glycine leads to complete loss of transport activity although the permease binds the high-affinity ligand p-nitrophenyl-α,D-galactopyranoside (NPG) normally. Moreover, replacement of Cys-154 with serine or valine yields permease with 10 or 30%, respectively, of the wild-type rate, indicating that although Cys-154 is needed for full activity, it is not mandatory. Brooker and Wilson (1986) then replaced Cys-176 or -234 with serine, and Menick *et al.* (1987a) replaced Cys-117, -333, or -353 and -355 with serine with little or no effect on activity. Therefore, out of a total of eight cysteinyl residues in the permease, only Cys-154 appears to be important for transport, but even this residue is not essential. Finally, experiments in which each of the cysteine mutants was purified and reconstituted (van Iwaarden *et al.*, 1991a) indicate that sulfhydryl–disulfide interconversion probably does not play a role in regulation of permease activity (Konings and Robillard, 1982; Robillard and Konings, 1982).

More recent studies (van Iwaarden *et al.*, 1991b) provide definitive support for the contention that cysteinyl residues in lac permease do not play a essential role in the mechanism of lac permease. When Cys-154 is replaced with valine and each of the other cysteinyl residues in lac permease is replaced with serine, "C-less" permease catalyzes active lactose transport at about 30% of the intial rate and at about 60% of the steady state level of accumulation of wild-type permease. Moreoever, active lactose transport in right-side-out vesicles containing C-less permease is not inactivated by NEM, in dramatic contrast to vesicles containing wild-type permease. In addition, the results reinforce the conclusion (van Iwaarden *et al.*, 1991a) that sulfhydryl–disulfide interconversion is probably not important for regulation of permease activity.

As stated above, a high-resolution structure is required to begin to determine the role of essential amino acid residues in the mechanism. However, it is also clear that in order to solve the mechanism, dynamic information at high resolution is required as well. In this respect, chemical labeling and spectroscopic approaches in which reactive cysteinyl residues are tagged with radioactive sulfhydryl reagents, spin labels (Altenbach *et al.*, 1989, 1990; Todd *et al.*, 1989), or fluorescent probes (Dornmair *et al.*, 1985) represent potentially powerful means for examining static and dynamic aspects of protein structure–function relationships at high resolution. A principal difficulty with the general approach, however, is the complexity resulting from the presence of multiple cysteinyl residues in most proteins, eight in the case of lac permease. Thus, in addition to the important conclusion that cysteinyl residues do not play a critical role in the mechanism of lac permease, the construction of a functional permease molecule devoid of cysteinyl residues, in analogy to permease devoid of tryptophanyl residues (Menezes *et al.*, 1990), provides the basis for an approach to the analysis of static and dynamic aspects of permease structure–function relationships. By using the *lacY* gene encoding C-less permease, for instance, it is now possible to design mutants in which an individual amino acid residue in putative hydrophilic or hydrophobic domains is replaced with a cysteinyl residue that can then be reacted specifically with either permeant or impermeant sulfhydryl reagents in right-side-out or inside-out membrane vesicles, followed by solubilization and immunoprecipitation. In addition, single cysteine mutants can be tagged with appropriately reactive electron paramagnetic resonant or fluorescent probes after solubilization and purification and studied spectroscopically after reconstitution. Finally, it should be possible to study proximity relationships between transmembrane domains by placing single cysteinyl residues in pairs of helical domains predicted to lie close to each other within the membrane. In these contexts, it is encouraging that each amino acid residue in putative transmembrane helices I, IX, and X, as well as hydrophilic domain 10 of C-less permease, has already been replaced with cysteine (see Fig. 1), and the great majority of the mutants exhibit highly significant transport activity (M. Sahin-Toth, J. Schweiger, B. Persson, and H. R. Kaback, unpublished information).

Parenthetically, it is also noteworthy that cysteine replacements for certain amino acid residues in C-less permease have yielded an interesting and unexpected result. As judged by replacements with phenylalanine, out of the 13 tyrosinyl residues in the permease, 4 appear to be important for lactose/H^+ symport, one of which is Tyr-26 in putative helix I (Roepe and Kaback, 1989). Strikingly, when Tyr-26 is replaced with cysteine, the permease exhibits wild-type activity. Clearly, therefore, it is the electronegativity of the oxygen or sulfur, respectively, in tyrosine or cysteine at position 26 that is important rather than bulk or hydrophobicity.

B. Pro Residues

Since prolyl residues are thought to play an important role in the structure and function of many proteins and because certain amino acid residues in putative helices IX and X of lac permease are important for lactose/H^+ symport and/or substrate recognition (see C. The Putative Charge Relay), Pro-327 in putative helix X was initially subjected to site-directed mutagenesis (Fig. 7) (Lolkema *et al.*, 1988). Permease with P327A catalyzes active transport in a manner indistinguishable from wild-type permease, permease with P327G exhibits about one-tenth the initial rate of wild-type permease but catalyzes lactose accumulation to the same steady state level as wild-type permease, and permease with P327L is inactive. Thus, there

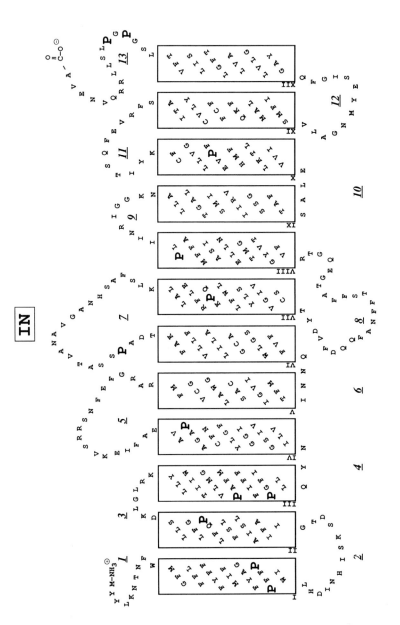

FIG. 7 Secondary structure model of lac permease showing the positions of proline residues (**P**). (From Consler *et al.*, 1991.)

is no relationship between permease activity and the helix-breaking (proline and glycine) or helix-making (alanine and leucine) properties of the residue at position 327, and the suggestion was made that it is a specific chemical property of the side chain at position 327, rather than cis/trans isomerization of Pro-327 or the presence of a kink, that is important. Subsequently, it was demonstrated (Roepe *et al.*, 1989; McKenna *et al.*, 1991) that the C-terminal tail of the permease, which contains proline at positions 403 and 405 (Fig. 7), can be deleted with no significant effect on either permease activity or the lifetime of the protein.

Recently, each remaining prolyl residue in the lac permease of *E. coli* at positions 28 (putative helix I), 31 (helix I), 61 (helix II), 89 (helix III), 97 (helix III), 123 (helix IV), 192 (putative hydrophilic region 7), 220 (helix VII), and 280 (helix VIII) (Fig. 7) was systematically replaced with glycine, alanine, or leucine (Consler *et al.*, 1991). With the exception of Pro-28, each prolyl residue can be replaced with glycine or alanine, and significant lac permease activity is retained. In contrast, when Pro-28 is replaced with glycine, alanine, or serine, lactose transport is abolished, but permease with Ser-28 binds NPG and catalyzes active transport of TDG. Replacement of Pro-28, -31, -123, -280 or -327 with leucine abolishes lactose transport, while replacement of Pro-61, -89, -97 or -220 with leucine has relatively minor effects. None of the alterations in permease activity is due to inability of the mutant proteins to insert into the membrane or to diminished lifetimes after insertion, since the concentration of each mutant permease in the membrane is comparable to that of wild-type permease as judged by immunological analyses. The results indicate that (1) Pro-28 is important for substrate binding and recognition; (2) it is primarily the hydrophobicity and/or size of the side chain at the other positions that is important for lac permease activity; and (3) neither cis/trans isomerization of prolyl residues nor the presence of kinks at these positions is important for membrane insertion, stability, or substrate translocation.

C. The Putative H$^+$ Relay

The most provocative and controversial findings from site-directed mutagenesis studies on lac permease began when the 4 histidinyl residues in the protein were mutagenized (Padan *et al.*, 1985; Püttner *et al.*, 1986, 1989). Replacement of His-35 and His-39 (putative hydrophilic domain; see Fig. 1) with arginine or replacement of His-205 (putative hydrophilic domain)

[2] Although initial experiments (Padan *et al.*, 1985) indicated that permease with Arg-205 is defective in transport, the mutant construct was found to have two additional mutations in the 5' end of the gene. When the secondary mutations are removed, Arg-205 permease exhibits normal activity (Püttner *et al.*, 1986).

with arginine[2], asparagine, or glutamine has no effect on active lactose transport, whereas replacement of His-322 (putative helix X) with Arg, Asn, Gln, or Lys causes dramatic loss of activity. Conversely, a permease mutant with a single His at position 322 exhibits properties identical to wild-type permease (Püttner and Kaback, 1988). Strikingly, H322R permease catalyzes downhill lactose influx at high substrate concentrations without concomitant H^+ translocation.

Efflux, exchange, and counterflow are useful for studying permease turnover because specific steps in the overall kinetic cycle can be delineated. Permease with arginine, asparagine, glutamine, or lysine in place of His-322 is markedly defective in all translocation reactions presumed to involve protonation or deprotonation (Fig. 8). Furthermore, the primary kinetic effect of $\Delta \bar{\mu}_{H^+}$ (i.e., a decrease in apparent K_m for lactose) is not observed. Interestingly, permeases with asparagine, glutamine, or lysine in place of His-322 catalyze downhill efflux, as well as influx, but both processes appear to occur without concomitant H^+ translocation (Püttner *et al.*, 1989).

Since His-322 may be directly involved in lactose-coupled H^+ translocation and this residue is located in putative helix X (Fig. 1), attention focused on Glu-325, which should be on the same face of helix X as His-322 and may be ion paired with this residue. In addition, structure–function studies on chymotrypsin (Blow *et al.*, 1969) and other serine proteases have led to the notion that acidic amino acid residues may function with His as components of a charge-relay system, a type of mechanism that might conceivably be related to H^+ translocation. For these reasons, Glu-325 was subjected to site-directed mutagenesis (Carrasco *et al.*, 1986, 1989). Permease with alanine, glutamine, valine, histidine, cysteine, or

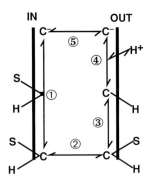

FIG. 8   Schematic representation of reactions involved in lactose efflux, exchange, and counterflow. C, Lac permease; S, substrate (lactose). The order of substrate and H^+ binding at the inner surface of the membrane is not implied.

tryptophan in place of Glu-325 catalyzes downhill influx of lactose without H^+ translocation, but does not catalyze either active transport or efflux. Remarkably, the rate of equilibrium exchange with the altered permeases is at least as great as that observed with wild-type permease. Moreover, permease mutated at position 325 catalyzes counterflow at the same rate and to the same extent as wild-type permease, but the internal concentration of $[^{14}C]$lactose is maintained for a prolonged period due to the defect in efflux. It is also noteworthy that permease mutated at position 325 catalyzes counterflow three to four times better than wild-type permease when the external lactose concentration is below the apparent K_m.

The results can be rationalized by the simple kinetic scheme shown in Fig. 8. Efflux down a concentration gradient is thought to consist of a minimum of five steps: (1) binding of substrate and H^+ on the inner surface of the membrane (order unspecified), (2) translocation of the ternary complex to the outer surface, (3) release of substrate, (4) release of H^+, and (5) return of the unloaded permease to the inner surface. Alternatively, exchange and counterflow with external lactose at saturating concentrations involve steps 1–3 only. Furthermore, release of H^+ appears to be rate limiting for the overall cycle (Viitanen et al., 1983).

All steps in the mechanism presumed to involve protonation or deprotonation appear to be blocked in the His-322 mutants. Therefore, it seems reasonable to suggest that protonation of His-322 may be involved in step 1. In contrast, replacement of Glu-325 results in a permease that is defective in all steps involving net H^+ translocation but catalyzes exchange and counterflow normally. Clearly, therefore, permease mutated at position 325 is probably blocked in step 4 (i.e., it is unable to lose H^+).

Experiments in which Glu-325 was replaced with asparatate have yielded unexpected results (P. D. Roepe, D. Mechling, L. Patel, and H. R. Kaback, unpublished information). Permease with Asp-325 is partly uncoupled and catalyzes symport about 30% as well as wild-type permease (Carrasco et al., 1989). The observation is not surprising, as the side chain containing the carboxylate is about 1.5 ÅA shorter in aspartate relative to glutamate. However, E325D permease catalyzes equilibrium exchange normally below pH 7.7, but as ambient pH is increased, exchange activity is progressively and reversibly inhibited with a midpoint at about pH 8.5, while equilibrium exchange with wild-type or E325A permease is unaffected by bulk-phase pH over a wide range. The findings indicate that translocation of the fully loaded permease does not tolerate a negative charge at position 325 and suggest that the carboxylate at position 325 may undergo protonation and deprotonation during lactose/H^+ symport. The observation that equilibrium exchange with wild-type permease is insensitive to pH over the same range is consistent with the notion that Glu-325 is strongly H bonded to His-322.

Replacement of Arg-302 with leucine, histidine, or lysine (putative helix IX; Fig. 1) yields permease with properties similar to those of H322R permease (Menick *et al.*, 1987b). In marked contrast, replacement of most of the other residues in putative helices IX and X of C-less permease has no significant effect on active lactose transport (M. Sahin-Toth and H. R. Kaback, unpublished information), thereby highlighting the specificity of Arg-302, His-322, and Glu-325 and providing further support for the contention that single amino acid changes do not indiscriminately cause conformational alterations. Furthermore, by molecular modeling of putative helices IX and X, it can be postulated that the guanidino group in Arg-302 may be sufficiently close to His-322 to participate in H bonding with the imidazole ring that, in turn, may be H bonded to the carboxylate of Glu-325 (Fig. 9). Minimally, therefore, the putative charge-relay in the permease would involve interactions between Arg-302, His-322, and Glu-325.

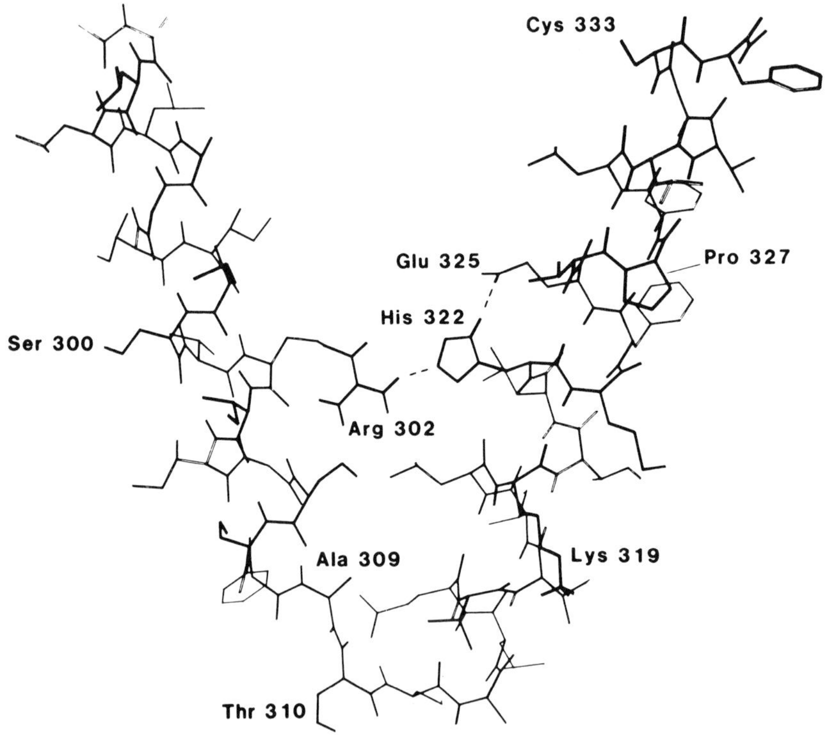

FIG. 9 Modified molecular model of putative helices IX and X in lac permease. The study was performed by Dr. Vincent Madison on an Evans-Sutherland computer on the basis of the hydropathy profile of Foster *et al.* (1983), except that Ala-309 and Thr-310 were transferred from helix IX to hydrophilic segment 10, which connects helices IX and X (see Fig. 1). See Menick *et al.* (1987b) for details.

As evidenced by binding studies with the high-affinity ligand NPG, permease mutated at position 325 binds with a K_d approximating that of wild-type permease (Carrasco et al., 1989). The finding is consistent with the observation that counterflow, a process that exhibits an apparent K_m similar to that observed for active transport, is intact in the mutants, but is in marked contrast to findings with permeases mutated at Arg-302 or His-322, which exhibit markedly decreased affinities (Menick et al., 1987b; Püttner et al., 1989). Therefore, it is tempting to speculate that the pathways for H^+ and lactose may overlap [i.e., that Arg-302 and His-322 may also be components of the substrate binding site in addition to being involved in H^+ translocation (see Collins et al., 1989; Franco et al., 1989, in addition)] and that protonation of His-322 may be required for high-affinity binding.

If Arg-302, His-322, and Glu-325 are sufficiently close to H bond and function as components of a charge-relay, the polarity, distance, and orientation between the residues should be critical (Lee et al., 1989). The importance of polarity between His-322 and Glu-325 was studied by interchanging the residues, and the modified permease is inactive in all modes of translocation. The effect of distance and/or orientation between His-322 and Glu-325 was investigated by interchanging Glu-325 with Val-326, thereby moving the carboxylate one residue around putative helix X. The resulting permease molecule is also completely inactive, and control mutations indicate that a glutamate residue at position 326 inactivates the permease. The wild-type orientation between histidine and glutamate was then restored by further mutation to introduce a histidinyl residue into position 323 or by interchanging Met-323 with His-322. The resulting permease molecules contain the wild-type histidine/glutamate orientation, but the putative histidine/glutamate ion pair is rotated about the helical axis by 100° relative to Arg-302 in putative helix IX. Both mutants are inactive with respect to all modes of translocation. The results provide support for the contention that the polarity between His-322 and Glu-325 and the geometric relationships between Arg-302, His-322, and Glu-325 are critical for permease activity. In addition, the results suggest that perturbation of the putative His-322/Glu-325 ion pair alone is insufficient to account for inactivation (i.e., Glu-322/His-325 should remain ion paired) and are consistent with the possible role of His-322 and Glu-325 as components of an H^+ relay.

Previous studies (Menick et al., 1987b) suggested that replacement of Lys-319, which should be on the same face of putative helix X as His-322 and Glu-325, with leucine has no effect on permease activity; however, this conclusion has been found to be incorrect (B. Persson, P. D. Roepe, L. Patel, J. Lee, and H. R. Kaback, unpublished information). Rather, K319L permease is defective in lactose/H^+ symport. The mutant is unable to catalyze lactose accumulation or efflux, but exhibits downhill lactose influx and catalyzes equilibrium exchange at about half the rate of wild-type permease. Unlike the Glu-325 mutants, which catalyze exchange at least as well as wild-type permease, K319L permease binds NPG poorly (see Collins *et al.*, 1989, in addition) and is defective in counterflow activity. The results suggest that Lys-319 is important for substrate recognition, as well as lactose-coupled H^+ translocation. In addition, it is noteworthy that the double mutant K319L/E325A catalyzes equilibrium exchange.

Although the notion that a type of charge-relay mechanism may be involved in lactose/H^+ symport is consistent with the findings discussed above, recent experiments question the notion that His-322 is obligatory for lactose-coupled H^+ translocation. From studies on permease mutants with tyrosine or phenylalanine in place of His-322, King and Wilson (1989a,b; 1990) conclude that sugar-dependent H^+ transport is observed, albeit with low efficiency, that melibiose efflux, in particular, remains coupled to H^+ translocation in the mutants, and that reactions involving exchange are limiting for lactose but not melibiose efflux, suggesting that slow exchange is substrate specific in the His-322 mutants. In addition, Brooker (1990) has shown that permease with valine in place of Ala-177 *and* asparagine in place of His-322 catalyzes lactose-dependent H^+ influx with a stoichiometry close to unity. Taken at face value, the results are difficult to reconcile with the contention that His-322 is obligatory for lactose-driven H^+ translocatin. However, it should be emphasized that *all* of the His-322 mutants isolated thus far are *grossly defective* with regard to sugar accumulation. Thus, whatever its precise role in the mechanism, a histidinyl residue at position 322 appears to be very important for β-galactoside accumulation against a concentration gradient.

On a broader level, H^+- and Na^+-coupled symport are conceptually and thermodynamically analogous, but it is unclear whether the two types of transport occur by the same general mechanism. Since melibiose (mel) permease uses H^+, Na^+, or Li^+ as the coupling cation depending on the sugar transported and, like lac permease, mel permease is inactivated by diethylpyrocarbonate, Pourcher *et al.* (1990) replaced each of the seven His residues in mel permease with arginine and demonstrated that His-94 alone is important for transport and binding. Therefore, mel permease and lac permease appear to require a single specific histidinyl residue for activity, although the residues are located at opposite ends of the respective molecules. Importantly, however, mel permease with H94R does not exhibit an uncoupled phenotype, and the defect observed may be due specifically to a loss in ability to bind substrate. As outlined above, certain data with lac permease are consistent with the idea that a type of H^+ relay may be operative. On the other hand, it is not obvious how the same mechanism can be directly involved in Na^+ symport unless nitrogen or oxygen atoms that can coordinate with Na^+ are present in the pathway in addition to the minimal structures necessary for a charge-relay or an "H^+-wire" of the type suggested by Onsager (1969). Despite certain marked kinetic differences between H^+- and Na^+-coupled sugar translocation, a few lines of evidence suggest that substrate-coupled Na^+ and H^+ translocation may occur by similar mechanism in mel permease: (1) NPG binding studies suggest that H^+ and Na^+ compete for a common binding site; (2) Na^+ is a competitive inhibitor of melibiose-coupled H^+ translocation; and (3) certain mutations in *melB* have been isolated and characterized that alter cation specificity.

Boyer (1988) has suggested that H_3O^+, rather than H^+, may be the symported species. Appropriately placed nitrogen or oxygen atoms in symporters like lac or mel permease could provide cation binding domains akin to those in the crown ethers or cryptates, both of which form coordination complexes with Na^+ and N_3O^+. In this context, however, some of the translocation reactions catalyzed by lac permease exhibit a significant D_2O effect (Viitanen *et al.*, 1983) which is not expected if coordination with H_3O^+ is the rate-limiting step in translocation (i.e., the mass of D_3O^+ is only 3/19 greater than H_3O^+). Furthermore, it is difficult to explain the behavior of the Glu-325 mutants with this type of model. In summary, therefore, although the contrast between H^+- and Na^+-coupled symport is of singular importance, the mechanistic relationship between the two is presently unclear.

VIII. Summary and Concluding Remarks

The lac permease, a hydrophobic polytopic protein from the membrane of *E. coli* that catalyzes the coupled translocation of β-galactosides and H^+, has been solubilized, purified, reconstituted into artificial phospholipid vesicles, and shown to be solely responsible for β-galactoside transport. The *lacY* gene that encodes the permease has been cloned and sequenced, and based on spectroscopic analyses of the purified protein and hydropathy profiling of the amino acid sequence a secondary structure model has been proposed in which the protein has 12 hydrophobic domains in α-helical configuration that traverse the membrane in zig-zag fashion connected by hydrophilic loops. Exclusive support for the 12-helix model has been obtained by analyzing a large number of lac permease–alkaline phosphatase (*lacY–phoA*) fusions. Experiments are discussed that suggest that the N terminus may be important for insertion of the permease into the membrane and that the last turn of the last helix in the permease is essential for proper folding and stability. Evidence is also presented indicating that topogenic determinants for the insertion and organization of the protein are distributed throughout the molecule. Remarkably, *lacY* gene fragments encoding certain N- and C-terminal portions of the permease produce functional complexes when the peptides are expressed independently. Although the permease appears to be functional as a monomer, certain paired in-frame deletions devoid of large transmembrane domains are able to complement functionally when expressed together in an appropriate host cell. The phenomenon seems to be specific for relatively large deletions separated by at least two putative transmembrane domains, and it is not observed with either point mutations or point deletions. Insertional mutagenesis has been used to demonstrate that most of the surface loops in the permease are not essential for function. Finally, site-directed mutagenesis is being utilized to delineate amino acid residues that are essential for activity, and it appears that relatively few residues are critically involved in the mechanism. The studies focus on the question of whether proton- and sodium-coupled transport occur by similar or different mechanisms. It has also been shown that cysteinyl residues, long thought to play a critical role in the mechanism of action of the permease, are not essential. A permease molecule devoid of cysteinyl residues has been constructed, shown to catalyze active lactose transport, and is currently being used for insertion of cysteinyl residues in strategic locations in the molecule in order to study static and dynamic aspects of permease structure and function. With current developments in molecular biology, it is now possible to manipulate proteins to an extent that was unimaginable only a few years ago. However, it has become apparent that without high-resolution structural information, which is particularly difficult to obtain with hydrophobic membrane proteins, interpretation of the results of many of the manipulations, particularly those related to the molecular mechanism of transport, will remain highly speculative.

References

Altenbach, C., Flitsch, S. L., Khorana, G., and Hubbell, W. L. (1989). *Biochemistry* **28**, 7806.

Altenbach, C., Marti, H., Khorana, H. G., and Hubbell, W. L. (1990). *Science* **248**, 1088.

Beyreuther, K., Bieseler, B., Ehring, R., and Müller-Hill, B. (1981). *In* "Methods in Protein Sequence Analysis" (M. Elzina, ed.), pp. 139–148. Humana Press, Clifton, New Jersey.

Bibi, E., and Kaback, H. R. (1990). *Proc. Natl. Acad. Sci. U.S.A.* **87**, 4325–4329.

Bibi, E., and Kaback, H. R. (1991). *Proc. Natl. Acad. Sci. U.S.A.* **89**, 1524.

Bibi, E., Verner, G., Chang, C.-Y., and Kaback, H. R. (1991). *Proc. Natl. Acad. Sci. U.S.A.* **88**, 7271.

Bieseler, B., Heinrich, P., and Beyreuther, C. (1985). *Ann. N.Y. Acad. Sci.* **456**, 309–325.

Blobel, G. (1980). *Proc. Natl. Acad. Sci. U.S.A.* **77**, 1496–1500.

Blow, D. M., Birktoft, J. J., and Hartley, B. S. (1969). *Nature (London)* **221**, 337–340.

Boyer, P. D. (1988). *Trends Biochem. Sci.* **13**, 5–7.
Brooker, R. J. (1990). *J. Biol. Chem.* **265**, 4155–4160.
Brooker, R. J., and Wilson, T. H. (1986). *J. Biol. Chem.* **261**, 11765.
Büchel, D. E., Gronenborn, B., and Mller-Hill, B. (1980). *Nature (London)* **283**, 541–545.
Calamia, J., and Manoil, C. (1990). *Proc. Natl. Acad. Sci. U.S.A.* **87**, 4937–4941.
Calamia, J., and Manoil, C. (1992). *J. Mol. Biol.,* in press.
Carrasco, N., Tahara, S. M., Patel, L., Goldkorn, T., and Kaback, H. R. (1982). *Proc. Natl. Acad. Sci. U.S.A.* **79**, 6894–6898.
Carrasco, N., Herzlinger, D., Mitchell, R., DeChiara, S., Danho, W., Gabriel, T. F., and Kaback, H. R. (1984a). *Proc. Natl. Acad. Sci. U.S.A.* **81**, 4672–4676.
Carrasco, N., Viitanen, P., Herzlinger, D., and Kaback, H. R. (1984b). *Biochemistry* **23**, 3681–3687.
Carrasco, N., Antes, L. M., Poonian, M. S., and Kaback, H. R. (1986). *Biochemistry* **25**, 4486–4488.
Carrasco, N., Püttner, I. B., Antes, L. M., Lee, J. A., Larigan, J. D., Lolkema, J. S., and Kaback, H. R. (1989). *Biochemistry* **28**, 2533–2539.
Collins, J. C., Permuth, S. F., and Brooker, R. J. (1989). *J. Biol. Chem.* **264**, 14698–14703.
Consler, T. G., Tsolas, O., and Kaback, H. R. (1991). *Biochemistry* **30**, 1291–1298.
Costello, M. J., Viitanen, P., Carrasco, N., Foster, D. L., and Kaback, H. R. (1984). *J. Biol. Chem.* **259**, 15570–15586.
Costello, M. J., Escaig, J., Matsushita, K., Viitanen, P. V., Menick, D. R., and Kaback, H. R. (1987). *J. Biol. Chem.* **262**, 17072–17082.
Danho, W., Makofske, R., Humeic, F., Gabriel, T. F., Carrasco, N., and Kaback, H. R. (1985). *Pept.: Struct. Funct., Proc. Am. Pept. Symp. 1985, 9th,* p. 59.
Deisenhofer, J., and Michel, H. (1989). *Science* **245**, 1463–1473.
Dornmair, K., Corin, A. S., Wright, J. K., and Jähnig, F. (1985). *EMBO J.* **4**, 3633–3638.
Engleman, D M., and Steitz, T. A. (1981). *Cell (Cambridge, Mass.)* **23**, 411–422.
Foster, D. L., Garcia, M. L., Newman, M. J., Patel, L., and Kaback, H. R. (1982). *Biochemistry* **21**, 5634–5638.
Foster, D. L., Boublik, M., and Kaback, H. R. (1983). *J. Biol. Chem.* **258**, 31–34.
Fox, C. F., and Kennedy, E. P. (1965). *Proc. Natl. Acad. Sci. U.S.A.* **51**, 891.
Franco, P. J., Eelkema, J. A., and Brooker, R. J. (1989). *J. Biol. Chem.* **264**, 15988–15992.
Friedlander, M., and Blobel, G. (1985). *Nature (London)* **318**, 338–343.
Goldkorn T., Rimon, G., and Kaback, H. R. (1983). *Proc. Natl. Acad. Sci. U.S.A.* **80**, 3322–3326.
Herzlinger, D., Viitanen, P., Carrasco, N., and Kaback, H. R. (1984). *Biochemistry* **23**, 3688–3693
Kaback, H. R. (1983). *J. Membr. Biol.* **76**, 95–112.
Kaback, H. R. (1986). *In* "Physiology of Membrane Disorders" (T. E. Andreoli, J. F. Hoffman, D. D. Fanestil, and S. G. Schultz, eds.), pp. 387–408. Plenum, New York.
Kaback, H. R. (1989). *Harvey Lect.* **83**, 77–103.
Kaback, H. R. (1990). *In* "The Bacteria" (T. A. Krulwich, ed.), Vol. 12, pp. 151–202. Academic Press, San Diego, California.
Kaback, H. R., and Barnes, E. M. (1971). *J. Biol. Chem.* **246**, 5523.
Kaback, H. R., and Patel, L. (1978). *Biochemistry* **17**, 1640.
Kaback, H. R., Bibi, E., and Roepe, P. D. (1990). *Trends Biochem. Sci.* **15**, 309.
Kaczorowski, G. J., Leblanc, G., and Kaback, H. R. (1980). *Proc. Natl. Acad. Sci. U.S.A.* **77**, 6319–6323.
Kennedy, E. P., Rumley, M. K., and Armstrong, J. B. (1974). *J. Biol. Chem.* **249**, 33.
King, S. C., and Wilson, T. H. (1989a). *J. Biol. Chem.* **264**, 7390–7394.
King. S. C., and Wilson, T. H. (1989b). *Biochim. Biophys. Acta* **982**, 253–264.

King, S. C., and Wilson, T. H. (1990). *J. Biol. Chem.* **265**, 3153–3160.
Konings, W. N., and Robillard, G. T. (1982). *Proc. Natl. Acad. Sci. U.S. A.* **79**, 5480.
Lee, J. A., Püttner, I. B., Carrasco, N., Antes, L. M., and Kaback, H. R. (1989). *Biochemistry* **28**, 2540–2544.
Li, J., and Tooth, P. (1987). *Biochemistry* **26**, 4816–4823.
Lolkema, J. S., Püttner, I. B., and Kaback, H. R. (1988). *Biochemistry* **27**, 8307–8310.
Manoil, C., and Beckwith, J. (1986). *Science* **233**, 1403–1408.
Matsushita, K., Patel, L., Gennis, R. B., and Kaback, H. R. (1983). *Proc. Natl. Acad. Sci. U.S.A.* **80**, 4889–4893.
McKenna, E., Hardy, D., Pastore, J. C., and Kaback, H. R. (1991). *Proc. Natl. Acad. Sci. U.S.A.* **88**, 2969–2973.
McKenna, E., Hardy, D., and Kaback, H. R. (1992). In preparation.
Menezes, M. E., Roepe, P. D., and Kaback, H. R. (1990). *Proc. Natl. Acad. Sci. U.S.A.* **87**, 1638.
Menick, D. R., Sarkar, H. K., Poonian, M. S., and Kaback, H. R. (1985). *Biochem. Biophys. Res. Commun.* **132**, 162.
Menick, D. R., Lee, J. A., Brooker, R. J., Wilson, T. H., and Kaback, H. R. (1987a). *Biochemistry* **26**, 1132.
Menick, D. R., Carrasco, N., Antes, L., Patel, L., and Kaback, H. R. (1987b). *Biochemistry* **26**, 6638–6644.
Neuhaus, J.-M., Soppa, J., Wright, J. K., Riede, I., Bocklage, H., Frank, R., and Overath, P. (1985). *FEBS Lett.* **185**, 83.
Newman, M. J., and Wilson, T. H. (1980). *J. Biol. Chem.* **255**, 10583–10586.
Newman, M. J., Foster, D. L., Wilson, T. H., and Kaback, H. R. (1981). *J. Biol. Chem.* **256**, 11804.
Onsager, L. (1969). *Science* **166**, 1359–1364.
Padan, E., Sarkar, H. K., Viitanen, P. V., Poonian, M. S., and Kaback, H. R. (1985). *Proc. Natl. Acad. Sci. U.S.A.* **82**, 6765–6768.
Page, M. G. P., and Rosenbusch, J. P. (1988). *J. Biol. Chem.* **263**, 15906–15914.
Popot, J. L., and de Vitry, C. (1990). *Annu. Rev. Biophys Biophys. Chem.* **19**, 369–403.
Popot, J. L., and Engleman, D. M. (1990). *In* "Protein Form and Function" (R. A. Bradshaw and M. Purton, eds.), p. 147. Elsevier, Cambridge.
Pourcher, T., Sarkar, H. K., Bassilana, M., Kaback, H. R., and Leblanc, G. (1990). *Proc. Natl. Acad. Sci. U.S.A.* **87**, 468–472.
Püttner, I. B., and Kaback, H. R. (1988). *Proc. Natl. Acad. Sci. U.S.A.* **85**, 1467–1471.
Püttner, I. B., Sarkar, H. K., Poonian, M. S., and Kaback, H. R. (1986). *Biochemistry* **25**, 4483–4485.
Püttner, I. B., Sarkar, H. K., Padan, E., Lolkema, J. S., and Kaback, H. R. (1989). *Biochemistry* **28**, 2525–2533.
Rapoport, T. A. (1986). *CRC Crit. Rev. Biochem.* **20**, 73–137.
Robillard, G. T., and Konings, W. N. (1982). *Eur. J. Biochem.* **127**, 597.
Roepe, P. D., and Kaback, H. R. (1989). *Biochemistry* **28**, 6127–6132.
Roepe, P. D., and Kaback, H. R. (1992). *CRC Crit. Rev. Biochem.* (in press).
Roepe, P. D., Zbar, R., Sarkar, H. K., and Kaback, H. R. (1989). *Proc. Natl. Acad. Sci. U.S.A.* **86**, 3992–3996.
Roepe, P. D., Consler, T. G., Menezes, M. E., and Kaback, H. R. (1990). *Res. Microbiol.* **141**, 290–308.
Sarkar, H. K., Viitanen, P. V., Poonian, M. S., and Kaback, H. R. (1986). *Biochemistry* **25**, 2778.
Seckler, R., and Wright, J. K. (1984). *Eur. J. Biochem.* **142**, 269–279.
Seckler, R., Wright, J. K., and Overath, P. (1983). *J. Biol. Chem.* **258**, 10817–10820.

Seckler, R., Möröy, T., Wright, J. K., and Overath, P. (1986). *Biochemistry* **25**, 2403–2409.
Singer, S. J., Maher, P. A., and Yaffe, M. P. (1987). *Proc. Natl. Acad. Sci. U.S.A.* **84**, 1960–1964.
Stochaj, U., and Ehring, R. (1987). *Eur. J. Biochem.* **163**, 653–658.
Stochaj, U., Bieseler, B., and Ehring, R. (1986). *Eur. J. Biochem.* **158**, 423–428.
Stochaj, U., Fritz, H.-J., Heibach, C., Markgraf, M., Schaewen, A. V., Sonnewald, U., and Ehring, R. (1988). *J. Bacteriol.* **170**, 2639–2645.
Teather, R. M., Müller-Hill, B., Abrutsch, U., Aichele, G., and Overath, P. (1978). *Mol. Gen. Genet.* **159**, 239–248.
Todd, A. P., Cong, J., Levinthal, C., Levinthal, W., and Hubbell, W. L. (1989). *Proteins: Struct. Funct. Genet.* **6**, 294.
Trumble, W. R., Viitanen, P. V., Sarkar, H. K., Poonian, M. S., and Kaback, H. R. (1984). *Biochem. Biophys. Res. Commun.* **119**, 860.
van Iwaarden, P., Menick, D. R., Kaback, H. R., and Konings, W. N. (1991a). *J. Biol. Chem.*, **266**, 15688.
van Iwaarden, P., Pastore, J. C., Konings, W. N., and Kaback, H. R. (1991b). *Biochemistry* **30**, 9595.
Viitanen, P., Garcia, M. L., Foster, D. L., Kaczorowski, G. J., and Kaback, H. R. (1983). *Biochemistry* **22**, 2531.
Viitanen, P., Garcia, M. L., and Kaback, H. R. (1984). *Proc. Natl. Acad. Sci. U.S.A.* **81**, 1629–1633.
Viitanen, P. V., Menick, D. R., Sarkar, H. K., Trumble, W. R., and Kaback, H. R. (1985). *Biochemistry* **24**, 7628.
Viitanen, P., Newman, M. J., Foster, D. L., Wilson, T. H., and Kaback, H. R. (1986). *In* "Methods in Enzymology" (S. Fleischer and B. Fleischer, eds.), Vol. **125**, p. 429. Academic Press, Orlando, Florida.
Vogel, H., Wright, J. K., and Jähnig, F. (1985). *EMBO J.* **4**, 3625–3631.
von Heijne, G., and Blomberg, C. (1979). *Eur. J. Biochem.* **97**, 175–181.
Wickner, W. T., and Lodish, H. F. (1985). *Science* **230**, 400–407.
Wright, J. K., Seckler, R., and Overath, P. (1986). *Annu. Rev. Biochem.* **55**, 255.
Wrubel, W., Stochaj, U., Sonnewald, U., Theres, C., and Ehring, R. (1990). *J. Bacteriol.* **172**, 5374–5381.

Group Translocation of Glucose and Other Carbohydrates by the Bacterial Phosphotransferase System

Bernhard Erni[1]
Institut für Biochemie, Universität Bern
CH-3012 Bern, Switzerland

I. Introduction to and Scope of This Article

Solute transport across a biological membrane and against a concentration gradient is an energy-requiring process. The driving force is provided by transmembrane ion gradients, a transmembrane electrical potential difference, or by the free energy of ATP hydrolysis. Bacteria utilize yet another energy-coupling mechanism to accumulate nutrients: group translocation. The hallmark of this mechanism is the coupling of solute translocation with chemical modification of the transported solute. Coupling is tight and chemical modification does not simply maintain the concentration gradient by continuously removing the substrate from the diffusion equilibrium. In facultative and obligate anaerobic bacteria a large number of sugars and sugar alcohols are translocated and simultaneously phosphorylated. Both processes are catalyzed by sugar-specific proteins in the cytoplasmic membrane known as "enzymes II." These enzymes II are the sugar-specific components of a more complex system known as the bacterial phosphotransferase system (PTS). It comprises in addition to the sugar-specific transporters a number of cytoplasmic proteins that sequentially transfer phosphoryl groups from phosphoenolpyruvate to the different transporters and that are important for allosteric regulation of catabolic enzymes, regulation of gene expression, and chemotaxis.

This article attempts to describe the structure and function of the transporters of the bacterial phosphotransferase system, in particular of those specific for mannitol, glucose, and mannose, which have been purified and for which biochemical data are available. Reviews with similar scope were

[1] Present address: Institute for Biochemistry, University of Bern, CH-3012 Bern, Switzerland.

127

published by Robillard and Lolkema (1988) and Erni (1989). For more comprehensive reviews on all aspects of the bacterial phosphotransferase system, readers are referred to the excellent articles by Meadow *et al.* (1990), Postma (1987), Postma and Lengeler (1985), and to a collection of papers edited by Danchin (1989). Aspects of metabolic regulation by the bacterial phosphotransferase system are discussed in great detail by Saier (1989a) and Saier *et al.* (1990). The function of the phosphotransferase system in chemotaxis has been reviewed by Lengeler and Vogler (1989) and by Taylor and Lengeler (1990).

II. The Bacterial Sugar Phosphotransferase System

A. Components

In 1964 Saul Roseman and collaborators (Kundig *et al.*, 1964) found that cell-free extracts of *Escherichia coli* catalyzed phosphorylation of glucose with phosphoenolpyruvate rather than ATP as the phosphoryl donor. Fractionation of the extract into membrane and cytoplasmic components revealed that both were necessary for phosphorylation of glucose. The two complementing activities were designated enzyme I and enzyme II. The cytoplasmic extract turned out to contain two complementing activities, which again were designated enzyme I and HPr (for heat-stable protein). Genetic and biochemical analysis revealed that the membrane fraction also contained several activities, which in contrast to the cytoplasmic components differed in their sugar specificity (Kundig and Roseman, 1971; Kundig, 1974). They were specified as enzymes II^{sugar}. Occasionally additional sugar-specific components could be found in the cytoplasmic fraction, which were designated enzymes III^{sugar}. Later, amino acid sequence comparisons showed that the "sugar-specific" enzymes II are functional subunits of transporter complexes but do not contain a sugar binding site of their own.

Figure 1 shows the different PTS proteins and how they interact. Selected examples from different bacteria are included to highlight the similarities and differences. The two cytoplasmic proteins, enzyme I and HPr of *E. coli,* are phosphoryl carrier proteins that sequentially transfer a phosphoryl group from phosphoenolpyruvate to the membrane-bound, sugar-specific transporters. Enzyme I is a dimer of identical subunits of M_r 63,489 (Saffen *et al.*, 1987; de Reuse and Danchin, 1988). In the presence of phosphoenolpyruvate enzyme I autophosphorylates at the N3 position of His-189 (Alpert *et al.*, 1985; Meadow *et al.*, 1990). Enzyme I can also be phosphorylated by acetate kinase with GTP, ATP, or acetyl phosphate as

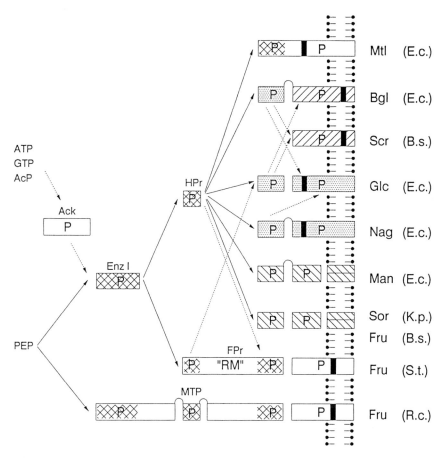

FIG. 1   The bacterial phosphotransferase system. Enzyme I (Enz I), HPr, FPr, MTP, Cytoplasmic proteins that sequentially transfer phosphoryl groups from phosphoenolpyruvate
to the sugar-specific transporters. Enzyme I can also be phosphorylated by acetate kinase
(Ack). Different types of membrane transporters (II^{sugar}) from different origins are lined up
on the right. The shading indicates colinear amino acid sequence similarities (from 30 to 70%
identity) within functionally homologous domains. The highly conserved sequence containing the phosphorylatable cysteine is marked by a solid bar. P, The phosphorylation sites that
are either histidines or cysteines. Phophoryl transfer between subunits is indicated by arrows. Broken arrows indicate examples of phosphoryl transfer in heterooligomeric complexes ("cross-talk"). The phosphoryl transfer between site 1 (left) and site 2 (right) within
the transporters is not explicitly shown. Arches indicate alanine–proline hinge peptides
connecting functional domains. The length of a box corresponds roughly to the molecular
weight of the protein, with the exception of FPr and MTP, which are smaller than suggested
by the drawing. Mtl (E.c.), II^{Mtl}; Bgl (E.c.), II^{Bgl}; Scr (B.s.), II^{Scr}; Glc (E.c.), II^{Glc}/III^{Glc}; Nag
(E.c.), II^{Nag}; Man (E.c.), II-M^{Man}/II-P^{Man}/III^{Man}; Sor (K.p.), II-M/II-A/III-B/III-F; Fru
(B.s.), P30/P28/P18/P16; Fru (S.t.), II^{Fru}; Fru (R.c.), II^{Fru}. E.c., *Escherichia coli;* B.s., *B.
subtilis;* K.p., *K. pneumoniae;* S.t., *S. typhimurium;* R.c., *R. capsulatus.*

phosphoryl donors (Fox *et al.*, 1986). The phosphoryl group is then transferred from phosphoenzyme I to either HPr or FPr. HPr carries the phosphoryl groups to the different sugar-specific transporters. The 9.1-kDa protein is phosphorylated at the N1 position of His-15 (Anderson *et al.*, 1971). For HPr, X-ray (El-Kabbani *et al.*, 1984; Delbaere *et al.*, 1989; Kapadia *et al.*, 1990) and two-dimensional nuclear magnetic resonance (NMR) structures are known (Klevit and Waygood, 1986; Wittekind *et al.*, 1989, 1990). *Salmonella typhimurium* expresses in addition to HPr a protein termed FPr, which contains an HPr-like domain and the enzyme III component of the fructose transporter as a single polypeptide chain (Geerse *et al.*, 1989). FPr is first phosphorylated by enzyme I at the HPr-like domain, from where the phosphoryl group is transferred intramolecularly to the enzyme III-like domain. FPr contains a third domain ("RM"), exhibiting 60% similarity to the consensus receiver module of two-component regulatory systems (Wu *et al.*, 1990). In *Rhodobacter capsulatus* the functional equivalents of enzyme I, HPr, and the enzyme III component of the fructose transporter are distinct structural domains contained in a single polypeptide chain termed MTP (Wu *et al.*, 1990).

Escherichia coli has between 10 and 12 PTS transporters with different and sometimes overlapping specificities for hexoses, hexitols, and disaccharides. All transporters consist of between one and four different protein subunits. The sum of the subunit molecular weights varies between 70,000 and 80,000. Independent of subunit composition and origin, all transporters have two phosphorylation sites that sequentially transfer a phosphoryl group from phospho-HPr to the sugar substrate. In the single-subunit transporters, the two sites are located on different protein domains. In the two-subunit transporters, the first phosphorylation site is located on the hydrophilic subunit (enzyme III) and the second on the membrane-bound subunit (enzyme II). In the three- and four-subunit transporters, the two sites are located either on two domains of a hydrophilic subunit or on two separate protein subunits. It appears that there are altogether four phosphorylation sites engaged in phosphoryl transfer between phosphoenolpyruvate and sugar.

B. Transphosphorylation

Biochemical and genetic experiments indicate that phosphoryl transfer occurs not only between sites 3 and 4 of the same transporter, but also between sites 3 and 4 belonging to different transporters (dashed arrows in Fig. 1). Similarly, phosphoryl transfer occurs between the HPr domain of FPr (site 2) and site 3 of other transporters (Geerse *et al.*, 1986; Chin *et al.*, 1987). The components of the PTS therefore constitute a network of interacting phosphoryl carrier proteins. A detailed analysis by Weigel *et al.* (1982) of thermodynamic and kinetic aspects of the phosphoryl transfer between phosphoenolpyruvate and III^{Glc} indicate that the phosphate transfer potentials of the PTS proteins are similar and very close to that of phosphoenolpyruvate. The phosphoryl transfer between III^{Glc} and II^{Glc} is also reversible (Erni, 1986). It is only the last step, phosphoryl transfer from the transporter to the sugar, that is associated with a large drop of free energy. From the intracellular concentration of enzyme I, HPr, and III^{Glc} (Postma, 1987) the concentration of phosphorylatable sites on PTS proteins can be estimated to be as high as 300 μM and thus to be of the same order if not higher than the concentration of free phosphoenol-pyruvate (50 μM in *E. coli* growing exponentially in the presence of glucose; Weigel *et al.*, 1982). In the presence of a limiting amount of phosphoenolpyruvate available for rephosphorylation of the system, the transporters with a higher affinity for phosphoryl carrier proteins will be phosphorylated preferentially. If they are actively engaged in sugar trans-location they will compete for the limited supply of phosphoryl donors and thus inhibit the uptake of sugars by transporters with lower affinity for a common phosphoryl donor (Scholte and Postma, 1981). The affinity of the different transporters for phospho-HPr would thus define a preference scale for the uptake of different PTS sugars by the bacterium.

C. Multiple Functions: Transport, Regulation, and Sensory Transduction

Transport and phosphorylation were the first PTS functions discovered and characterized (Kundig *et al.*, 1964). They will be described in greater detail below. However, the study of mutants soon revealed that the PTS is also involved in regulation of metabolic processes. The aspects of regula-tion by the PTS have been reviewed in great detail (Saier, 1989a; Saier *et al.*, 1990) and are mentioned here for the sake of completeness only.

The hydrophilic subunit III^{Glc} of the glucose transporter regulates the activity of several target enzymes by stoichiometric binding. Unphos-phorylated III^{Glc} binds to non-PTS transporters (e.g., for lactose or malt-ose) and thereby inhibits their activity as shown in greatest detail for the LacY-dependent transport of lactose (Osumi and Saier, 1982; Nelson *et al.*, 1983; Misko *et al.*, 1987; Mitchell *et al.*, 1987; Wilson *et al.*, 1990; Dean *et al.*, 1990). The ratio of III^{Glc} to phospho-III^{Glc} increases during transport of glucose or other PTS sugars because III^{Glc} is continuously dephos-phorylated by II^{Glc} engaged in glucose transport or is inadequately re-phosphorylated due to other PTS transporters draining phosphoryl groups from HPr (Scholte and Postma, 1981). When the concentration of PTS substrates drops, the steady state concentration of phospho-III^{Glc} increases and the inhibition of other uptake systems is relieved. Free III^{Glc} allosterically inhibits glycerol kinase (Postma *et al.*, 1984) and other catabolic enzymes might also be regulated by III^{Glc} or phospho-III^{Glc}. There is circumstantial evidence that phospho-III^{Glc} stimulates adenyl cyclase (Liberman *et al.*, 1986). Cyclic AMP generated by adenyl cyclase binds to the catabolite activator protein and permits the expression of operons that are otherwise repressed as long as glucose and to a lesser degree other PTS sugars are available to the cell. To be effective III^{Glc} must be present in stoichiometric amounts relative to the regulated target enzymes. However, there are more target molecules than III^{Glc} in a cell. Binding studies with III^{Glc} and the lactose transporter (LacY) have shown that binding is strongly stimulated in the presence of lactose, suggesting that only those enzymes for which a substrate is available become targets for inhibition by III^{Glc}. Such cooperativity for binding of substrate and of the allosteric inhibitor might reduce the need for larger amounts of III^{Glc} (reviewed in Saier, 1989a). An unexpected solution to match a non-PTS target protein with a PTS regulator is realized in *Streptococcus thermophilus*. In this microorganism the lactose–proton symporter consists of a 12-helix transmembrane domain with a fused C-terminal domain similar to III^{Glc} of *E. coli*. The allosteric PTS regulator is thus an integral part of a non-PTS target (Poolman *et al.*, 1989).

HPr in gram-positive but not in gram-negative bacteria can function as a protein kinase and thus regulate target enzymes by phosphorylation. In addition, an HPr-specific protein kinase can inactivate HPr by phosphorylation at a seryl residue (for a review, see Reizer *et al.*, 1988, 1989).

Some of the transporters are finally involved in feedback regulation of their own expression. The transporter specific for β-glucosides (II^{Bgl}), regulates the expression of enzymes involved in the catabolism of β-glucosides. II^{Bgl} reversibly phosphorylates an antiterminator protein (BglG) necessary for the transcription of the β-glucoside operon. In the absence of a sugar substrate II^{Bgl} is expressed constitutively in low amounts, which are sufficient to phosphorylate and thus inactivate the antiterminator protein. Transcription of the *bgl* operon is aborted prematurely. In the presence of a substrate, the porter transfers phosphoryl groups to the substrate rather than to the antiterminator. The dephosphorylated antiterminator. The dephosphorylated antiterminator permits transcription of the *bgl* operon, including gene *bglF* encoding II^{Bgl} (Amster-Choder *et al.*, 1989; Amster-Choder and Wright, 1990; Houman *et al.*, 1990; Schnetz and Rak, 1990). A similar mechanism regulates the operons specific for uptake and metabolism of sucrose in *Bacillus subtilis* (Débarbouillé *et al.*, 1990; Zukowski *et al.*, 1990).

The RM domain of FPr displays sequence similarity with the receiver module consensus sequences of two-component regulatory systems (Wu et al., 1990; Kofoid and Parkinson, 1988). The II^{Glc} and III^{Glc} subunits of the glucose transporter also have marginal similarity with the consensus sequences characteristic of two-component regulatory systems (Kofoid and Parkinson, 1988). This family of proteins, which includes membrane receptors and cytoplasmic proteins, is involved in intracellular signaling and regulation of metabolic processes. They control enzyme activity and gene expression by protein phosphorylation at histidyl and aspartyl residues (e.g., in chemotaxis and osmoregulation; for a review, see Stock et al., 1989). The presence of related sequences in PTS proteins suggests that components of the PTS might participate in other regulatory circuits that are based on reversible transphosphorylation of proteins.

The soluble phosphoryl carriers as well as the transporters of the PTS are required for chemotaxis toward PTS sugars (Adler and Epstein, 1974; Niwano and Taylor, 1982; Lengeler and Vogler, 1989; Taylor and Lengeler, 1990). In contrast to the classical chemotaxis receptors, where binding of the stimulus elicits a chemotactic response, chemotactic stimulation by PTS substrates is inseparably linked to transport and phosphorylation. Grübl et al. (1990) describe an HPr-deficient strain that can transport PTS sugars due to complementation by FPr (Fig. 1) but no longer responds chemotactically to PTS sugars. They further characterize HPr point mutations with unbalanced impairment of transphosphorylation (transport) and chemotaxis functions. They propose that HPr is the component at the interface between PTS and the general chemotactic signal transducing Che proteins. An increase in substrate concentration could, e.g., cause a transient decrease in the ratio of phosphorylated to dephosphorylated HPr. Rephosphorylation would reset the system to prestimulus activity and thus result in adaptation to the new stimulus concentration. It is not known how HPr interacts with components of the general chemotaxis proteins (Hess et al., 1988a,b) and whether signal transduction is affected by noncovalent binding of PTS components to chemotaxis proteins or by transphosphorylation.

D. Genetic Organization and Transcriptional Regulation

The PTS comprises over a dozen mutually interacting proteins, the genes of which are spread all over the bacterial chromosome. In E. coli the genes ptsH and ptsI for HPr and enzyme I, respectively, constitute an operon at 52 min. Immediately downstream of this operon but only partly coregulated with it is crr, the gene encoding the cytoplasmic subunit (III^{Glc}) of the glucose transporter (de Reuse and Danchin, 1988). Gene ptsG encoding the II^{Glc} subunit maps at 24 min. In B. subtilis, in contrast, the genes for the glucose transporter, enzyme I, and HPr are arranged adjacent to each other in a sequence that directly reflects the sequence in which they functionally interact (Gonzy-Tréboul *et al.*, 1989; Sutrina *et al.*, 1990). The other sugar-specific membrane transporters are encoded in operons together with other enzymes required for the catabolism of the respective sugar.

Transcription of PTS genes appears to be regulated by different mechanisms depending on the particular operon. All the genes that had been looked for are under catabolite control, as revealed by physiological measurements and/or by CAP binding consensus sequences in the promotor region (de Reuse and Danchin, 1988, Peri *et al.*, 1990; Plumbridge, 1990; Buhr, 1990).

Expression of the *bgl* operon for the catabolism of β-glucosides is regulated by an antiterminator protein, the activity of which is regulated by phosphorylation/dephosphorylation as described above. De Reuse and Danchin (1991) showed that glucose stimulates the transcription of the *ptsHI* operon (encoding enzyme I and HPr) approximately threefold in a II^{Glc}-dependent way, and that overexpression of II^{Glc} by itself has the same effect. It appears that dephosphorylated II^{Glc} is activating some as yet unknown regulator of *ptsHI* transcription. The operon for transport and catabolism of sorbose is regulated by protein SorC, which acts as a repressor in the absence of the inducer L-sorbose and as an activator in its presence (Wöhrl *et al.*, 1990). The repressor of the *nag* operon (*N*-acetylglucosamine transporter) appears also to modulate the transcription of the *ptsLPM* (*manXYZ*) gene encoding the mannose transporter (Vogler and Lengeler, 1989). This last example indicates that the different PTS transport systems might influence each other at the level of gene expression in addition to mutual interaction at the level of protein function, as will be described in Section III,D below.

III. Structure and Function of the PTS Transporters

A. Amino Acid Sequences and Hydropathy

More than a dozen genes of gram-positive and gram-negative origin encoding PTS transporters (enzymes II) have been sequenced and with each additional sequence more comprehensive sequence comparisons have been made (Saier, 1989b; Lengeler, 1990; Lengeler *et al.*, 1990; Wu and Saier, 1990; Meadow *et al.*, 1990). The results from these comparisons can be summarized as follows: The PTS permeases are composed of between one and four different polypeptide subunits. The sum molecular weights of the functional complexes are between 70,000 and 80,000. The sequence similarity between the different transporters varies from not recognizable to 70% amino acid sequence identity. There is a clear correspondence between the subunits of the complex transporters and protein domains in the single-polypeptide transporters (Fig. 1). According to amino acid sequence and subunit composition the different transporters can be assigned to four major types. To type I belong the two-domain single-subunit (enzyme II- enzyme III fused) transporters for N-acetylglucosamine (II^{Nag}) and the two-subunit (II^{Glc}/III^{Glc}) transporter for glucose of $E.$ $coli.$ The amino acid sequences of the two transporters are colinear and about 40% identical (Peri and Waygood, 1988; Rogers et $al.,$ 1988). The mannitol transporter (II^{Mtl}) resembles II^{Nag} with respect to the hydropathy profile and the linear arrangement of the functional domains, but the two transporters share only limited sequence similarities (Lee and Saier, 1983; Erni and Zanolari, 1986). The type II transporters are related to group I, displaying up to 40% sequence identity. However, they differ from type I inasmuch as the most highly conserved sequence containing the phosphorylatable cysteinyl residue (black bars in Fig. 1) is not at the C-terminal but at the N-terminal end of the polypeptide chain. To type II belong the single-subunit (enzyme II - enzyme III fused) transporters for β-glucoside of $E.$ $coli$ (Bramley and Kornberg, 1987; Schnetz et $al.,$ 1987) and for glucose of $B.$ $subtilis$(Sutrina et $al.,$ 1990). The sucrose transporters of $B.$ $subtilis$ (Fouet et $al.,$ 1987) and of $Klebsiella$ $pneumoniae$ (Lengeler et $al.,$ 1990) also belong to type II, but they do not have an enzyme III domain/subunit of their own. They depend on the enzyme III domain of another PTS transporter for functional complementation. Sutrina et $al.$ (1990) showed that the sucrose transporter in $B.$ $subtilis$ is phosphorylated by the enzyme III domain of the glucose transporter. Type III is represented by the fructose transporters of $E.$ $coli$ and $R.$ $capsulatus$ (Prior and Kornberg, 1988; Wu and Saier, 1990). The two transporters are 55% identical but display little similarity with the transporters belonging to type I. It appears that in contrast to type I and type II transporters the hydrophilic domain of the type III transporters is entirely N-terminal. Type IV includes the three-subunit mannose transporter of $E.$ $coli$ (Erni et $al.,$ 1987), and the four-subunit transporters for fructose of $B.$ $subtilis$ and for sorbose of $K.$ $pneumoniae.$ They have up to 70% identical amino acid sequences (Martin-Verstraete et $al.,$ 1990; Wöhrl and Lengeler, 1990; Lengeler, 1990) but have no sequence similarity with any other transporter. Type IV transporters also differ radically from all other transporters inasmuch as both phosphorylation sites are histidines (Erni et $al.,$ 1989) and none of the conserved cysteines appears to be essential (E. Rhiel and B. Erni, unpublished results).

Figure 2 proposes a hypothetical model of a "consensus structure"

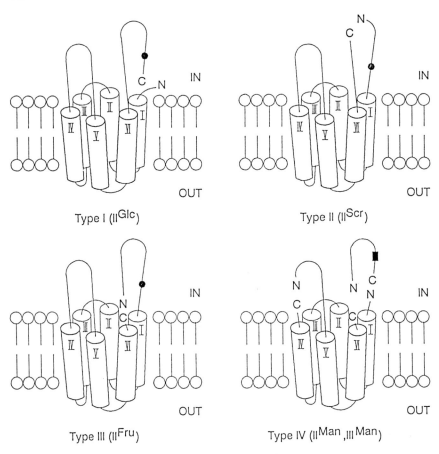

FIG. 2 Speculative model of the "consensus" structure (topology) underlying the different types of PTS transporters. Putative transmembrane helices are numbered I–VI. The two cytoplasmic loops are not drawn to scale. Phosphorylation site 2 is indicated as a dot when a cysteine, and as a rectangle when a histidine. N, N terminus; C, C terminus of the protein (subunits).

common to all transporters. It is based on the assumption that PTS transporters function similarly and therefore must have similar overall structures in spite of divergent amino acid sequences. The consensus structure is envisaged to consist of a bundle of an even number, most likely six, membrane spanning helices and a cytoplasmic domain composed of two longer hydrophilic loops. Loop 1 is predicted to span helices IV and V, loop 2 helices VI and I. The different types are cyclically permuted forms of this consensus structure, with the N- and C-terminal ends of the polypeptide chain in loop 2. In type I transporters (II^{Glc} and II^{Mtl}) loop

2 is formed by the C-terminal hydrophilic end of the polypeptide chain. In type II (II^{Bgl} and II^{Scr}) the N-terminal and C-terminal ends together form loop 2, and in type III (II^{Fru}) the loop is formed by the N-terminal end alone. With some imagination even type IV transporters (consisting of three and four subunits) can be fitted into this scheme. The six-helix bundle would be formed by two rather than by a single polypeptide subunit. The first subunit (e.g., $II\text{-}P^{Man}$ of the mannose transporter) comprises helices I–IV. Of the second subunit ($II\text{-}M^{Man}$) the N-terminal half forms loop 1 and the C-terminal half forms the two helices V and VI. An independent hydrophilic subunit (P18, $III\text{-}B^{Sor}$, and the P20 domain of III^{Man}) is the structural and functional equivalent of the hydrophilic loop 2. This topology would predict that the C terminus of helix IV and the N terminus of loop 1 of the transmembrane subunits are exposed on the same face of the membrane and possibly close to each other. To test this, the two subunits, $II\text{-}P^{Man}$ and $II\text{-}M^{Man}$, were fused by site-directed deletion of the stop codon between genes ptsP and ptsM. Cells expressing this fusion protein continue to ferment (transport) mannose. The specific activity of the fusion is approximately 10% of wild type. The fusion protein has a reduced affinity for the III^{Man} subunit and membrane assembly is compromised when the fusion is overexpressed (E. Rhiel and B. Erni, unpublished results).

The domains/subunits containing the first phosphorylation site (enzyme III) have been omitted from the drawings in Fig. 2. They are independent subunits (e.g., III^{Glc}, P16, SorF, FPr, MTP) or are structurally distinct domains fused to the C terminus of the transmembrane domain (e.g., in II^{Bgl}, II^{Nag}, II^{Mtl}). In the mannose transporter the P13 domain containing the first phosphorylation site is fused to the N-terminal end of the P20 domain.

It must be emphasized once more that the model proposed in Fig. 2 is highly speculative. However, it could be tested by cyclically permuting gene fragments within individual transporters and exchanging gene fragments between different transporters.

B. Active Sites of the PTS Transporters

1. Phosphorylation Sites

The soluble phosphoryl carrier proteins and the sugar-specific transporters are transiently phosphorylated. Histidines were identified as the phosphorylated residues of the soluble PTS proteins, enzyme I, HPr, and III^{Glc}, by sequencing of phosphorylated peptides (Anderson et al., 1971; Alpert et al., 1985; Kalbitzer et al., 1982; Dörschug et al., 1984).

Transient phoshorylation of the transporters was first inferred by Begley *et al.* (1982), who observed net inversion of configuration of the chiral phosphoryl group during transfer from phosphoenolpyruvate to glucose. With enzyme I, HPr, and III^{Glc} being phosphorylated, II^{Glc} had to be the fourth transiently phosphorylated protein intermediate. Using the same technique Mueller *et al.* (1990) demonstrated that the mannitol transporter II^{Mtl} must be phosphorylated twice. The two phosphorylation sites of the mannitol transporter were identified by Pas and Robillard (1988) as His-554 and Cys-384. His-554 is the first phosphorylation site that is phosphory-lated by phospho-HPr. Similarly, histidine-containing sequences are found in the III^{Glc} subunit of the glucose transporter and in the other transporters belonging to types I, II, and III. Cys-384 is the second phos-phorylation site. It accepts the phosphoryl group from His-554. The find-ing of a phosphorylated cysteine is consistent with the results of Nuoffer *et al.* (1988), who systematically mutagenized the cysteines of the glucose transporter and found that out of three cysteines only Cys-421 is essen-tial for function. A ^{32}P-labeled dodecapeptide containing Cys-421 could by isolated from phospho-II^{Glc} (M. Meins, personal communication). Schnetz *et al.* (1990) found that Cys-24 of the β-glucoside transporter is essential for phosphorylation and transport. The amino acid sequence, including the cysteine corresponding to Cys-384 of II^{Mtl}, Cys-421 of II^{Glc}, and Cys-24 of II^{Bgl}, is the most highly conserved motif recognizable in all transporters of types I, II, and III. The mannose transporter also contains two phosphorylation sites, which, however, radically differ from those just described. Both are histidines and are in sequence contexts that have no similarities with the sites of type I–III transporters. His-10, the first phos-phorylation site, is located on the P13 domain, His-175 on the P20 domain of III^{Man} (Erni *et al.*, 1989). The two histidines are conserved in the cytoplasmic subunits of the other type IV transporters, the fructose and sorbose transporters of *B. subtilis* and *K. pneumoniae,* respectively (Martin-Verstraete *et al.*, 1990; J. Lengeler, personal communication). As no phosphorylated intermediates of the membrane-bound II-P^{Man} and II-M^{Man} subunits of the mannose transporter could be detected, it is likely that these subunits are not phosphorylated. Cysteine-37 of II-P^{Man} and Cys-47 of II-M^{Man} are both conserved in the two related transporters for fructose and sorbose but are both dispensable. The C37S mutation of the mannose transporter has 100% of wild-type activity. The C47E mutation retains 15% activity, reflecting a loss of protein stability (folding and assembly) rather than of catalytic activity (E. Rhiel and B. Erni, unpub-lished results). Although the amino acid sequences around Cys-16 of II-P^{Man} and Cys-384 of II^{Mtl} are strikingly similar (Pas and Robillard, 1988; Sutrina *et al.*, 1990) this cysteine is missing in the homologous region of the fructose and sorbose transporters. As III^{Man} does not contain cysteines either, it is likely that sulfhydryl groups do not play a role in phosphorylation and transport of sugars by the type IV transporters. By systematically mutagenizing additional histidines, one particular histidine could be identified in III^{Glc} (His-76), III^{Man} (His-86), and II^{Bgl} (His-306) that appears to be important for phosphoryl transfer between sites 1 and 2 but not for phosphorylation of site 1 by phospho-HPr (Erni *et al.*, 1989; Presper *et al.*, 1989; Schnetz *et al.*, 1990). However, while His-76 of III^{Glc} and His-86 of III^{Man} are in the same domain with phosphorylation site 1, His-306 is in the transmembrane domain of II^{Bgl}, and it is not clear whether these histidines are mechanistically equivalent. The H86N and the H175N mutants of III^{Man} compete with wild-type III^{Man} for binding to the II^{Man} subunits, and when overexpressed are negative dominant. H10Q can complement the H86N and the H175N mutations, indicating that phosphoryl transfer can occur between different III^{Man} subunits. In contrast, H86N and H175N are noncomplementing, which suggests that His-86 and His-175 form a functional unit only if on the same III^{Man} subunit but not if on different subunits of a dimer (Erni *et al.*, 1989, and unpublished results).

2. Substrate Binding Sites

The apparent K_m of solute transport by PTS transporters is in the micromolar range (Stock *et al.*, 1982). The substrate binding sites are located in the transmembrane part of the transporters. Deletion of the C-terminal hydrophilic domain of the mannitol transporter containing both phosphorylation sites afforded a truncated protein that remained membrane inserted, bound mannitol with the same affinity as the intact protein (Grisafi *et al.*, 1989), and catalyzed slow, passive mannitol transport (Lolkema *et al.*, 1990). These results indicate that phosphorylation of the transporter does not affect the affinity of the transporter for the substrate. A hybrid transporter consisting of the N-terminal, hydrophobic end of II^{Glc} (residues 1–386) and the C-terminal hydrophilic end of II^{Nag} (residues 369–648) was fully active as a glucose-specific transporter (Hummel *et al.*, 1992). The C-terminal domain of II^{Nag} included both phosphorylation sites (His-569 and Cys-412), which suggests that Cys-412 is not part of the sugar recognition site although it must interact with the sugar substrate during transphosphorylation. In order to further localize the protein regions of the glucose transporter involved in substrate translocation, mutant forms of the II^{Glc} subunit were selected that could no longer efficiently transport extracellular α-methylglucoside but continued to phosphorylate glucose generated intracellularly from maltose (Buhr *et al.*, 1992; Nuoffer *et al.*, 1988). Mutations affording this phenotype were found clustered in three regions of the hydrophobic N-terminal part of II^{Glc} (Met-17, between residues 149–157, and between residues 339–343). However, it is not yet clear whether the three regions are involved in substrate translocation only, in substrate binding, or in both.

It appears that the PTS transporters not only phosphorylate solutes in transit from the outside to the inside but also substrates generated on the inside (e.g., glucose generated from disaccharides; Thompson and Chassy, 1985; Thompson, 1987; Nuoffer et al., 1988). This could be explained if the "sugar binding sites" of the transporters were alternatively accessible from either the periplasmic or the cytoplasmic side of the membrane, whereby only substrates bound on the cytoplasmic side could be phosphorylated (Robillard and Lolkema, 1988; Erni, 1990; Lolkema et al., 1990). It is conceivable that phosphorylation of the substrate lowers its affinity for the binding site and thus effects substrate release. The observed low efficiency of transporter-dependent phosphate exchange between substrate and substrate phosphate at equilibrium (transphosphorylation; Saier et al., 1977) could be due to the very low affinity of the transporter for the sugar phosphate.

Coupling between transport and phosphorylation appears to be tight in wild-type transporters, and protein phosphorylation/dephosphorylation might be a means of generating the conformational change that makes the substrate binding site move. Nevertheless, mutant glucose transporters have been found in S. typhimurium that catalyze facilitated diffusion uncoupled from phosphorylation (Postma, 1981). They had a 1000-fold higher K_m than wild-type II^{Glc}. Secondary mutants with only a 20-fold increased K_m could be isolated from them in a glucose-limited chemostat (Ruijter et al., 1990). It could be that these uncoupled mutants can release the translocated solute without phosphorylation because of the reduced affinity for the substrate.

C. Domain Structure

The amino acid sequence similarities on the one hand and the variable subunit compositions on the other hand suggest that PTS transporters consist of structurally and functionally independent protein domains. The domain structure is particularly obvious in III^{Man}, II^{Nag}, II^{Bgl}, and MTP, where the domains are connected by alanine–proline-rich hinges (Fig. 1; Erni et al., 1987; Wu et al., 1990) and in the glucose transporter of B. subtilis, where the enzyme III and enzyme II domains are connected by a Q linker (Sutrina et al., 1990). Both linkers contain characteristic polypeptide sequences present in multidomain proteins of both prokaryotic and eukaryotic origin (Erni, 1989; Wooton and Drummond, 1989). Moreover, highly similar transporters exist as multidomain forms in one bacterial species and as multisubunit proteins in another (Fig. 1). Gene fusion and gene splitting experiments have confirmed that interacting subunits can be fused and that multidomain proteins can be split with surprisingly little effect on enzymatic activity. The III^{Glc} subunit could be fused to the C terminus of II^{Glc} using alanine–proline-rich hinges of different length. All fusions functioned *in vivo*. The *in vitro* activities differed slightly with the different lengths of the spacers. The fusion protein had a catalytic advantage at low protein concentration, where diffusion-controlled interaction between nonfused III^{Glc} and II^{Glc} became rate limiting. Enzyme I and HPr of *E. coli* when properly fused through an alanine—proline-rich linker functioned *in vivo*, and after purification had the same activity as a 1:1 mixture of purified enzyme I and HPr. The transphosphorylation rate was limited by HPr but it could be increased by adding extra HPr. It therefore appears that the fusion between HPr and enzyme I is flexible enough to allow the additional binding of free HPr to the active center of enzyme I (T. Schunk and B. Erni, unpublished results). Similarly, a hybrid protein consisting of the N-terminal hydrophobic domain of II^{Glc} (residues 1–386) and the C-terminal hydrophilic end of II^{Nag} (residues 369–648) was fully active, displaying specificity for glucose (Hummel *et al.*, 1992). The reciprocal fusion, however, was inactive. The III^{Man} subunit of the mannose transporter could be split into its two domains, P13 and P20, which complement each other *in vivo*. Their *in vitro* activity was 2% of that of intact III^{Man}, reflecting that the transphosphorylation rate became diffusion controlled, once the domains were no longer covalently associated with each other (Erni *et al.*, 1989). It is noteworthy that the split domains could be crystallized whereas the intact protein with its floppy hinge could not (Génovésio-Taverne *et al.*, 1990).

II^{Mtl} in membrane vesicles could be proteolytically split into a cytoplasmic portion and a membrane-embedded part. The cytoplasmic domain could be phosphorylated by HPr and the membrane-embedded domain bound mannitol with high affinity (Grisafi *et al.*, 1989). The cytoplasmic and the membrane-embedded domains of II^{Mtl} were also expressed as separate polypeptides *in vivo*. They functionally complemented each other depending on the precise position of gene interruption. White and Jacobson (1990) observed complementation between overlapping fragments comprising amino acids 1–500 and 379–637 but not between nonoverlapping fragments 1–377 and 379–637. In contrast, van Weeghel *et al.* (1991a) obtained good complementation between the nonoverlapping fragments 1–347 and 348–637. The cytoplasmic domain of II^{Mtl} itself consists of two subdomains (348–489 and 490–637) containing the two phosphorylation sites (Cys-384 and His-554). The C-terminal subdomain (490–637) is a stable polypeptide that can complement the full-length H554A II^{Mtl} mutant (van Weeghel et al., 1991b). It remains to be demonstrated whether the 490–637 subdomain can also complement the truncated 1–489 subdomain.

D. Oligomeric Structure

Three transporters, the mannitol, glucose, and mannose transporters, have been purified so far (Jacobson et al., 1983; Robillard and Blaauw, 1987; Meins et al., 1988; Erni and Zanolari, 1985). Of all three transporters homodimeric forms have been characterized. Sedimentation equilibrium centrifugation and chemical cross-linking of the glucose transporter showed that detergent-solubilized II^{Glc} is a dimer of two identical subunits (Erni, 1986). An intersubunit disulfide bridge involving the phosphory-latable cysteines is formed on air oxidation of the purified protein. The cross-linked dimer is inactive but can be reactivated by reduction. The noncovalent interaction between the subunits in the dimer is weak. II^{Glc} sediments as a dimer in sucrose gradients at 4°C but as a monomer at 15°C (Meins et al., 1988). The III^{Man} subunit of the mannose transporter forms homodimers as shown by sedimentation equilibrium centrifugation and sodium dodecylsulfate (SDS) polyacrylamide gel electrophoresis (Erni and Zanolari, 1985; Erni et al., 1987). The subunit contacts are established between the N-terminal P13 domain (Erni et al., 1989). The stability of the dimer against dissociation by SDS is increased by phosphorylation of the subunits. The dimer is stable at pH above 7.0 but labile below pH 6, indicating that a (phosphorylated) histidyl residue is important for the stabilization of the dimer. The mutant H10Q of III^{Man} does not form dimers on polyacrylamide gels and equilibrium sedimentation indicates a monomer–dimer equilibrium (B. Erni and A. Lustig, unpublished results). It appears that the monomer interaction is labile and strongly dependent on the conditions under which an experiment is performed. The subunit stoichiometry of the two membrane-bound subunits $II-P^{Man}$ and $II-M^{Man}$ appears to be 1:1 as concluded from the fusion experiments referred to above. Analytical ultracentrifugation experiments of the purified $II-P^{Man}/II-M^{Man}$ complex failed due to physical heterogeneity of the purified protein. Both dimeric and monomeric forms of the mannitol transporter have been observed by gel filtration and on polyacrylamide gels (Roossien and Robillard, 1984; Stephan and Jacobson, 1986; Khandekar and Ja-cobson, 1989). A nonlinear relation between II^{Mtl} activity and protein concentration was described by Stephan and Jacobson (1986). It cannot be excluded, however, that the low activity at very low protein concentration was due to nonspecific adsorption of the protein to the reaction vessel (and subsequent denaturation) rather than to dissociation of dimers. Lolkema and Robillard (1990) concluded that II^{Mtl} forms a cooperative dimer, with the dimer having higher activity than the monomer. While the *in vitro* experiments mentioned above were generally interpreted to mean that dimerization is important for sugar translocation and phosphorylation, a number of *in vivo* complementation experiments indicate that dimerization is important for phosphoryl transfer between sites 1 and 2 within the transporters. Two different His mutants of III^{Man}, H10Q and H175N, which are inactive by themselves, regain activity when mixed together *in vitro* (Erni *et al.*, 1989). van Weeghel *et al.* (1991c) showed similar *in vitro* complementation between the II^{Mtl} mutants, H554A lacking the first phosphorylation site and C384S lacking the second phosphorylation site. Complementation is possible because of phosphoryl transfer between mutant monomers in a heterodimer. In addition to this intragenic complementation, intergenic complementation was observed (Fig. 1, dashed arrows). The absence of III^{Glc}, the hydrophilic subunit of the glucose transporter containing the first phosphorylation site, could be compensated by either II^{Nag} or II^{Bgl} (Vogler *et al.*, 1988; Schnetz *et al.*, 1990). Vice versa, III^{Glc} could complement II^{Nag} and II^{Bgl} lacking phosphorylation sites 1 due to deletions of or mutations in their C-terminal domains (Vogler and Lengeler, 1988; Schnetz *et al.*, 1990). These *in vivo* results suggest that heterooligomers between PTS transporters of different sugar specificity must occur and that they permit the efficient distribution and equilibration of phosphoryl groups between phosphorylation sites 1 and 2 belonging to different transporters.

What then is the function of dimerization? Is it important for sugar translocation because the two subunits together form a single translocation pathway? Or is it important for the "horizontal" phosphoryl exchange between the different transporters? There is probably no answer yet to this question because sugar translocation and phosphorylation are difficult to assay independently of phosphoryl transfer between sites 1 and 2. However, it appears unlikely that dimerization can be important for both sugar translocation and transphosphorylation. It is difficult to determine what the sugar specificity of a heterodimer would be, if each monomer contributed to the sugar translocation pathway. And a fast exchange of subunits between heterodimers (for intersite transphosphorylation) and homodimers (for sugar translocation) is also not very likely. The intersubunit contact sites are in the hydrophilic domains of III^{Man}, II^{Mtl}, and II^{Glc}, where the phosphorylation sites are also localized. Dimer interaction in III^{Man} is mediated by the protein domain containing phosphorylation site 1 (P13) while the second domain does not dimerize (Erni *et al.*, 1989). An intersubunit disulfide bridge is rapidly established between the two phosphorylatable Cys-421 in the purified glucose transporter (Meins *et al.*, 1988), indicating close contact between the hydrophilic domains. Similarly, a region between phosphorylation sites 1 and 2 of II^{Mtl} appears to be important for dimer formation and productive phosphory transfer (White and Jacobson, 1990). Taken together, these results could mean that dimerization is a means of regulating the flow of phosphoryl groups between the transporters and thus of regulating transport activity.

IV. Perspectives

The sugar transporters of the bacterial phosphotransferase system are a still growing group of bacterial membrane proteins. They are built of characteristic, structurally related modules that are similar enough to cross-react functionally with each other and to be recombined to form hybrid proteins. The modular design directly reflects the dual function of the transporters: catalysis of solute translocation coupled with solute phosphorylation and intracellular signal transduction. The bacterial phosphotransferase system is thus an exquisite model for how apparently independent cellular processes like transport, sensory stimulation, regulation of metabolic pathways, and gene expression are mechanistically linked with each other. The phosphorylation function is convenient for assaying the catalytic activity without a need to reconstitute the transporter into proteoliposomes. Reconstitution is, nevertheless, possible as demonstrated for the mannitol transporter by Elferink et al. (1990). The modular structure allows the protein to be broken down into smaller units that are more amenable for structural characterization. By judiciously combining elements from the different transporters, it might eventually become possible to engineer PTS transporters with novel substrate specificities.

Acknowledgments

I would like to thank all the colleagues who communicated their results to me before publication. Work in my laboratory was supported by Grants Er 147/1-1 and Er 147/2-1 from the Deutsche Forschungsgemeinschaft.

References

Adler, J., and Epstein, W. (1974). Proc. Natl. Acad. Sci. U.S.A. 71, 2895–2899.
Alpert C. A., Frank, R., Stüber, K., Deutscher, J., and Hengstenberg, W. (1985). Biochemistry 24, 959–964.
Amster-Choder, O., and Wright, A. (1990). Science 249, 540–542.

Amster-Choder, O., Houman, F., and Wright, A. (1989). *Cell (Cambridge, Mass.)* **58**, 847–855.

Anderson, B., Weigel, N., Kundig, W., and Roseman, S. (1971). *J. Biol. Chem.* **246**, 7023–7033.

Begley, G. S., Hansen, D. E., Jacobson, G. R., and Knowles, J. R. (1982). *Biochemistry* **21**, 5552–5556.

Bramley, H. F., and Kornberg, H. L. (1987). *Proc. Natl. Acad. Sci. U.S.A.* **84**, 4777–4780.

Buhr, A., Daniels, G. A., and Erni, B. (1992). *J. Biol. Chem.* **267**, 3847–3851.

Chin, A. M., Feucht, B. U., and Saier, M. H. (1987). *J. Bacteriol.* **169**, 897–899.

Danchin, A., ed. (1989). *FEMS Microbiol. Rev.* **63.**

Dean, D. A., Reizer, J., Nikaido, H., and Saier, M. H. (1990). *J. Biol. Chem.* **265**, 21005–21010.

Débarbouillé, M., Arnaud, M., Fouet, A., Klier, A., and Rapoport, G. (1990). *J. Bacteriol.* **172**, 3966–3973.

Delbaere, L. T. J., Vandonselaar, M., Quail, J. W., Waygood, E. B., and Lee, J. S. (1989). *J. Biol. Chem.* **264**, 18645–18646.

de Reuse, H., and Danchin, A. (1988). *J. Bacteriol.* **170**, 3827–2837.

de Reuse, H., and Danchin, A. (1991). *J. Bacteriol.* **173**, 727–733.

Dörschug, M., Frank, R., Kalbitzer, H. R., Hengstenberg, W., and Deutscher, J. (1984). *Eur. J. Biochem.* **144**, 113–119.

Elferink, M. G. L., Driessen, A. J. M., and Robillard, G. T. (1990). *J. Bacteriol.* **172**, 7119–7125.

El-Kabbani, O. A. L., Waygood, E. B., Brayer, G. D., and Delbaere, L. T. J. (1984). *Acta Crystallogr. Sect. A* **A40**, Suppl., C-35.

Erni, B. (1986). *Biochemistry* **25**, 305–312.

Erni, B. (1989). *FEMS Microbiol. Rev.* **63**, 13–24.

Erni, B. (1990). *Res. Microbiol.* **141**, 360–364.

Erni, B., and Zanolar, B. (1985). *J. Biol. Chem.* **260**, 15495–15503.

Erni, B., and Zanolari, B. (1986). *J. Bol. Chem.* **261**, 16398–16403.

Erni, B., Zanolari, B., and Kocher, H. P. (1987). *J. Biol. Chem.* **262**, 5238–5247.

Erni, B., Zanolari, B., Graff, P., and Kocher, H. P. (1989). *J. Biol. Chem.* **264**, 18733–18741.

Fouet, A., Arnaud, M., Klier, A., and Rapoport, G. (1987). *Proc. Natl. Acad. Sci. U.S.A.* **84**, 8733–8777.

Fox, D. K., Meadow, N. D., and Roseman, S. (1986). *J. Biol. Chem.* **261**, 13498–13503.

Geerse, R. H., Ruig, C. R., Schuitema, A. R. J., and Postma, P. W. (1986). *Mol. Gen. Genet.* **203**, 435–444.

Geerse, R. H., Izzo, F., and Postma, P. W. (1989). *Mol. Gen. Genet.* **216**, 517–525.

Génovésio-Taverne, J. C., Sauder, U., Pauptit, R. A., Jansonius, J. N., and Erni, B. (1990). *J. Mol. Biol.* **216**, 515–517.

Gonzy-Tréboul, G., Zagorec, M., Rain-Guion, M. C., and Steinmetz, M. (1989). *Mol. Microbiol.* **3**, 103–112.

Grisafi, P. L., Scholle, A., Sugiyama, J., Briggs, C., Jacobson, G. R., and Lengeler, J. W. (1989). *J. Bacteriol.* **171**, 2719–2727.

Grübl, G. A., Vogler, A. P., and Lengeler, J. W. (1990). *J. Bacteriol.* **172**, 5871–5876.

Hess, J. F., Oosawa, N. K., and Simon, M. I. (1988a). *Cell (Cambridge, Mass.)* **53**, 79–87.

Hess, J. F., Bourret, R. B., and Simon, M. I. (1988b). *Nature (London)* **336**, 139–143.

Houman, F., Diaz-Torres, M. R., and Wright, A. (1990). *Cell (Cambridge, Mass.)* **62**, 1153–1163.

Hummel, U., Nuoffer, C., Zanolari, B., and Erni, B. (1992). *Protein Science* (in press).

Jacobson, G. R., Lee, C. A., Leonard, J. E., and Saier, M. H. (1983). *J. Biol. Chem.* **258**, 10748–10756.

Kalbitzer, H. R., Hengstenberg, W., Rosch, P., Muss, P., Bernsmann, P., Engelmann, R., Dörschug, M., and Deutscher, J. (1982). *Biochemistry* **21**, 2879–2885.
Kapadia, G., Reizer, J., Sutrina, S. L., Saier, M. H., Reddy, P., and Herzberg, O. (1990). *J. Mol. Biol.* **212**, 1–2.
Khandekar, S. S., and Jacobson, G. R. (1989). *J. Cell. Biochem.* **39**, 207–216.
Klevit, R. E., and Waygood, E. B. (1986). *Biochemistry* **25**, 7774–7781.
Kofoid, E. C., and Parkinson, J. S. (1988). *Proc. Natl. Acad. Sci. U.S.A.* **85**, 4981–4985.
Kundig, W. (1974). *J. Supramol. Struct.* **2**, 695–714.
Kundig, W., and Roseman, S. (1971). *J. Biol. Chem.* **246**, 1407–1418.
Kundig, W., Gosh, S., and Roseman, S. (1964). *Proc. Natl. Acad. Sci. U.S.A.* **52**, 1067–1074.
Lee, C. A., and Saier, M. H. (1983). *J. Biol. Chem.* **253**, 10761–10767.
Lengeler, J. W. (1990). *Biochim. Biophys. Acta* **1018**, 155–159.
Lengeler, J. W., and Vogler, A. P. (1989). *FEMS Microbiol. Rev.* **63**, 81–93.
Lengeler, J. W., Titgemeyer, F., Vogler, A. P., and Wöhrl, B. M. (1990). *Philos. Trans. R. Soc. London, Ser. B* **326**, 489–504.
Liberman, E., Saffen, D., Roseman, S., and Peterkofsky, A. (1986). *Biochem. Biophys. Res. Commun.* **141**, 1138–1144.
Lolkema, J. S., and Robillard, G. T. (1990). *Biochemistry* **29**, 10120–10125.
Lolkema, J. S., Dijkstra, D. S., ten Hoeve-Duurkens, R. H., and Robillard, G. T. (1990). *Biochemistry* **29**, 10659–10663.
Martin-Verstraete, I., Débarbouillé, M., Klier, A., and Rapoport, G. (1990). *J. Mol. Biol.* **214**, 657–671.
Meadow, N. D., Fox, D. K., and Roseman, S. (1990). *Annu. Rev. Biochem.* **59**, 497–542.
Meins, M., Zanolari, B., Rosenbusch, J. P., and Erni, B. (1988). *J. Biol. Chem.* **263**, 12986–12993.
Misko, T. P., Mitchell, W. J., Meadow, N. D., and Roseman, S. (1987). *J. Biol. Chem.* **262**, 16261–16266.
Mitchell, W. J., Saffen, D. W., and Roseman, S. (1987). *J. Biol. Chem.* **262**, 16254–16260.
Mueller, E. G., Khandekar, S. S., Knowles, J. R., and Jacobson, G. R. (1990). *Biochemistry* **29**, 6892–6896.
Nelson, S. O., Wright, J. K., and Postma, P. W. (1983). *EMBO J.* **2**, 715–720.
Niwano, M., and Taylor, B. L. (1982). *Proc. Natl. Acad. Sci. U.S.A.* **79**, 11–15.
Nuoffer, C., Zanolari, B., and Erni, B. (1988). *J. Biol. Chem.* **263**, 6647–6655.
Osumi, T., and Saier, M. H. (1982). *Proc. Natl. Acad. Sci. U.S.A.* **79**, 1457–1461.
Pas, H. H., and Robillard, G. T. (1988). *Biochemistry* **27**, 5835–5839.
Peri, K. G., and Waygood, E. B. (1988). *Biochemistry* **27**, 6054–6061.
Peri, K. G., Goldie, H., and Waywood, E. B. (1990). *Biochem. Cell. Biol.* **68**, 123–137.
Plumbridge, J. A. (1990). *J. Bacteriol.* **172**, 2728–2735.
Poolman, B., Royer, T. J., Mainzer, S. E., and Schmidt, B. F. (1989). *J. Bacteriol.* **171**, 244–253.
Postma, P. W. (1981). *J. Bacteriol.* **147**, 382–389.
Postma, P. W. (1987). *In* "*Escherichia coli* and *Salmonella typhimurium:* Cellular and Molecular Biology" (F. C. Neidhardt, J. L. Ingraham, K. B. Low, B. Magasanik, M. Schaechte, and H. E. Umbarger, eds.), pp. 127–141. Am Soc. Microbiol. Washington, D.C.
Postma, P. W., and Lengeler, J. (1985). *Microbiol. Rev.* **49**, 232–269.
Postma, P. W., Epstein, W., Schuitema, A. R. J., and Nelson, S. O. (1984). *J. Bacteriol.* **158**, 351–353.
Presper, K. A., Wong, C. Y., Liu, L., Meadow, N. D., and Roseman, S. (1989). *Proc. Natl. Acad. Sci. U.S.A.* **86**, 4052–4055.

Prior, T. I., and Kornberg, H. L. (1988). *J. Gen. Microbiol.* **134**, 2757–2768.

Reizer, J., Saier, M. H., Deutscher, J., Grenier, F., Thompson, J., and Hengstenberg, W. (1988). *CRC Crit. Rev. Microbiol.* **15**, 297–338.

Reizer, J., Sutrina, S. L., Saier, M. H., Stewart, G. C., Peterkofsky, A., and Reddy, P. (1989). *EMBO J.* **8**, 2110–2120.

Robillard, G. T., and Blaauw, M. (1987). *Biochemistry* **26**, 5796–5803.

Robillard, G. T., and Lolkema, J. S. (1988). *Biochim. Biophys. Acta* **947**, 493–520.

Rogers, M. J., Ohgi, T., Plumbridge, J., and Söll, D. (1988). *Gene* **62**, 197–207.

Roossien, F. F., and Robillard, G. T. (1984). *Biochemistry* **23**, 5682–5685.

Ruijter, G. J. G., Postma, P. W., and van Dam, K. (1990). *J. Bacteriol.* **172**, 4783–4789.

Saffen, D. W., Presper, K. A., Doering, T. L., and Roseman, S. (1987). *J. Biol. Chem.* **262**, 16241–16253.

Saier, M. H. (1989a). *Microbiol. Rev.* **53**, 109–120.

Saier, M. H. (1989b). *FEMS Microbiol. Rev.* **63**, 183–192.

Saier, M. H., Cox, D. F., and Moczydlowski, E. G. (1977). *J. Biol. Chem.* **252**, 8908–8916.

Saier, M. H., Wu, L. F., and Reizer, J. (1990). *Trends Biochem. Sci.* **15**, 391–395.

Schnetz, K., and Rak, B. (1990). *Proc. Natl. Acad. Sci. U.S.A.* **87**, 5074–5078.

Schnetz, K., Toloczyki, C., and Rak, B. (1987). *J. Bacteriol.* **169**, 2579–2590.

Schnetz, K., Sutrina, S. L., Saier, M. H., and Rak, B. (1990). *J. Biol. Chem.* **265**, 13464–13471.

Scholte, B. J., and Postma, P. W. (1981). *Eur. J. Biochem.* **114**, 51–58.

Stephan, M. M., and Jacobson, G. R. (1986). *Biochemistry* **25**, 4046–4051.

Stock, J. B., Waygood, E. B., Meadow, N. D., Postma, P. W., and Roseman, S. (1982). *J. Biol. Chem.* **257**, 1453–14552.

Stock, J. B., Ninfa, A. J., and Stock, A. M. (1989). *Microbiol. Rev.* **53**, 450–490.

Sutrina, S. L., Reddy, P., Saier, M. H., and Reizer, J. (1990). *J. Biol. Chem.* **265**, 18581–18589.

Taylor, B. L., and Lengeler, J. W. (1990). *In* "Membrane Transport and Information Storage," pp. 69–90. Alan R. Liss, New York.

Thompson, J. (1987). *FEMS Microbiol. Rev.* **46**, 221–231.

Thompson, J., and Chassy, B. M. (1985). *J. Bacterial* **162**, 224–234.

van Weeghel, R. P., Meyer, G., Pas, H. H., Keck, W., and Robillard, G. T. (1991a). *Biochemistry,* **30**, 9478–9485.

van Weeghel, R. P., Meyer, G., Keck, W., and Robillard, G. T. (1991b). *Biochemistry* **30**, 1774–1779.

van Weeghel, R. P., van der Hoek, Y. Y., Pas, H. H., Elferink, M., Keck, W., and Robillard, G. T. (1991c). *Biochemistry* **30**, 1768–1773.

Vogler, A. P., and Lengeler, J. W. (1988). *Mol. Gen. Genet.* **213**, 175–178.

Vogler, A. P., and Lengeler, J. W. (1989). *Mol. Gen. Genet.* **219**, 97–105.

Vogler, A. P., Broekhuizen, C. P., Schuitema, A., Lengeler, J. W., and Postma, P. W. (1988). *Mol. Microbiol.* **2**, 719–726.

Weigel, N., Kukuruzinska, M. A., Nakazawa, A., Waygood, E. B., and Roseman, S. (1982). *J. Biol. Chem.* **257**, 14477–14491

White, D. W., and Jacobson, G. R. (1990). *J. Bacteriol.* **172**, 1509–1515.

Wilson, T. H., Yunker, P. L., and Hansen, C. L. (1990). *Biochim. Biophys. Acta* **1029**, 113–116.

Wittekind, M., Reizer, J., Deutscher, J., Saier, M. H., and Klevit, R. E. (1989). *Biochemistry* **28**, 9908–9912.

Wittekind, M., Reizer, J., and Klevit, R. E. (1990). *Biochemistry* **29** 7191–7200.

Wöhrl, B. M., and Lengeler, J. W. (1990). *Mol. Microbiol.* **4**, 1557–1565.

Wöhrl, B. M., Wehmeier, U. F., and Lengeler, J. W. (1990). *Mol. Gen. Genet.* **224,** 193–200.
Wootton, J. C., and Drummond, M. H. (1989). *Protein Eng* **2,** 535–543.
Wu, L. F., and Saier, M. H. (1990). *J. Bacteriol.* **172,** 7167–7178.
Wu, L. F., Tomich, J. M., and Saier, M. H. (1990). *J. Mol. Biol.* **213,** 687–703.
Zukowski, M. M., Miller, L., Cosgwell, P., Chen, K., Aymerich, S., and Steinmetz, M. (1990). *Gene* **90,** 153–155.

Sugar–Cation Symport Systems in Bacteria*

Peter J. F. Henderson, Stephen A. Baldwin, Michael T. Cairns, Bambos M. Charalambous, H. Claire Dent, Frank Gunn, Wei-Jun Liang, Valerie A. Lucas, Giles E. Martin, Terry P. McDonald, Brian J. McKeown, Jennifer A. R. Muiry, Kathleen R. Petro, Paul E. Roberts, Karolyn P. Shatwell, Glenn Smith, and Christopher G. Tate
Department of Biochemistry, University of Cambridge, Cambridge CB2 1QW, England

I. Introduction

In 1963 P. Mitchell suggested that the accumulation of sugars by microbial cells might be linked to a transmembrane electrochemical gradient of protons. This hypothesis accounted for the respiration dependence and uncoupler sensitivity of the lactose transport system of *Escherichia coli* (Kennedy, 1970), and was analogous to Mitchell's theory that a transmembrane electrochemical gradient of protons was the fundamental high-energy intermediate in oxidative and photosynthetic phosphorylation (Mitchell, 1961, 1963). The idea required that an individual transport system catalyze the simultaneous translocation of protons with a sugar molecule, "symport," or the experimentally indistinguishable "antiport" of hydroxyl ions. Energy derived from respiration or ATP hydrolysis is "stored" as the electrochemical gradient of protons (Fig. 1), and can therefore drive accumulation of the nutrient against a concentration gradient (Mitchell, 1963, 1973). However, this brilliant prediction remained untested until 1970, when I. West devised experimental conditions in which the movement of lactose or substrate analogs into *E. coli* cells containing the lactose transport protein (Lac Y) evoked an alkaline pH change showing proton movement in the same direction (West, 1970; West and Mitchell, 1972, 1973). Since then the expression of the *lacY* gene has been amplified (Teather *et al.*, 1978), the protein has been solubilized and

* This article is dedicated to Peter Mitchell, whose ideas stimulated the characterization of bacterial sugar–cation symport systems.

purified (Newman and Wilson, 1980; Newman *et al.*, 1981), its structure has been explored by immunological and chemical methods (Carrasco *et al.*, 1986; Page and Rosenbusch, 1988; Kaback, 1989), and its molecular mechanism has been investigated by mutagenesis (Kaback, 1987, 1989; Roepe *et al.*, 1990; Brooker, 1990). The structure and activity of LacY are reviewed in detail in this volume by H. R. Kaback.

More recently, other cation-linked transport systems for L-arabinose, D-xylose, D-galactose, L-fucose, L-rhamnose, and raffinose[1] have been discovered in *E. coli* and several of the Enterobacteriaceae, for D-glucose in cyanobacteria, and for lactose in streptococci. Remarkably, some of these proteins are closely related to glucose transport proteins occurring in eukaryotes as diverse as yeasts, algae, plants, protozoa, and mammals, including humans (Maiden *et al.*, 1987; Baldwin and Henderson, 1989; Henderson and Maiden, 1990; Henderson, 1990, 1991), while others have primary sequences that are not recognizably related at all, despite the similarity of their function. The glucose–Na^+ transporter of rabbits and humans is different in primary sequence from all the other sugar transporters characterized, but it is homologous to the proline–Na^+ and pantothenate–Na^+ transporters of *E. coli* (Hediger *et al.*, 1987, 1989; Jackowski and Alix 1990; Turk *et al.*, 1991).

The relationship between prokaryote and eukaryote sugar transport proteins is one theme of this article, which will describe selected properties of the bacterial sugar–H^+ transport proteins—their substrate specificites, the effects of certain inhibitors, their genes, and the amino acid sequences of the proteins. The recurrence of certain motifs in the primary sequences is highlighted, together with the presence or absence of residues that might be expected to interact with the sugars, the proton, or the inhibitors. A similarity in structure is proposed between these proteins and others that transport the apparently unrelated substrates citrate, α-ketoglutarate, quaternary ammonium compounds, quinolones, and tetracycline and a unifying two-dimensional model is presented for the structure of all the homologous proteins in the membrane.

Additional mechanisms exist for energizing the transport of sugars and other nutrients into *E. coli,* and for expelling toxins, for example the phosphotransferase systems, ATP-dependent binding protein systems, hexosephosphate/P_i antiports and ATP-dependent arsenate/arsenite extrusion. These are the subjects of other reviews (G. F.-L. Ames and B. Erni, this volume; see also Henderson, 1986; Furlong, 1987; Rosen, 1990; Lengeler *et al.*, 1990; Kornberg, 1990; Quiocho, 1990; Higgins *et al.*, 1990a; Maloney, 1990a,b). For the reader who wishes to follow the progress of the sugar molecules into the metabolic pathways after the initial transport event, reviews by Cooper (1986) and Lin (1987) are highly recommended.

---

[1] The D- or L-configurations of sugars will be omitted unless ambiguity could arise.

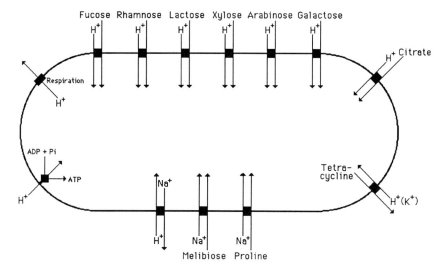

FIG. 1   An illustration of ion-linked transport systems in *Escherichia coli*. The large oval represents the cytoplasmic membrane of the microorganism. A transmembrane electrochemical gradient of protons is generated by respiration, depicted on the top left. This gradient is used for the energization of ATP synthesis (left) and transport as follows. A series of different sugar–H$^+$ symport proteins (top) catalyze accumulation of their individual substrate at the expense of the proton gradient. A similar mechanism operates for citrate uptake (top right) and for some other nutrients (not shown). By contrast, a tetracycline–H$^+$ antiport protein (bottom right) catalyzes efflux of tetracycline at the expense of the proton gradient. Similarly, an Na$^+$–H$^+$ antiporter catalyzes efflux of Na$^+$ at the expense of the proton gradient (bottom). The resulting transmembrane electrochemical gradient of sodium ions is used to energize melibiose uptake or proline uptake via individual substrate–Na$^+$ symport proteins (bottom). References are given in the text.

II. The Number of Proton-Linked Sugar Transport Systems in Bacteria

A. Experimental Criteria

1. Measurement of Sugar–H$^+$ Symport

A proton-linked sugar transport system can be most readily identified in the following type of experiment, in which uphill proton movement is measured in response to an inwardly directed substrate gradient. A concentrated suspension of bacteria is depleted of endogenous substrate and ATP, and resuspended under anaerobic conditions to prevent respiration. A relatively high concentration (1–5 mM) of sugar is added. If the sugar is a substrate for a sugar–H$^+$ symport system present in the cells (these have usually been grown in the presence of a suitable inducer), it diffuses inward down the concentration gradient. This results in an alkaline pH change in the medium reflecting the movement of protons with the sugar. The experimental details are described by West (1970) and by Henderson and Macpherson (1986). Quite often the initial alkaline pH change, which is rarely greater than 0.1 pH units, is obscured by subsequent metabolism of the sugar to acidic end products. Where this is the case the use of appropriate nonmetabolized sugar analogs (see Section IV, F below), or of mutants impaired in a metabolic enzyme, can facilitate detection of the pH change (Henderson and Macpherson, 1986).

2. Measurement of Respiration-Energized Transport in Vesicles

The second method requires the manufacture of right side-out subcellular vesicles by the method of Kaback (1972, 1974). The use of vesicles has several advantages: the binding protein and phosphotransferase types of transport system are not operative; the metabolic enzymes are virtually absent; and vesicles are much more sensitive than intact cells to ionophore and protonophore reagents that modify the electrochemical gradient of protons across the membrane. Transport of radioisotope-labeled sugars is energized by respiration (ascorbate plus tetramethylphenylenediamine is most effective; Kaback, 1974). No net pH change occurs during respiration-energized transport, but the sensitivity to protonophores and ionophores reveals its proton-linked nature. Sugar-promoted pH changes can be observed in such vesicles under deenergized conditions (Patel *et al.*, 1982; Horne and Henderson, 1983), but the amounts of vesicles required and their relative "leakiness" to protons means that this assay is more conveniently practised on intact cells.

3. Other Assays

Sugar–H^+ symport proteins will also catalyze facilitated diffusion of radioisotope-labeled sugar across the membrane, either into or out of the cells/vesicles (net influx or efflux, respectively; Viitanen *et al.*, 1986; Wilson *et al.*, 1986). Various counterflow experiments can be devised, in which the protein catalyzes exchange of the same or different labeled/ unlabeled substrates across the membrane, each at the same or different concentrations (Viitanen *et al.*, 1986; Walmsley, 1988). A particularly useful variant of this is the "overshoot" experiment (Wilson *et al.*, 1986; Viitanen *et al.*, 1986; Stein, 1986), because it is very sensitive and therefore valuable in reconstitution and purification assays, where it is not convenient to energize transport and only small amounts of transport protein may be available.

4. Value of Different Assay Methods

The existence of a repertoire of assay methods is invaluable, because it constitutes a series of "partial" reactions with which to examine the activities of mutants. In the absence of a means of determining the three-dimensional structure of the protein it is essential to show that this has not been seriously compromised in a mutant by showing that at least one of the assayable activities is unimpaired while characterizing those others that have been changed (see, e.g., Kaback, 1987; Roepe *et al.*, 1990).

Unfortunately, these assays are often performed in a semiquantitative way, expressing activities as percentages of those measured in wild-type controls. Measurements of reproducibility are not always given, and advantage is not taken of computerized methods of calculating best-fit values of kinetic "constants" and their standard deviations (Cleland, 1979; Henderson, 1985). It is always desirable to extrapolate the determination of K_m and k_{cat} values to conditions where all substrates, i.e., the sugar *and* the cation, are saturating (Cleland, 1979; Fromm, 1979). It is essential to measure k_{cat} and not just V_{max}, especially in mutants or recombinants where the level of expression of a protein may well have changed by a factor of two or more. Furthermore, the value of an "apparent" K_m for one substrate determined at a less-than-saturating concentration of the other is an arbitrary number, which will vary in an *apparently* unsystematic way when, for example, comparing mutants and wild type, if account is not taken of the steady state mechanism of the overall process, i.e., ordered or random addition and leaving of reactants, with steps at thermodynamic equilibrium or not (see, e.g., Severin *et al.*, 1989). Substrate specificities are best expressed quantitatively as the k_{cat}/K_m ratio (Fersht, 1985). The reasons for this are discussed by King and Wilson (1990a).

B. Different Sugar–H⁺ Transport Systems in *Escherichia coli*

Using the first two assay methods described above six different sugar–H⁺ systems were identified in *E. coli*. Their physiological substrates are lactose (LacY; West, 1970), galactose (GalP; Henderson *et al.*, 1977), xylose (XylE; Lam *et al.*, 1980), arabinose (AraE; Daruwalla *et al.*, 1981), fucose (FucP: Bradley *et al.*, 1987), and rhamnose (RhaT; Muiry, 1989). It seems likely that the raffinose transport system (Aslanidis *et al.*, 1990) will also turn out to be of this type in view of its homology to the lactose–H^+ symporter, although its cation selectivity is not yet characterized. Since the melibiose–Na^+ symporter can transport H^+ (Wilson et al., 1986; Wilson and Wilson, 1987; Leblanc et al., 1989) it is also included in this class; the glucuronide transport system should perhaps be included because of its similarity to the melibiose–Na^+ transporter, although there is no direct evidence of its cation selectivity. Some of the properties of all these proteins are summarized in Table I.

Further evidence indicating that each system requires a different protein (or proteins) resulted from the mapping of the genes at different locations on the E. coli chromosome (Table I). In addition each one had a different and characteristic specificity for a range of sugars and their structural analogs (Table II), as determined by the assays described above and discussed in Section IV,F.

TABLE I

Properties of Sugar Transport Proteins in Bacteria

Substrate	Organism	True M_r	Apparent M_r	Amino acids	Gene location (min)[a]	Inhibitors			K_m^{app} (μM)[d]
						NEM	CB[b]	Fsk[c]	
D-Glucose	Synechocystis PCC6803	49,743	—[e]	468	—	—	—	—	—
D-Glucose	Zymomonas mobilis	50,199	—	473	—	—	—	—	—
D-Xylose	Escherichia coli	53,603	39,000	491	91	Yes	No	No	70–170
L-Arabinose	Escherichia coli	51,683	37,000	472	61	Yes	Yes	No	60–320
D-Galactose	Escherichia coli	50,983	37,000	464	64	Yes	Yes	Yes	50–450
Lactose	Escherichia coli	46,502	30,000	417	8	Yes	No	No	50–900
Raffinose	Escherichia coli	46,700	—	425	—	—	—	—	1,000
Melibiose	Escherichia coli	52,202	39,000	469	92	Yes	—	—	300[f]
Glucuronide	Escherichia coli	49,916	—	456	36	—	—	—	132
Lactose	Streptococcus thermophilus	69,454	—	634	—	—	—	—	—
L-Fucose	Escherichia coli	47,773	32,000	439	60	No	No	No	18–38
L-Rhamnose	Salmonella typhimurium	37,390	27,000	344	88	No	No	No	16–60

[a] Location of the gene on the chromosome of E. coli.
[b] CB, Cytochalasin B.
[c] Fsk, Forskolin.
[d] The K_m for sugar has not necessarily been determined in the presence of saturating cosubstrate.
[e] —, The information is not yet known.
[f] The value depends on the identity of the cation cosubstrate.

TABLE II
Molecular Recognition by Sugar Transporters

Xylose–H+ symporter (E. coli)[a]	Arabinose–H+ symporter (E. coli)[b]	Galactose–H+ symporter (E. coli)[c]	Glucose transporter (erythrocyte)[d]	Glucose transporter (adipocyte)[e]	Fucose–H+ symporter (E. coli)[f,g]	Rhamnose–H+ symporter (E. coli)[g]
D-Xylose	L-Arabinose	D-Glucose	2-Deoxy-D-glucose	6-F-D-Galactose	L-Fucose	L-Rhamnose
	5-CH3-L-Arabinose (D-Fucose)	2-Deoxy-D-glucose	D-Glucose	D-Glucose	L-Galactose	L-Mannose
	5-FCH2-L-Arabinose (6-F-D-Galactose)	D-Galactose	6-Deoxy-D-glucose	6-Deoxy-D-glucose	D-Arabinose	L-Lyxose
	D-Xylose	6-F-D-Glucose	D-Mannose	2-Deoxy-D-galactose		
		6-F-D-Galactose	D-Galactose	D-Galactose		
		6-Deoxy-D-glucose	2-Deoxy-D-galactose	6-Deoxy-D-galactose		
		D-Talose	D-Xylose	D-Talose		
		2-Deoxy-D-galactose	L-Arabinose	3-Deoxy-D-glucose		
		D-Mannose	6-Deoxy-D-galactose	D-Xylose		
		6-Deoxy-D-galactose (D-Fucose)		L-Arabinose		
		D-Xylose				

[a] Davis (1986).
[b] Petro (1988).
[c] Henderson et al.(1977); Horne (1980); Henderson and Macpherson (1986).
[d] Lefevre (1961); Barnett et al. (1973).
[e] Rees and Holman (1981).
[f] Bradley et al. (1987).
[g] Muiry (1989).

C. Sugar–H⁺ Transport Systems in Other Bacteria

The occurrence of proton-linked sugar transport systems in different species of bacteria has not been systematically investigated. Consequently, their distribution *appears* to be rather capricious. The following summarizes the current state of knowledge, but it is expected that many more examples will soon be identified. In view of the occurrence of related proteins in organisms as diverse as the cyanobacteria, yeasts, plants, and mammals (see Section III) these proteins are of considerable general interest, and the acquisition of more information on their distribution in bacteria and higher organisms is needed in order to expose their evolutionary relationships.

1. Enterobacteriaceae

The lactose–H⁺ transporter occurs in *E. coli* and in species of *Klebsiella* (Buvinger and Riley, 1985; McMorrow *et al.*, 1988), but not in *Salmonella typhimurium*. Its evolution and distribution may be more related to the need to detoxify environmental galactosides rather than utilize them for growth (Lin, 1987). Rhamnose–H⁺ symport activity was found in *E. coli, S. typhimurium, Klebsiella* species, and *Erwinia* species, and both fucose–H⁺ and galactose–H⁺ activity occurred in the first three of these, whereas xylose–H⁺ symport was detected only in *E. coli* (T. P. McDonald, B. J. McKeown, and P. J. F. Henderson, unpublished observations). DNA homologous at high stringency to the *E. coli araE* gene encoding the arabinose–H⁺ symporter was readily detected in *S. typhimurium, Klebsiella* and *Enterobacter* species, but not in *Erwinia* or *Hafnia* (Charalambous *et al.*, 1989; G. E. Martin, B. Charalambous, and P. J. F. Henderson, unpublished observations). The sequence of the putative *araE* gene from *Klebsiella pneumoniae* was determined and found to encode an open reading frame, the predicted amino acid sequence of which is very similar to that of the *E. coli* AraE protein (B. Charalambous, T. P. McDonald, and K. P. Shatwell, unpublished observations). Charalambous *et al.* (1989) describe a combination of biochemical and genetical strategies by which sugar–H⁺ transport proteins can be detected in any organism.

2. Lactic Acid Bacteria

Dairy lactic acid bacteria utilize lactose as a carbon source. Uptake is effected by a phosphotransferase system, or in some species, such as *Streptococcus thermophilus, Lactobacillus bulgaricus, L. helveticus, L. brevis,* and *L. buchneri,* by a proton-linked mechanism (Hickey *et al.,* 1986). The gene for one of these, from *Strep. thermophilus,* has been cloned and sequenced. It resembles the melibiose–Na^+ and glucuronide transporters of *E. coli* with an additional C-terminal region like the enzyme III of phosphotransferase systems (Poolman *et al.*, 1989), and so is remarkably different from both the *E. coli* and the yeast lactose transporters.

3. Cyanobacteria

While most cyanobacteria are photosynthetic a few species can grow chemo- or photoheterotrophically on sugars as carbon source (Rippka *et al.*, 1979). Proton-linked glucose transporters were identified in several species, e.g., *Synechocystis, Nostoc,* and *Plectonema* (Raboy and Padan, 1978; Beauclerk and Smith, 1978; Flores and Schmetterer, 1986; Joset *et al.*, 1988). The gene from *Synechocystis* has been sequenced (Zhang *et al.*, 1989) and is similar to the *E. coli* xylose–H^+, arabinose–H^+, and galactose–H^+ proteins and the eukaryote sugar transporters described in the next section.

III. Sugar Transport Systems in Eukaryotes

A. *Chlorella kessleri*

Glucose–H^+ symport activity has long been reported (Komor and Tanner, 1974) in species of the alga *Chlorella vulgaris,* now renamed *Chlorella kessleri,* and the cDNA encoding the inducible hexose–H^+ symporter has been cloned and sequenced (Sauer and Tanner, 1989). The cDNA clone was isolated by differential screening of a cDNA library prepared from glucose-induced cells. The predicted open reading frame encodes a protein of 533 amino acids with about 30% identity to the bacterial and eukaryote sugar transporters that comprise the homologous family in Table III. The molecular weight of 57,445 was higher than the value of 47,000 determined from sodium dodecyl sulfate (SDS)-polyacrylamide gel electrophoresis (Sauer, 1986), a discrepancy that is also typical of bacterial sugar transport proteins (see Section X,E).

Direct proof that the *Chlorella* gene encoded glucose–H^+ symport activity was obtained by expressing it in yeast (*Schizosaccharomyces pombe*). Unlike the indigenous yeast transporter the *Chlorella* gene encoded an energy-dependent accumulation of 3-*O*-methylglucose or 6-deoxyglucose which, in deenergized conditions, catalyzed net accumulation of sugar in response to an imposed pH gradient (Sauer *et al.*, 1990a). The K_m values for glucose, 6-deoxyglucose, and 3-*O*-methylglucose were

TABLE III

Homologous Sugar Transport Proteins in Different Organisms

Sugar	Organism	Ref.
D-Glucose	*Synechocystis* PCC6803	Zhang *et al.* (1989)
D-Glucose	*Zymomonas mobilis*	Barnell *et al.* (1991)
D-Xylose	*Escherichia coli*	Maiden *et al.* (1987)
L-Arabinose	*Escherichia coli*	Maiden *et al.* (1987)
D-Galactose	*Escherichia coli*	Henderson (1990)
D-Galactose	*Saccharomyces cerevisiae*	Szkutnicka *et al.* (1989)
D-Glucose	*Saccharomyces cerevisiae*	Celenza *et al.* (1988)
Maltose	*Saccharomyces carlsbergensis*	Yao *et al.* (1989)
	Saccharomyces cerevisiae	Cheng and Michels (1989)
Lactose	*Kluyveromyces lactis*	Chang and Dickson (1988)
Quinate	*Neurospora crassa*	Geever *et al.* (1989)
D-Glucose	*Chlorella kessleri*	Sauer and Tanner (1989)
D-Glucose	*Arabidopsis thaliana*	Sauer *et al.* (1990b)
D-Glucose	*Leishmania donovanii*	Zilberstein *et al.* (1986)
	Leishmania enriettii	B. R. Cairns *et al.* (1989)
D-Glucose	Human hepatoma/erythrocyte	Mueckler *et al.* (1985)
	Rat brain	Birnbaum *et al.* (1986)
D-Glucose	Rat liver	Thorens *et al.* (1988)
	Human liver	Fukumoto *et al.* (1988)
D-Glucose	Rat adipocyte/heart/muscle	James *et al.* (1989)
(insulin	Mouse adipocyte/heart/muscle	Birnbaum (1989)
regulated)		Kaestner *et al.* (1989)
D-Glucose	Human small intestine	Gould and Bell (1990)
D-Glucose	Human fetal muscle	Gould and Bell (1990)

the same as measured in *Chlorella,* and so was the specificity for glucose, fructose, galactose, xylose, and D-arabinose (Sauer *et al.,* 1990a).

B. *Arabidopsis thaliana*

An energy-linked glucose transport protein has been identified in *Arabidopsis* by an elegant application of modern molecular biology (Sauer *et al.,* 1990b). A ^{32}P-labeled DNA probe from the coding region of the cDNA clone encoding the *C. kessleri* glucose–H[+] symporter was used to screen a genomic library of *Arabidopsis* DNA in λ EMBL4 (Sauer *et al.,* 1990b). An *Eco*RI fragment that strongly hybridized to the probe was derived from a positive EMBL4 clone and sequenced. The *Eco*RI fragment was 4447 bp long and contained the putative sequence of a gene interrupted by three introns. The locations of the introns were verified by sequencing the cDNA isolated by screening with a 400-bp fragment derived from a putative exon in the genomic clone. The predicted open reading frame contained 552 amino acids of M_r 57,581 (Sauer et al., 1990b), the sequence of which was 47.1% identical to the C. kessleri hexose–H^+ transporter, 31.6% identical to the E. coli arabinose transporter, and 28–30% identical to other prokaryote and eukaryote transporters in the homologous family in Table III. There was no apparent similarity to the other sugar–H^+ symport proteins for lactose, melibiose, fucose, or rhamnose in bacteria (Table I), nor with the mammalian glucose–Na^+ transporter. The conclusion that the Arabidopsis DNA encoded a sugar transport protein was confirmed by expressing it in S. pombe, where it caused highly significant rates of accumulation of glucose, 3-O-methylglucoside, galactose, and possibly fructose, but not sucrose (Sauer et al., 1990b). Unlike the indigenous yeast transporter the Arabidopsis gene encoded an energy-dependent sugar transport and is probably therefore a glucose–H^+ symport protein, although this was not confirmed by direct measurements.

This work opens the way to characterizing and cloning other sugar transport genes from plants, if low-stringency hybridization can detect related sequences in libraries of plant DNA. Both glucose and sucrose transporters are of profound importance in the distribution of energy resources around plant tissues (Madore and Lucas, 1987). The properties of a sucrose–H^+ symport activity in Beta vulgaris (sugar beet) have been reviewed (Bush, 1990), and a sucrose–H^+ transporter found in the cotyledons of Ricinus (Hutchings, 1978a,b; Daie, 1989) may also be a member of the same family. A useful way to test this would be to determine sensitivity of the sugar transport to cytochalasin B and to forskolin (see Section IV,B,C). However, the glucose–H^+ symport activities reported in plant cell membranes of Zea mays (maize) coleoptiles (Rausch et al., 1989), and in a cell line cultured from Chenopodium rubrum ("fat hen" or "goosefoot"; Gogarten and Bentrup, 1989), are more likely to be related to the glucose–Na^+ transporter of mammalian cells than to the homologous family in Table III, since the former is reported to cross-react with antibodies raised to the glucose–Na^+ protein and the latter is reported to be sensitive to phlorhizin (Gogarten and Bentrup, 1989; Rausch et al., 1989).

C. Leishmania

By the criterion of reaction with cytochalasin B (see Section IV,B) the protozoan parasite Leishmania donovani is likely to contain a glucose transport protein (Zilberstein et al., 1986; Baly and Horuk, 1988) related to the homologous family of sugar transporters (Zilberstein and Dwyer, 1985; Table III). There is also a developmentally regulated gene cloned and sequenced from *Leishmania enriettii* that has not yet been shown actually to catalyze transport (B. R. Cairns *et al.*, 1989). This is obviously homologous to the sugar transporters in Table III, but some of the motifs very highly conserved in all the others are missing or different in the amino acid sequence deducted from the DNA sequence of B. R. Cairns *et al.* (1989). It is expressed primarily in the insect or promastigote stage of the parasite life cycle. At least two isoforms exist, which differ only in the N-terminal hydrophilic domains (Stack *et al.*, 1990). This is different from the structural isoforms found in mammalian cells, where the differences are scattered throughout the sequences (see M. J. Birnbaum in this volume). It will be interesting to see if there are more of the isoforms in *Leishmania,* whether they have different kinetic and/or regulatory properties, and even whether their gene products are found in different subcellular locations.

D. Yeast Transport Proteins for Glucose, Galactose, Lactose, and Maltose

The occurrence of proton-linked and facilitated diffusion transport systems for sugars in various species of yeast is well established (Eddy, 1982; Kotyk, 1983; van Dijken and Scheffers, 1986; Romano, 1986; van Leeuwen *et al.*, 1991). The comprehensive paper by Kruckeberg and Bisson (1990) discusses the genetical, structural, and physiological significance of the glucose transporters in particular, and their conclusions are summarized here together with information from the previous papers.

There are now five genetically and biochemically distinct sugar transport systems in yeasts for which sequence information is available: lactose transporter (LAC12; Chang and Dickson, 1988), maltose transporter (MAL61; Cheng and Michels, 1989; Yao *et al.*, 1989), galactose transporter (GAL2; Szkutnicka *et al.*, 1989), and three glucose transporters (SNF3, HXT2, and HXT1; Celenza *et al.*, 1988; Kruckeberg and Bisson, 1990; Lewis and Bisson, 1990). All are members of the homologous group in Table III, but they all have "extra" regions upstream and downstream of the "sugar transporter" part. For example, in SNF3 these constitute 85 residues at the N terminus and 303 residues at the C terminus. These extra regions are probably involved with membrane targeting, regulation of activity, regulation of transcription/translation, or other special functions (Kruckeberg and Bisson, 1990; Marshall-Carlson *et al.*, 1990).

All the yeast sugar transporters are predicted to have 12 membrane-spanning segments. However, for HXT2, transmembrane domains 1, 2, 3, 4, 5, 7, and 12 are more strongly predicted to be β sheet than α helix, whereas α helix is more likely for transmembrane domains 6, 8, 9, and 10 (Kruckeberg and Bisson, 1990). Of course the algorithms for predicting secondary structure, especially of membrane proteins, are unlikely to be reliable (von Heijne, 1988). There are two consensus sites in HXT2 at which N-linked glycosylation may occur, at residues 82 and 299, which are predicted to be located on the extracellular face of the cytoplasmic membrane. The first of these is located between transmembrane domains 1 and 2, corresponding to the location of predicted glycosylation sites in all the mammalian GLUT transporters; the one in GLUT1 has been shown to exist on the extracellular side of the membrane (Carruthers, 1990). Also, a leucine zipper motif was observed in or near transmembrane domain 2 in HXT2, where it is also proposed to occur in the sequences of SNF3, GAL2, and several mammalian transporters of group I. Whether this is involved in the formation of homodimers or in the interaction of two proteins such as transporter and kinase remains to be established. In yeast high-affinity glucose transport depends on glucose kinase activity, possibly by direct interaction of the proteins (Kruckeberg and Bisson, 1990).

The HXT2 transporter is 65% identical to the GAL2 transporter, and 21–31% identical to other members of the group I family. Thus, genetical manipulation of the sequences of the first two, combined with quantitation of substrate binding, may be a useful experimental approach to elucidate molecular features of the sugar recognition process. The alignment of all these sequences is an excellent basis for the design of site-directed mutagenesis experiments, but so far it has not been possible unequivocally to identify residue(s) associated with proton translocation or with sugar recognition. However, Marshall-Carlson and co-workers (1990) isolated mutants in the *SNF3* gene impaired in sugar transport function. Two of the mutations, Gly-112→Asp and Gly-153→Arg introduced charged residues into predicted transmembrane domain 1 and transmembrane domain 2, respectively, with the apparent consequence that the protein failed to localize to the cell membrane. A third mutation changed Val-402→Leu in predicted transmembrane domain 8; this relatively conservative change did not affect localization to the membrane, but did affect transport (Marshall-Carlson *et al.*, 1990). Analyses of a series of C-terminal deletions and fusions to *lacZ* showed that the C-terminal region is important, but not essential, for transport function (Marshall-Carlson *et al.*, 1990). Further *in vivo* selection of yeast mutants altered in cation and/or sugar recognition is expected to be a powerful method of exploring transport, membrane targeting, and regulatory functions of yeast transporters.

In *Saccharomyces cerevisiae* at least three different sequenced genes are involved in glucose transport, *SNF3, HXT2,* and *HXT1* (Celenza *et al.*, 1988; Kruckeberg and Bisson, 1990; Lewis and Bisson, 1990). The *SNF3* gene is required for expression of high-affinity transport by facilitated diffusion, and its expression is repressed by high glucose concentrations. Kruckeberg and Bisson (1990) conclude that *HXT2* encodes a high-affinity transporter. Loss of either *SNF3* or *HXT2* leads to a deficiency in high-affinity glucose transport. However, a low level of high-affinity glucose uptake was evident in a double-null *SNF3 HXT2* mutant; this may be a consequence of the expression of one or more additional homologs of the glucose transporters, such as the *HXT1* gene, which has been isolated, and other similar genes detected by low-stringency Southern blotting.

It therefore seems likely that *S. cerevisiae*, at least, has a panoply of glucose transporters expressed and regulated in a highly controlled way in response to changes in environmental sugar concentration. Hence the proposed regulatory role of the ''extra'' amino acid sequences (see above). It is also perhaps significant that there are two consensus cAMP-dependent protein kinase phosphorylation sites at residues 266 and 539 in the HXT2 protein, predicted to be on the cytoplasmic side of the membrane. These and other observations are of particular importance for elucidating the physiology of the fermentation process as yeast grows and adapts from very high to very low sugar concentrations in its environment.

In view of the success in expressing *Chlorella* and *Arabidopsis* glucose transport proteins in *S. pombe* it would seem to be relatively straightforward to express the yeast sugar transporters in the same host. Provided that the expression of the relevant host cell sugar transport is relatively low, or that the appropriate transport-negative mutations are present, this could prove to be a convenient route for future investigations of both functional and structural features of the yeast transport proteins.

IV. Inhibitors of Proton-Linked Sugar Transport

Inhibitors have proved to be of great value in the characterization of transport proteins. For example, the combination of a covalent modifier, the sulfhydryl reagent *N*-ethylmaleimide, and an alternative substrate, ''thio-di-galactoside'' (TDG), which protected the LacY protein much more effectively against *N*-ethylmaleimide than the normal substrate lactose, enabled the first identification of a transport protein (Fox and Kennedy, 1965; Carter *et al.*, 1968; Jones and Kennedy, 1969; Kennedy, 1970). Subsequently, Kacszorowski *et al.* (1980) found that 4-nitrophenyl-α-D-galactoside (NPG) reacted covalently with the LacY protein when the two were exposed to UV light, and conditions were found that achieved reasonable specificity. This convenient labeling method was invaluable for assaying the protein during purification (Newman *et al.*, 1981; Viitanen *et al.*, 1986; Kaback, 1986, 1989). Furthermore, NPG is a tight-binding reversible ligand that can be used to quantitate the amount of both LacY and MelB proteins and to test the integrity of their sugar binding sites in mutants (see, e.g., Damiano-Forano et al., 1986; Overath et al., 1987; Kaback, 1987; Leblanc et al., 1989; Roepe et al., 1990).

Alternative substrates serve first to characterize the specificity of a transport protein, which is especially useful when comparing a homologous series of proteins, like those in Table III, or in comparing wild-type and mutant proteins (see Section IV,F below). They can also be useful to isolate transport from further metabolism. For example, D-talose, D-fucose, and to a lesser extent 2-deoxy-D-galactose are substrates for GalP, but not for galactokinase, the first enzyme of galactose metabolism; D-fucose is a substrate for AraE, but not for arabinose isomerase, the first enzyme of arabinose metabolism; L-galactose and D-arabinose are substrates for FucP, but not for L-fucose isomerase, the first enzyme of L-fucose metabolism; L-mannose and L-lyxose are substrates for RhaT, and are relatively poor substrates for the L-rhamnose isomerase, the first enzyme of L-rhamnose metabolism. Some sugar analogs are gratuitous inducers of transport that are often more effective than the natural substrate; thus, D-fucose is an inducer of the expression of the *galP* gene, but not the *araE* gene. Some are relatively poor substrates that are nevertheless more effective than the natural one for protecting the protein against covalent modification; this appears to be the case for 6-deoxyglucose, which is a better agent for protecting XylE against *N*-ethylmaleimide than xylose, although only the latter is a substrate for transport.

6-Deoxy-D-glucose, methyl-β-D-galactoside, cytochalasin B, and forskolin appear to be dead-end inhibitors for some of the sugar–H^+ transporters, but not all (see Section IV,B–E). Their availability will be important for exploiting kinetic measurements of the bacterial transporter activities to the level that has already been valuable in mechanistic studies of the mammalian glucose transporter (Deves and Krupka, 1978; Lowe and Walmsley, 1986; Walmsley, 1988).

In the rest of this section compounds are briefly described that are likely to be of similar value in future investigations of the relatively novel series of transporters listed in Table III.

A. *N*-Ethylmaleimide

When incubated with a sulfhydryl reagent, such as 1 mM *N*-ethylmaleimide for 15 min at 25°C, the *E. coli* proteins for arabinose–H^+, xylose–H^+, and galactose–H^+ transport lose 70–90% of their activity (Macpherson et al., 1981, 1983; Henderson and Macpherson, 1986; Davis, 1986). In general, the appropriate sugar substrates protect these proteins against

N-ethylmaleimide, except that methyl-β-D-galactoside, which is not a substrate for GalP (Henderson *et al.*, 1977), does protect it against *N*-ethylmaleimide, and 6-deoxyglucose, which is not a substrate for XylE, does protect it against *N*-ethylmaleimide. However, the fucose–H^+ or rhamnose–H^+ transport proteins are insensitive to *N*-ethylmaleimide under the same conditions (P. J. F. Henderson, T. P. McDonald, and S. A. Bradley, unpublished data).

Substrate protection against inhibition by, or reaction with, *N*-ethylmaleimide is also a property of the lactose–H^+, melibiose–Na^+, and proline–Na^+ symporters (Fox and Kennedy, 1965; Carter *et al.*, 1968; Hanada *et al.*, 1985; Damiano-Forano *et al.*, 1986), although the amino acid sequences of their proteins are different (see Sections VI,F and G).

The glucose transporter of human erythrocytes is also inhibited by sulfhydryl reagents (Bloch, 1974). However, in this case the presence of D-glucose *accelerates* the rate of inactivation by, e.g., *N*-ethylmaleimide (Rampal and Jung, 1987), whereas maltose or cytochalasin B protect the protein against *N*-ethylmaleimide (reviewed by Carruthers, 1990).

The implication of all these results is that the environment of a sulfhydryl group is affected by the binding of a sugar to the transport protein. It is not easy to identify the residue(s) involved in proteins expressed at wild-type levels. However, since there are no cysteine residues in the N-terminal half of the arabinose–H^+ transport protein (AraE), the region interacting with *N*-ethylmaleimide must be in the C-terminal half of this protein and, by implication, in all the homologous proteins. Furthermore, AraE, and the galactose–H^+ symporter (GalP), have been overexpressed to the point where reaction with *N*-ethylmaleimide can be measured directly (V. A. Lucas, H. C. Dent, and P. J. F. Henderson, unpublished data). Inhibition is protected against by the appropriate substrates, and yields approximate estimates of stoichiometry indicating that not all of the cysteine residues in each protein are available to the reagent. Directed mutagenesis of each of the three cysteine residues in GalP to a serine revealed that only Cys-374 is involved in substrate-protected reaction with *N*-ethylmaleimide (T. P. McDonald and P. J. F. Henderson, unpublished data). In the model of the folding of the protein in the membrane (see, e.g., Fig. 1) this is located on the inside of the membrane between predicted transmembrane domains 10 and 11. Perhaps this indicates the location of (part of) the sugar binding site. All the GalP mutants were capable of transporting galactose. It is interesting that the substrate-protected residue in the LacY protein that reacted with *N*-ethylmaleimide is located in a very different place in its proposed model, just inside transmembrane domain 4 (Menick *et al.*, 1987; Roepe *et al.*, 1990).

B. Cytochalasin B

Cytochalasin B is a fungal product that inhibits mammalian glucose transport proteins (Jung and Rampal, 1977) and binds covalently to them when
irradiated with UV light (Carter-Su et al., 1982; Shanahan, 1982; Baly and
Horuk, 1988). It inhibited the transport of arabinose by AraE into vesicles
made from appropriate induced strains of E. coli (T. P. McDonald and
P. J. F. Henderson, unpublished data). Binding of radioactive cytochalasin
B to the amplified arabinose transport protein (AraE) could be measured
directly, revealing a dissociation constant of 0.7–1.1 μM (Petro, 1988); the
cytochalasin B was displaced competitively by substrate, with a dissociation constant for arabinose of 20–30 mM (Petro, 1988).

Similar results were obtained with GalP (Cairns et al., 1991), except that
the dissociation constant for cytochalasin B was 3–7 μM, and for galactose
it was about 9 mM (P. E. Roberts and P. J. F. Henderson, unpublished
data). Thus, cytochalasin B bound much more tightly to both AraE and
GalP than substrate, reminiscent of the situation for a transition state
analog (Fersht, 1985). For both proteins the pattern of displacement of
cytochalasin B by different sugars was consistent with the substrate specificity obtained in transport assays, with the exception that methyl-β-D-
galactoside was an effective protecting agent but not a substrate for GalP,
and that 6-deoxy-D-glucose was an effective protecting agent, but not a
substrate, for AraE.

These results reinforce the conclusion that the prokaryote and eukaryote sugar transport proteins are structurally similar. The corollary is that
all transport proteins that interact with cytochalasin B are likely to be
related. The glucose transport protein of L. donovani (Zilberstein et al.,
1986; Baly and Horuk, 1988) is therefore one of this family of transport
proteins (Table III). Since the fucose–H$^+$, rhamnose–H$^+$, melibiose–
(Na$^+$, H$^+$), and lactose–H$^+$ transporters are not susceptible to cytochalasin B it might be concluded that they are not structurally related to this
family. However, this suggestion must be treated with caution because
attempts to inhibit or photolabel the XylE transporter with cytochalasin B
have failed, despite its homology to the proteins in Table III.

When vesicles from appropriately induced strains of E. coli or S. ty-
phimurium are irradiated with UV light in the presence of [^{3}H]cytochalasin
B, the label becomes covalently attached to the AraE or GalP proteins
(Petro, 1988; M. T. Cairns et al., 1989, 1991; Charalambous et al., 1989; G.
Smith, unpublished data). Substrates and inhibitors protect the proteins
against labeling with cytochalasin B.

Cytochalasin B will be an invaluable tool for explaining the differences
in specificity between these sugar transport proteins, for labeling and

assaying the proteins in purification experiments, and for the determination of their structures and molecular mechanisms.

C. Forskolin

The antibiotic forskolin inhibits mammalian glucose transporters, in addition to numerous other effects (Laurenza et al., 1989). By exploiting a high-affinity photolabel/inhibitor derivative of forskolin, 3-[^{125}I]iodo-4-azido-phenethylamino-7-O-succinyldeacetylforskolin, Wadzinski et al. (1990) came to the conclusion that it probably labels transmembrane domain 10 in the mammalian glucose transporter. The photolabeling was protected against by D-glucose and by cytochalasin B, which implies that the antibiotics are interacting at or near (part of) the sugar binding site. The forskolin analog also labeled a bacterial membrane protein; the pattern of protection by sugars indicated that this is probably the GalP transporter (Wadzinski et al., 1990; G. E. Martin, M. F. Shanahan, K. B. Seamon, and P. J. F. Henderson, unpublished results). We have demonstrated that forskolin inhibits transport activity of GalP by measuring sugar uptake in subcellular vesicles of E. coli. However, the transport activity of AraE or XylE is unaffected by the presence of forskolin. Thus, the AraE protein appears to bind cytochalasin B, but not forskolin. There are a variety of structural analogs of forskolin available (Robbins et al., 1991), which are likely to be of particular value in future explorations of the architecture of the sugar binding site in the GalP protein.

D. 6-Deoxy-D-glucose

This sugar analog (also called 5-methyl-D-xylose) was not a substrate for XylE in the sugar–H$^+$ symport assay, although it did inhibit the transport of xylose, the natural substrate (Davis, 1986; B. J. McKeown, unpublished data). 6-Deoxy-D-glucose was also more effective than xylose in protecting XylE against reaction with N-ethylmaleimide (E. O. Davis and P. J. F. Henderson, unpublished data). Although it was not a substrate for AraE in the sugar–H$^+$ symport assay, 6-deoxy-D-glucose did protect AraE against cytochalasin B. In contrast, 6-deoxy-D-glucose was a good substrate for GalP (Table II).

These observations imply that 6-deoxy-D-glucose either binds to the normal (external?) sugar binding site of XylE and AraE, but fails to trigger the events leading to sugar–H$^+$ symport, or that it binds to a second (internal?) site that may also interact with cytochalasin B (Walmsley, 1988).

E. Methyl-β-D-galactoside

Rotman *et al.* (1968) originally showed that methyl-β-D-galactoside is a substrate for one of the galactose transport systems of *E. coli,* but not for the other. When the GalP and MglP transport systems were separated by genetical manipulation, it turned out that methyl-β-D-galactoside is a substrate for the latter, but not for the former (Henderson *et al.*, 1977, 1983b). However, it was still very effective in protecting the GalP protein against reaction with *N*-ethylmaleimide (Henderson *et al.*, 1983b). Furthermore, methyl-β-D-galactoside again appeared to be an effective agent in protecting the GalP protein against reaction with cytochalasin B (Cairns *et al.*, 1991). These observations imply that, like the interaction of 6-deoxy-D-glucose with AraE, methyl-β-D-galactoside is a dead-end inhibitor of GalP, i.e., it may bind to the sugar site without triggering the events leading to sugar–H^+ symport, or to a second site.

F. Alternative Substrates

Alternative substrates are listed in Table II in the order of their affinities for each transport protein as reported in the literature. The following observations can be made.

Of the homologous transporters, XylE is most selective for its substrate, followed by AraE and then GalP and the mammalian sugar transporters. One may speculate that if evolution started with the glucose transporter, then the changes in primary sequence may first have effected discrimination against the 6-OH position (not the 6-HOCH$_2$) to achieve a pentose transporter (AraE) and then enhanced recognition of the precise 4-OH configuration to achieve a xylose transporter (XylE). Analysis of the recognition of each hydroxyl group in the hexose molecule has led to models of the sugar recognition site in the glucose transporters of mammals (Barnett *et al.*, 1973; Rees and Holman, 1981; Walmsley, 1988). These may be equally relevant to the binding of sugars by the bacterial protein, GalP.

Future mutagenesis experiments may succeed in associating changes in just a few amino acid residues in each protein with the changes in sugar recognition. This is an important goal in the elucidation of the relationship of structure to function of these transport proteins.

Individual sugar analogs have proved to be of particular value. 2-Deoxy-D-galactose, D-talose, and D-fucose are all substrates for GalP but not for the first metabolic enzyme, galactokinase, to any significant extent. Thus, their radioisotope-labeled forms are useful for isolating transport from metabolism in transport assays, and the unlabeled sugars give maximal alkaline pH changes in the direct sugar–H^+ symport assays without the complication of acid production due to metabolism (Henderson and Macpherson, 1986). Furthermore, D-fucose is a gratuitous inducer of all the *gal* operons, useful for maintaining high levels of expression in cultures of *E. coli,* and probably other enterobacteria. D-Fucose is also a valuable, nonmetabolizable substrate of the AraE transporter (Henderson and Macpherson, 1986).

L-Rhamnose (6-deoxy-L-mannose) and L-fucose (6-deoxy-L-galactose) can be regarded as either 6-deoxy-hexoses or 5-methyl-pentoses. The L-rhamnose–H^+ or L-fucose–H^+ transporters of *E. coli* accept the corresponding hexoses or pentoses (Table II), albeit with reduced affinity, so loss of the normal substrate methyl group or substitution of an –OH group in it does not prevent binding (Bradley *et al.,* 1987; Muiry, 1989). However, each transport protein recognizes the correct orientation of the –OH at the C-2, C-3, or C-4 position in the pyranose ring and so a sugar differing in only one of these configurations is not a substrate (Muiry, 1989). Thus, the L-rhamnose transporter accepts L-rhamnose, L-mannose, and L-lyxose, but not L-fucose, L-galactose, and D-arabinose, which are substrates of the L-fucose transporter (Muiry, 1989). Each of the substrate analogs is metabolized relatively slowly, and has consequently proved very useful in optimization of sugar–H^+ symport assays. None of them appear to be gratuitous inducers of the *E. coli* operons. Since the predicted primary sequences of the RhaT and FucP transport proteins are not homologous to each other or to any of the other sugar–H^+ symporters (see Sections VI,G and H) comparative studies are unlikely to yield insights into the nature of the substrate recognition process by the two L-deoxy-sugar transporters.

V. The Locations of Genes Encoding Sugar–H^+ Transport Proteins on the Chromosome of *Escherichia coli*

An important step in cloning and sequencing a transport protein of *E. coli* is to establish the location of the corresponding gene on the chromosome. This can be achieved by first isolating a mutant impaired in the relevant transport activity. Random mutagenesis of *E. coli* is conveniently accomplished by insertion of specially constructed transposons or phages into the genome (Jones-Mortimer and Henderson, 1986). These usually contain antibiotic-resistance or other markers to facilitate selection of cells into which they have become incorporated. It is then necessary to devise selection and screening procedures, which show that a particular transport system has been mutated. Among these are the appearance of sugar-resistant phenotypes in strains carrying *fda*, *fdp*, or *ppc*, the acquisition of resistance to toxic sugar analogs, and the disappearance of sugar–H^+ symport activity (Henderson and Kornberg, 1975; Jones-Mortimer and Henderson, 1986).

One mutant was isolated that contained Mu d(Ap^r) phage inserted into the *xylE* gene, which encodes the xylose–H^+ symport protein; this facilitated the location of the gene, because it was much easier to score the ampicillin-resistance phenotype on solid media than the presence or absence of the transport activity (Davis *et al.*, 1984). Classical genetic techniques of interrupted conjugation and phage P1-mediated transduction were then used to locate the ampicillin-resistance determinant, and consequently the *xylE* gene, at 91.4 min on the *E. coli* chromosome. Another useful mutagenic phage is λ p*lac* Mu (Bremer *et al.*, 1984; Silhavy *et al.*, 1984), which was exploited to clone the *xylE*, *araE*, and *galP* genes located at 91.4, 61.2, and 63.7 min, respectively (Davis and Henderson, 1987; Maiden *et al.*, 1988; P. E. Roberts, D. C. M. Moore, and P. J. F. Henderson, unpublished data).

The positions of a number of relevant sugar transport genes on the *E. coli* chromosome are summarized in Fig. 2. The transport genes can be located more precisely by comparison of the positions of restriction sites in their genes and in flanking DNA with a restriction map of the entire *E. coli* genome (Kohara *et al.*, 1987).

Some transport genes occur in operons where they are adjacent to the genes encoding the enzymes for metabolism of the sugar and to the genetic elements regulating gene expression. The best known case of this type is the contiguous *lac* operon (Beckwith and Zipser, 1970). Similarly, the gene encoding the fucose–H^+ transporter is part of a contiguous fucose operon at 60 min (Fig. 2; Lu and Lin, 1989) and the gene encoding the rhamnose–H^+ transporter is part of a contiguous rhamnose operon at 88 min (Fig. 2; Tate *et al.*, 1992). Both of these operons also contain the genes for the metabolic enzymes and regulatory proteins (Tobin and Schleif, 1987; Lu and Lin, 1989; Badia *et al.*, 1989).

In other cases the transport gene is entirely separate from the metabolism genes. For example, the *araE* gene encoding the arabinose–H^+ transporter is at 61 min, distant from the genes for the enzymes of arabinose metabolism at 0 min, from the associated regulatory region, also at 0 min, and from the genes for the binding protein transport system at 45 min (Fig. 2; Bachmann, 1987). Similarly, the *xylE* gene for the xylose–H^+ transporter is at 91 min (Davis *et al.*, 1984), separate from the other genes encoding the enzymes of xylose metabolism, the regulatory region, and the binding protein xylose transport system; these are clustered together at about 80 min (Bachmann, 1987; Sumiya, 1989). The *galP* gene for galactose–H^+

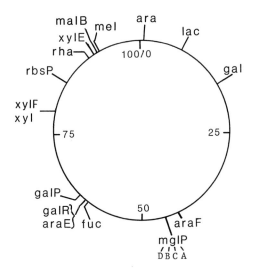

FIG. 2   Locations of some genes encoding enzymes of sugar transport and metabolism on the gene linkage map of *Escherichia coli*. The circle represents the chromosome of *E. coli*, which is calibrated arbitrarily 0–100 min from the index mark on the inside at the top. Details are given by Bachmann (1987) and the nomenclature is as follows: *ara* contains genes for regulation of transcription and for the enzymes of metabolism of arabinose; *lac* contains genes for regulation of transcription, lactose–H^+ transport, and metabolism of lactose; *gal* contains genes for metabolism of galactose; *araF* contains three genes encoding proteins for ATP-energized binding protein-mediated transport of arabinose; *mgl* contains genes (probably three) for ATP-energized binding protein-mediated transport of galactose; *fuc* contains genes for regulation of transcription, fucose–H^+ transport, and metabolism of fucose; *araE* encodes the arabinose–H^+ symport protein; *galR* produces a repressor protein for regulation of transcription of *gal, mgl,* and *galP; galP* encodes the galactose–H^+ symport protein; *xyl* contains the genes for internal metabolism of xylose and for regulation of transcription; *xylF* contains genes (probably three) for ATP-energized binding protein-mediated transport of xylose; *rbs* contains genes for regulation of transcription, ATP-energized binding protein-mediated transport (three genes), and metabolism of ribose; *rha* contains genes for regulation of transcription, rhamnose–H^+ transport (probably one gene), and metabolism of rhamnose; *xylE* encodes the xylose–H^+ symport protein; *malB* contains five genes involved in ATP-energized binding protein-mediated transport of maltose; and *mel* contains a gene for melibiose–Na^+ transport and one for melibiose metabolism. References are in the text.

transport is located at about 64 min (Riordan and Kornberg, 1977; P. E. Roberts, D. C. M. Moore, and P. J. F. Henderson, unpublished data), and is again separate from the genes encoding the metabolic enzymes at 17 min, the *galR* regulatory region at 61 min, and the genes for the galactose binding protein transport system at 45 min (Fig. 2; Bachmann, 1987).

   However discontinuous the operon, the expression of all genes involved in the transport and metabolism of a particular sugar is highly coordinated.

The presence of the sugar in the environment derepresses expression of all the genes by factors of 10 to several hundred (Buttin, 1968), although this may be attenuated by catabolite repression (Saier, 1985; Postma and Lengeler, 1985; Lengeler *et al.*, 1990). To date, only the expression of the *ara* operons has been studied in any detail (Kolodrubetz and Schleif, 1981; Kosiba and Schleif, 1982; Stoner and Schleif, 1983), and it is still unclear precisely how the transcription of the *araE* gene for transport is coordinated with the expression of the other *ara* operons. The physiological and/or evolutionary significance of this complex organisation of the genetic elements involved in sugar transport, metabolism, and the regulation of gene expression is as yet unclear.

The gene for each of the transport proteins in Table I has been cloned and sequenced. In the case of the lactose–H^+ transporter it has been shown unequivocally that only a single gene product is necessary for activity (Kaback, 1989). For the other sugar–H^+ symporters no evidence for the involvement of more than one protein has been found, although the possibility has not been completely eliminated.

VI. The Primary Sequences of Sugar–H^+ Transport Proteins Determined from DNA Sequences of the Genes

A. The Arabinose–H^+ Transport Protein of *Escherichia coli*

The sequence of the 472 amino acids in the arabinose–H^+ (AraE) transport protein of *E. coli* was predicted from the DNA sequence of the gene cloned from a specialized transducing phage (Maiden *et al.*, 1987, 1988). When the overexpressed AraE protein was isolated by electroelution from an SDS-polyacrylamide gel, little amino acid sequence could be determined from sensitive Edman degradation of the intact protein. However, clear sequences corresponding to those predicted from the DNA sequence were obtained from peptides isolated after the protein was subjected to partial digestion with various proteases. It seems that the N terminus of the protein is blocked, possibly by retention of the *N*-formyl-methionine, a situation that does not appear to operate for the XylE and GalP proteins (see Sections VI,B and C). The matching of the predicted and determined sequences provided an unequivocal identification of the overexpressed AraE protein, and indicated that it had not undergone any posttranslational modification, despite its appearance in SDS-polyacrylamide gels at molecular weights around 36,000, compared with the value of 51,683 predicted from the DNA sequence. The protein reacted covalently with [³H] cytochalasin B (Section IV,B) when photoactivated with UV light. The appropriate substrates (Table I) and 6-deoxy-D-glucose protected the protein against this labeling.

B. The Xylose–H⁺ Transport Protein of *Escherichia coli*

The sequence of the 491 amino acids in the xylose–H⁺ (XylE) transport protein of *E. coli* was predicted from the DNA sequence of the *xylE* gene cloned from a specialized transducing phage (Davis and Henderson, 1987). When the overexpressed XylE protein was isolated by electroelution from an SDS-polyacrylamide gel, the amino acid sequence of the N-terminal 33 residues could be determined by Edman degradation and it corresponded to that predicted from the DNA sequence. The protein migrated in SDS-polyacrylamide gels at molecular weights around 35,000 to 38,000, compared with the value of 53,607 predicted from the DNA sequence. The XylE protein did not react covalently with [³H] cytochalasin B (Section IV,B) when photoactivated with UV light, under conditions where the AraE protein (Section VI,A) or the GalP protein (Section VI,C) did do so.

C. The Galactose (Glucose)–H⁺ Transport Protein of *Escherichia coli*

The gene *galP* encoding the *E. coli* galactose–H⁺ transporter (GalP) has been cloned, sequenced, and its expression amplified (P. E. Roberts, D. C. M. Moore, and P. J. F. Henderson, unpublished data). It is predicted to contain 464 amino acids, of which the N-terminal 30 residues have been confirmed by Edman degradation of the intact protein purified by electroelution from SDS-polyacrylamide gels. There was no evidence of modification of the N terminus. The protein was expressed at high levels sufficient for digestion by proteases to obtain derived peptides, the N-terminal sequences of which corresponded to those predicted from the DNA sequence. The intact protein migrated in SDS-polyacrylamide gels at molecular weights around 34,000 to 36,000, compared with the value of 50,913 predicted from the DNA sequence. It reacted covalently with [³H] cyto-chalasin B (Section IV,B) or with 3-[¹²⁵I]iodo-4-azido-phenethylamino-7-*O*-succinyldeacetylforskolin (Section IV,C) when photoactivated with UV light. The appropriate substrates (Table I) protected the protein against this labeling.

D. Homologies between the Monosaccharide Transport
Proteins of *Escherichia coli*

The AraE and XylE pentose transport proteins are homologous, with 141 identical residues out of 472 and 491, respectively, in the aligned sequences (Fig. 4; Maiden *et al.*, 1987). There are additional conservative substitutions throughout the proteins, so that nearly 40% of the residues can be regarded as highly conserved (Maiden *et al.*, 1987; Baldwin and Henderson, 1989). In contrast there is little homology in the DNA sequences of the genes (Maiden, 1987).

The AraE pentose transporter and the GalP hexose transporter are very similar, with 64% identical amino acid residues throughout the aligned sequences, and additional conservative substitutions (P. E. Roberts, D. C. M. Moore, and P. J. F. Henderson, unpublished observations). The GalP and XylE proteins are only 33% identical. Since GalP and AraE are therefore more closely related to each other than they are to XylE, which is reflected also in the similarity of the DNA sequences of the *galP* and *araE* genes, they perhaps resulted from a more recent gene duplication and subsequent divergence (Doolittle, 1981, 1990; von Heijne, 1987).

The substrate specificity of GalP is more similar to that of the mammalian glucose transporter than to those of AraE or XylE (Table II). GalP interacts with both cytochalasin B and forskolin, AraE interacts with cytochalasin B only, and XylE does not interact with either (cf. Section IV). Therefore, AraE and GalP are two proteins of relatively similar amino acid sequence with different substrate and inhibitor specificities, and GalP and the mammalian erythrocyte glucose transporter are of relatively different sequences with similar substrate and inhibitor specificities (cf. Table II).

The XylE protein is 42% identical and 62% similar to a protein that probably effects facilitated diffusion of glucose into the gram-negative bacterium, *Zymomonas mobilis* (Barnell *et al.*, 1991). In general the AraE, XylE, and GalP proteins exhibit 22–35% levels of identity with the homologous proteins from other organisms listed in Table III. It seems likely that sequence comparisons (Fig. 3) will eventually identify regions of all the proteins, and even individual amino acids, that contribute to molecular recognition of substrates and inhibitors.

E. The Lactose–H⁺ Transport Protein of *Escherichia coli*

There is no obvious homology between the primary sequence of the lactose–H^+ transport protein (LacY) of *E. coli* (Buchel *et al.*, 1980) and the sequences of any of the other sugar–cation transporters except the equivalent protein in *K. pneumoniae* (Buvinger and Riley, 1985; McMorrow *et al.*, 1988) and the raffinose transporter of *E. coli* (Aslanidis *et al.*, 1990). Nevertheless, LacY is also predicted to have 12 membrane-spanning segments (Foster *et al.*, 1983; Kaback, 1989), and it is possible that the 3-dimensional shape of the LacY protein will turn out to be similar to that of other sugar transport proteins. There are several precedents for similarity of tertiary structure between proteins that are not homologous at the primary sequence level (Fersht, 1985; Quiocho, 1986, 1990). Studies of the lactose–H$^+$ transport protein lead the field and have been extensively reviewed elsewhere (Kaback, 1986, 1987, 1989, and this volume; Brooker, 1990), and will not therefore be considered further here.

F. The Melibiose—Cation Transport Protein of *Escherichia coli*

1. Primary Sequence of the MelB Protein

The sequence of 469 amino acids in the melibiose–cation transport protein of *E. coli* (Yazyu *et al.*, 1984) has 96 identities to the 491 residues of the "membrane domain" of the lactose transport protein of *Strep. thermophilus* (Poolman *et al.*, 1989) and 101 identities with the glucuronide transport protein of *E. coli* (W.-J. Liang, D. M. Wilson, P. J. F. Henderson, and R. A. Jefferson, unpublished data). There is no obvious homology to any other sequenced transport protein. Reviews favor a structural model that predicts 12 transmembrane domains for the MelB protein (Leblanc *et al.*, 1989; Botfield *et al.*, 1990; but see Yazyu *et al.*, 1984, who suggest 10–11 transmembrane domains), and between transmembrane domains 2 and 3 and between 8 and 9 are sequences similar to the RXGRR motif found in the transporters of Table III (discussed in Section VIID). All these sugar transport proteins are of similar size (Table I) and display similar, anomalous behavior in SDS-polyacrylamide gels (see Sections VI,A–E,G,H and X,E). Despite the differences in primary sequence one can speculate that the melibiose, lactose, and glucuronide transporters may therefore have a three-dimensional structure similar to many other transport proteins that are not thought to be related at the moment.

2. Interdependence of Sugar and Cation Specificity

The nature of the sugar substrate determines which cation is the preferred cosubstrate of the MelB protein (Wilson and Wilson, 1987). For melibiose transport Na$^+$ or H$^+$ is utilized, whereas Na$^+$ or Li$^+$ is utilized for β-

galactoside transport and Na^+, H^+, or Li^+ is utilized for α-galactoside transport. Also, the kinetic constants for the transport of melibiose depend on the chemical identity of the coupling cation (Bassilana *et al.*, 1985; Pourcher *et al.*, 1990a). Leblanc *et al.* (1989) concluded that when the protein catalyzes melibiose–H^+ cotransport the transmembrane proton motive force essentially affects the transport constant K_t, whereas when the same carrier catalyzes melibiose–Na^+ cotransport the transmembrane sodium motive force selectively modifies the transport V_{max}. These observations are interpreted in terms of a structural interaction between the sugar and cation binding sites in the protein (Leblanc *et al.*, 1989; Botfield *et al.*, 1990), but it is important also to take kinetic considerations into account (see Section XII,C below).

3. Mutations That Alter Sugar and/or Cation Specificity

Yazyu *et al.* (1985) isolated a Li^+-resistant mutant of *E. coli* in which a point mutation altered Pro-122 to a serine residue. The carrier lost the ability to cotransport H^+ with melibiose and showed an absolute requirement for Na^+ or Li^+. Botfield and Wilson (1988) isolated 23 independent TMG-resistant mutants altered in their sugar specificity. Most also displayed altered recognition for cations. The sites of mutation were clustered in four regions of the protein (Botfield and Wilson, 1988), Asp-15 through Ile-18 (I), Tyr-116 through Pro-122 (IV), Val-342 through Ile-348 (X), and Ala-364 through Gly-374 (hydrophilic region before XI). The roman numerals in parentheses indicate the putative transmembrane domain in a 12-transmembrane domain model (Botfield *et al.*, 1990). The last two regions are of particular interest because their positions correspond to regions of LacY and the homologous sugar transporters that mutagenesis or sequence alignments implicate in the mechanism of sugar binding and/or transport (see Section VII).

Site-directed mutagenesis of Glu-361 to Gly, Asp, or Ala reduced transport activity to 2–6% of normal. Since the K_t for TMG transport was normal or subnormal and the binding of p-nitrophenyl-α-D-galactoside was as tight or tighter than normal it seemed that the V_{max} was affected by these changes. Each of the seven His residues was changed to Arg, and only His-94 was found to be essential for transport activity (Pourcher *et al.*, 1990c). N-Terminal deletions indicated that most of the first 11 residues were required for insertion of the protein into the membrane. Deletion of only a small part of the C-terminal end of the protein, which is thought to be located on the cytoplasmic side of the membrane (Botfield and Wilson, 1989b), led to loss of activity (Botfield and Wilson, 1989a). A T7 polymerase/promoter expression system has been used to identify the MelB protein on SDS-polyacrylamide gels (Pourcher et al., 1990b).

G. The L-Fucose—H$^+$ Transport Protein of Escherichia coli

Lu and Lin (1989) briefly reported the DNA sequence of the fucose operon of E. coli. By comparison with the complementation map of the genes they identified one of the open reading frames in the translated sequence (439 amino acids, predicted M_r 47,773) as that of the fucose transport protein (FucP). The predicted amino acid sequence is very hydrophobic, characteristic of a membrane protein, but it does not exhibit homology with any of the known transport proteins, nor is it easy to predict the number of transmembrane domains. The FucP protein has been overexpressed by placing the gene under the control of the λ P_L promoter; the N-terminal sequence of the purified protein was identical to the sequence predicted from the DNA sequence (F. Gunn, C. G. Tate, and P. J. F. Henderson, unpublished data). The FucP protein migrates anomalously on SDS-polyacrylamide gel electrophoresis (apparent M_r 32,000), analogous to the behavior of other integral membrane proteins (Section X,E). Its further characterization is in progress.

H. The L-Rhamnose—H$^+$ Transport Protein of Escherichia coli and Salmonella typhimurium

The DNA sequence of the rhaT gene, which encodes the rhamnose–H$^+$ symporter, has been determined for the genes isolated from S. typhimurium (Muiry, 1989) and E. coli (Tate et al., 1992). Both genes encoded proteins of 344 amino acids (M_r 37,400) that were predicted to have 10 transmembrane domains. The RhaT protein showed no homology to any sequenced transport protein in protein sequence databases, and the topology of the RhaT protein is completely different from that predicted for the other sugar–H$^+$ symporters (Fig. 5). The rhaT gene has been expressed under the control of the λ P_L promoter (Tate et al., 1992); this gave increased expression of the RhaT protein sufficient to identify the [^{35}S] Met-labeled protein, by SDS-polyacrylamide gel electrophoresis (apparent M_r 27,000) of membranes purified from maxicells (Section X,B). Like other transport proteins, the apparent molecular weight of the RhaT protein on SDS-polyacrylamide gel electrophoresis is considerably lower than the molecular weight predicted from the protein sequence (Section X,E). Further investigations on the structure of the RhaT protein are currently in progress.

I. The Glucose–H^+ Transport Protein of *Synechocystis*

Zhang *et al.* (1989) selected a clone containing the glucose transport protein of *Synechocystis* PCC6803 by complementation of a fructose-resistant mutant (cf. Flores and Schmetterer, 1986). Of the two reading frames in the sequenced DNA in this clone one encoded a protein predicted to contain 468 amino acids. This could be readily aligned with the other proteins listed in Table III, showing 25–39% identity and 46–60% homology (Zhang *et al.*, 1989). It was most similar to the *E. coli* xylose–H^+ symporter (XylE).

J. The Primary Sequences of Mammalian Glucose Transport Proteins

Mueckler *et al.* (1985) cloned and sequenced cDNA encoding a facilitated diffusion glucose transport protein from the human hepatoma HepG2 cell line. Of its 492 residues 24–31% were identical to those in the aligned AraE, XylE, and GalP sugar–H^+ transporters from *E. coli* (Maiden *et al.*, 1987; Fig. 3), and, taking into account conservative substitutions, each bacterial protein was as similar to the human protein as they were to each other. Immunological and peptide sequence data indicated that the hepatoma protein was very similar, if not identical, to the intensively studied glucose transporter of human erythrocytes (Wheeler and Hinkle, 1985; Baly and Horuk, 1988; Walmsley, 1988; Baldwin and Henderson, 1989; Baldwin, 1990; Carruthers, 1990). The glucose transporter of rat brain was virtually identical to the human hepatoma/erythrocyte ones, with only 7 different residues out of 492 (Birnbaum *et al.*, 1986).

Since then at least five different mammalian glucose transport proteins have been identified (Table III; Kayano *et al.*, 1988; Fukumoto *et al.*, 1989; Thorens *et al.*, 1988; James *et al.*, 1989; Birnbaum, 1989; Carruthers, 1990; Gould and Bell, 1990). They are about 55–65% identical to each other. Their structural differences must account for their different kinetic properties, hormone sensitivities, and inhibitor susceptibilities (Ciaraldi *et al.*, 1986; Baly and Horuk, 1988), which are presumably allied to differences in the physiological roles of glucose transport in each tissue (Nordlie, 1985; Baly and Horuk, 1988; Kasanicki and Pilch, 1990). They are expressed differently in each tissue (M. J. Birnbaum and B. Thorens, this volume).

The mammalian glucose transport proteins are homologous to the sugar transport proteins found in cyanobacteria, bacteria, yeasts, algae, plants, and protozoa (22–33% identity; Table III), and it seems likely that they also occur in the great diversity of intervening multicellular life forms. The

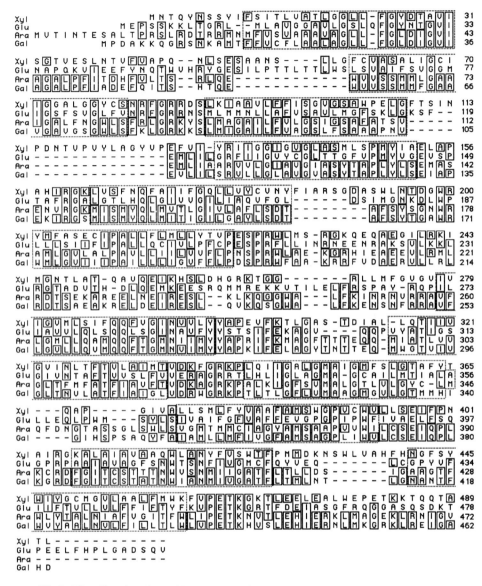

FIG. 3   The aligned amino acid sequences of homologous sugar transport proteins from bacteria and eukaryotes. The single-letter code for amino acids is employed for the following transport proteins: Xyl, xylose–H⁺ symporter of *E. coli;* Ara, arabinose–H⁺ symporter of *E. coli;* Gal, galactose–H⁺ symporter of *E. coli;* Glu, glucose transporter of a human hepatoma cell line. The alignments have been modified slightly from the individual published ones (references in Table III and text). The dotted rectangles enclose hydrophobic regions.

sequences of different glucose transporters have been compared by Henderson (1990, 1990b).

VII. Features of the Aligned Primary Sequences of the Homologous Sugar Transport Proteins

A. Twelve Transmembrane Domains

When the hydropathic profile of each of the homologous sugar transport proteins is analyzed independently there appear to be 12 hydrophobic regions, predicted to be transmembrane domains (references in Table III). In some cases uncertainty over the predicted position of these regions is overcome when the sequences are aligned according to the amino acid identities. One example of this is loop 11 of the AraE protein (Baldwin and Henderson, 1989; Fig. 4). Some faint evidence of duplication in the two halves of each protein (Figs. 3–5) suggests that the membrane-spanning regions should be regarded as two groups of six (Maiden et al., 1987).

The predicted occurrence of 12 transmembrane domains is not restricted to the homologous sugar transport proteins listed in Table III. It is also the case for the lactose–H^+ (Foster et al., 1983), glucose 6-phosphate/P_i (Friedrich and Kadner, 1987), glycerol 3-phosphate/P_i (Eiglmeier et al., 1987), proline–Na^+ (Nakao et al., 1987), melibiose–Na^+ (Leblanc et al., 1989), and glucuronide transporters of E. coli, the plasmid- or transposon-encoded citrate and tetracycline transporters (see Section VIII) of bacteria, and the uracil transporter of yeast (Jund et al., 1988). Perhaps this is a general property of transport proteins (cf. Kartner and Ling, 1989; Riordan et al., 1989; Higgins, 1989; Higgins et al., 1990b; Kane et al., 1990; Pacholocyk et al., 1991). There are exceptions, for example the rhamnose–H^+ transporter of S. typhimurium and E. coli (Muiry, 1989; see Section VI,H).

B. A Central Hydrophilic Region

In all the homologous sugar transport proteins there is a central hydrophilic region of 55–70 residues strongly predicted to comprise a high proportion of α helix (Figs. 3 and 4). This presumably plays an essential role in the maintenance of structure and function. In the erythrocyte glucose transporter it may be identified with a protease-sensitive region that is cleaved by trypsin only when it is applied to the cytoplasmic face of the membrane (Baldwin et al., 1980; Fig. 2 in Baldwin and Henderson, 1989). This may imply that the region is cytoplasmic in all the transporters.

Periplasmic side

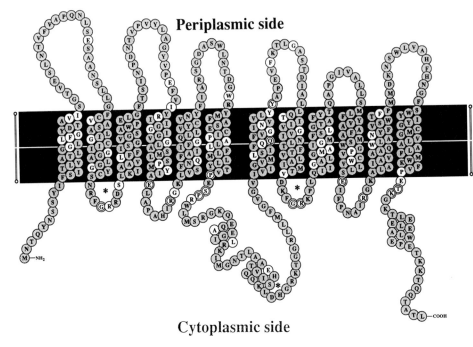

Cytoplasmic side

FIG. 4 Proposed orientation of the xylose–H$^+$ symport protein of *Escherichia coli* in the cytoplasmic membrane. The model is based on the alignment of 12 putative α-helical transmembrane domains with those in the mammalian glucose transporter (Mueckler *et al.*, 1985; Baldwin and Henderson, 1989). Some additional features are as follows. Residues absolutely conserved in all the aligned group I sugar transporters are indicated by a clear double circle; highly conserved residues are indicated by a clear circle. The highly conserved positions of predicted β turns in hydrophilic domains are asterisked. The short region between transmembrane domains 2 and 3 on the cytoplasmic side of the membrane is deliberately drawn to emphasize its similarity to the corresponding region between transmembrane domains 8 and 9. In the central and carboxy-terminal hydrophilic regions on the cytoplasmic side of the membrane there are regions of predicted helix that are conserved, usually more extensively, in all the group I transporters. (This figure was prepared by B. J. McKeown.)

FIG. 5 Proposed orientations of the arabinose–H$^+$ symporter, the citrate–H$^+$ symporter, and the tetracycline–H$^+$ antiporter in the cytoplasmic membrane of bacterial cells. The models are based on predictions of membrane-spanning regions and α-helix content and on the alignment with the glucose transporter (Eisenberg *et al.*, 1984; Mueckler *et al.*, 1985; von Heijne, 1987; Eckert and Beck, 1989). The 12 putative membrane-spanning segments, usually of 21 residues, are depicted as rectangles with their end residues numbered from the N terminus. The rectangles outside the membrane represent predicted α helices. The single residues indicated are usually conserved in all the aligned sugar transporters (Figs. 3 and 4), and some recur at equivalent positions in the two halves of the molecule, as discussed in text.

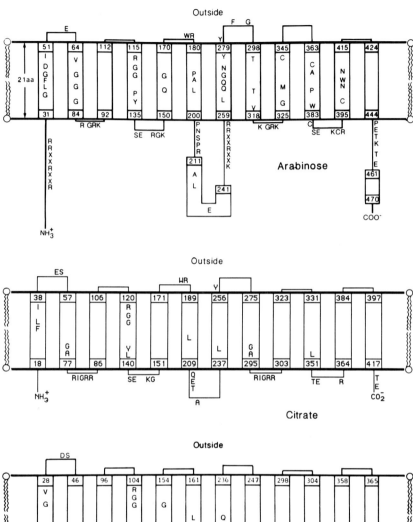

Arabinose

Citrate

Tetracycline

C. Significance of the Conserved Amino Acid Residues

When all the homologous sugar transporters are aligned there are 14 absolutely conserved residues and about 150 conservative substitutions (Fig. 4). These residues may be important features of the common structure and mechanism of these proteins. Analyses (Maiden, 1987; Kruckeberg and Bisson, 1990; P. E. Roberts and P. J. F. Henderson, unpublished data) showed that the most commonly conserved residues were glycine, proline, arginine, and glutamate. The significance of the first two probably reflects the unique spatial properties of their side chains (von Heijne, 1987). The second two are probably involved in structurally important charge interactions. The conserved glutamate residues are also of interest because of their potential interaction with protons in the sugar–H^+ symporters. However, so far there is no obvious relationship between the presence or absence of a particular glutamate (or aspartate) and the capacity for cation symport. This highlights the need to be certain whether a particular protein transports cations or not. The invariant residues in the aligned sequences occur predominantly in the hydrophilic regions predicted to be on the cytoplasmic side of the membrane (Fig. 4).

The bacterial transporters, unlike the mammalian transporters, are energized by the cotransport of a proton with the sugar; hence, the proton-translocating residues might be conserved only in the bacterial proteins. Likely candidates for such residues are histidine, glutamate, and aspartate. However, despite the proposed involvement of histidine in proton translocation by the functionally similar lactose–H^+ transporter of *E. coli* (Kaback, 1989; Kaback, 1990; but see also King and Wilson, 1989b), no histidine residues are conserved in the bacterial monosaccharide proteins (Fig. 3; Baldwin and Henderson, 1989). Furthermore, directed mutagenesis of all seven histidine residues in the melibiose–Na^+ transport protein failed to associate any of them with proton translocation even though alterations in His-94, but not any of the other histidine residues, did profoundly affect transport competence (Pourcher *et al.,* 1990c).

Aspartate, asparagine, glutamate, and glutamine residues are involved in the H-bonded interaction of soluble proteins with sugar molecules in their binding sites (Quiocho, 1986, 1990). It is therefore of interest that these residues are highly conserved in some locations predicted to be within transmembrane domains (loops 1, 6, 7, 8, and 10; see Figs. 3 and 4). These are therefore candidates for site-directed mutagenesis experiments to evaluate their possible roles in sugar binding.

In view of their inhibition by sulfhydryl reagents (above), it is interesting that no cysteine residues are conserved in the aligned transporters (Maiden *et al.,* 1987). Such inhibition may stem from steric hindrance of substrate binding or translocation, rather than from an essential mechanistic role of –SH groups. It happens that there are no cysteine residues in the N-terminal half of the AraE protein, so the substrate-protected residue(s) must be in the C-terminal part of AraE and, by extension, of the other proteins. Consistently, site-directed mutagenesis showed that only Cys-374 of the three Cys residues in GalP reacted with N-ethylmaleimide; Cys-374 is in the C-terminal half of the protein, predicted to be between transmembrane domains 10 and 11 (Fig. 3). This corresponds to Cys-384 in the AraE protein (Fig. 3). Walmsley (1988) has suggested that a reagent-sensitive cysteine residue in the glucose transporter may be at position 347, or possibly 417, in the C-terminal half of the protein. Directed mutagenesis could confirm the location of such residues, as it has in the case of the lactose–H^+ transporter—both peptide labeling/mapping and site-directed mutagenesis experiments identified Cys-148 as the residue susceptible to N-ethylmaleimide in LacY (Beyreuther et $al.$, 1981; Menick et $al.$, 1987). However, this is predicted to occur in the N-terminal half of the protein (Kaback, 1986, 1989).

D. The RXGRR Motif

Between the putative transmembrane domains 2 and 3, and between 8 and 9, is a sequence motif RXGRR (Fig. 6). The first and last R may be replaced by K, X is usually an amino acid with a large hydrophobic side chain, and the motif is often preceded by D or N. It is predicted to form a β turn, linking the adjacent helices. The positively charged side chains may interact with the head groups of lipids. The finding of this motif and the occurrence of 12 hydrophobic regions in a novel sequence are initial clues that a protein belongs to this homologous group. A quite similar motif is found in the corresponding regions of the LacY and MelB proteins (Maiden et $al.$, 1987), despite their differences in primary sequence.

E. The PESPR and PETK Motifs

These occur after the ends of the putative transmembrane domains 6 and 12, respectively, in most of the monosaccharide transporters (Fig. 7). They are also found in the lactose and maltose transporters of yeasts (Chang and Dickson, 1988; Cheng and Michels, 1989), but not in the citrate, tetracycline, or their related transporters (see Section VIII). In the AraE and GalP proteins the E in the first motif is changed to N or D, and the G in most glucose transporters after the second motif is not conserved. Again, site-directed mutagenesis experiments need to be undertaken to evaluate the possible significance of these differences in relation to substrate speci-

		Helix 2		Helix 3
Human	GLUT1	LSVAIFSVGGMIGSFSVGLFV	-NRFGRRNS	MLMMNLLAFVSAVLMGFSKLG
Rabbit	GLUT1	LSVAIFSVGGMIGSFSVGLFV	-NRFGRRNS	MLMMNLLAFVSAVLMGFSKLA
Rat	GLUT1	LSVAIFSVGGMIGSFSVGLFV	-NRFGRRNS	MLMMNLLAFVSAVLMGFSKLG
Pig	GLUT1	LSVAIFSVGGMIGSFSVGLFV	-NRFGRRNS	MLMMNLLAFISAVLMGFSKLG
Mouse	GLUT1	LSVAIFSVGGMIGSFSVGLFV	-NRFGRRNS	MLMMNLLAFVAAVLMGFSKLG
Human	GLUT2	LSVSSFAVGGMTASFFGGWLG	-DTLGRIKA	MLVANILSLVGALLMGFSKLG
Rat	GLUT2	LSVSSFAVGGMVASFFGGWLG	-DKLGRIKA	MLAANSLSLTGALLMGCSKFG
Human	GLUT3	LSVAIFSVGGMIGSFSVGLFV	-NRFGRRNS	MLIVNLLAVTGGCFMGLCKVA
Human	GLUT4	LSVAIFSVGGMISSFLIGIIS	-QWLGRKRA	MLVNNVLAVLGGSLMGLANAA
Rat	GLUT4	LSVAIFSVGGMISSFLIGIIS	-QWLGRKRA	MLANNVLAVLGGALMGLANAA
Mouse	GLUT4	LSVAIFSVGGMISSFLIGIIS	-QWLGRKRA	MLANNVLAVLGGALMGLANAV
Human	GLUT5	VTVSMFPFGGFIGSLLVGPLV	-NKFGRKGA	LLFNNIFSIVPAILMGCSRGA
Yeast	SNF3	ILVSFLSLGTFFGALTAPFIS	-DSYGRKPT	IIFSTIFIFSIGNSLQVGAGG
Yeast	GAL2	LIVAIFNIGCAFGGIILSKGG	-DMYGRKKG	LSIVVSVYIVGIII-QIASIN
Yeast	MAL61	LCL-CYMAGEIVGLQVTGPSV	-DYMGNRYT	LIMALFFLAAFIFILYFCKSL
Yeast	LAC12	LVFSIFNVGQICGAFFVPLM-	-DWKGRKPA	ILIGCLGVVIGAIISSLTTT-
E. coli	AraE	WVVSSMMLGAAIGALFNGWLS	-FRLGRKYS	LMAGAILFVLGSIGSAFATSV
E. coli	XylE	FCVASALIGCIIGGALGGYCS	-NRFGRRDS	LKIAAVLFFISGVGSAWPELG
E. coli	GalP	WVVSSMMFGAAVGAVGSGWLS	-FKLGRKKS	LMIGAILFVAGSLFSAAAPNV
E. coli	TET	VLLALYALMQFLCAPVLGALS	-DRFGRRPV	LLASLLGATIDYAIMATTPVL
E. coli	CIT	AVFGSGFLMRPIGAVVLGAYI	-DRIGRRKG	LMITLAIMGCGTLLIALVPGY
Chlorella		LFVSSLFLAGLVSCLFASWIT	-RNWGRKVT	MGIGGAFFVAGGLVNAFAQDM
Arabidopsis		MFTSSLYLAALISSLVASTVT	-RKFGRRLS	MLFGGILFCAGALINGFAKHV
Synechocystis		LSVSLALLGSALGAFGAGPIA	-DRHGRIKT	MILAAVLFTLSSIGSGLPFTI
Leishmania		IFAGSMIAGCLIGSVFAGPLA	-SKIGARLS	FLLVGLVGVVASVMYHASCAA
Neurospora qa-y		NIVSVYQAGAFFGCLFAYATS	-YFLGRRKS	LIAFSVVFIIGAAIMLAADGQ
Aspergillus qutD		NIVSLYQRGAFFGALFAYPIG	-HFWGRRWG	LMFSALIFFLGAGMMLGANGD
E. coli	LacY	IIFAAISLFSLLFQPLFGLLS	-DKLGLRKY	LLWIITGMLVMFAPFFIFIFG
K. pneumoniae		IVFSCISLFAIIFQPVFGLIS	-DKLGLRKH	LLWTITILLILFAPFFIFVFS
E. coli	MelB	VARIWDAINDPIMGWIVNATR	-SRWGKFKP	WILIGTLANSVILFLLFSAHL

		Helix 8		Helix 9
Human	GLUT1	VYATIGSGIVNTAFTVVSLFV	VERAGRRTLH	LIGLGLAGMAGCAILMTIALAL
Rabbit	GLUT1	VYATIGSGIVNTAFTVVSLFV	VERAGRRTLH	LIGLGLAGMAACAVLMTIALAL
Rat	GLUT1	VYATIGSGIVNTAFTVVSLFV	VERAGRRTLH	LIGLGLAGMAGCAVLMTIALAL
Pig	GLUT1	VYATIGSGIVNTAFTVVSLFV	VERAGRRTLH	LIGLGLAGMAGCAVLMTIALAL
Mouse	GLUT1	VYATIGSGIVNTAFTVVSLFV	VERAGRRTLH	LIGLGLAGMAGCAVLMTIALAL
Human	GLUT2	VYATIGVGAVNMVFTAVSVFL	VEKAGRRSLF	LIGMSGMFVCAIFMSVGLVL
Rat	GLUT2	VYATIGVGAINMIFTAVSVLL	VEKAGRRTLF	LAGMIGMFFCAVFMSLGLVL
Human	GLUT3	IYATIGAGVVNTIFTVVSLFL	VERAGRRTLH	MIGLGGMAFCSTLMTVSLLL
Human	GLUT4	AYATIGAGVVNTVFTLVSVLL	VERAGRRTLH	LLGLGAGMCGCAILMTVALLL
Rat	GLUT4	AYATIGAGVVNTVFTLVSVLL	VERAGQRTLH	LLGLGAGMCGCAILMTVALLL
Mouse	GLUT4	AYATIGAGVVNTVFTLVSVLL	VERAGRRTLH	LLGLGAGMCGCAILMTVALLL
Human	GLUT5	QYVTAGTGAVNVVMTFCAVFV	VELLGRRLLL	LLGFSICLIACCVLTAALAL
Yeast	SNF3	YLVSFITYAVNVVFNVPGLFF	VEFFGRRKVL	VVGGVIMTIANFVAIVGCS
Yeast	GAL2	FETSIVGVVNFASTFFSLWT	VENLGRRKCL	LLGAATMMACMVIYASVGVT
Yeast	MAL61	FTFSIIQYCLGIAATFVSWWA	SKYCGRFDLY	AFGLAFQAIMFFIIGGLGCS
Yeast	LAC12	VLMNGVYSIVTWISSICGAFF	IDKIGRREGF	-LGSISGAALALTGLSICTA
E. coli	AraE	MIATLVVGLTFMFATFIAVFT	VDKAGRKPAL	KIGFSVMALGTLVLGYCLMQ
E. coli	XylE	LLQTIIVGVINLTFTVLAIMT	VDKFGRKPLQ	IIGALGMAIGMFSLGTAFYT
E. coli	GalP	MWGTVIVGLTNVLATFIAIGL	VDRWGRKPTL	TLGFLVMAAGMGVLGTMMHI
E. coli	TET	LSLAVFGILHALAQAFVTGPA	TKRFGEKQAI	IAG--MAADALGYVLLAFAT
E. coli	CIT	LVVTMLVGISNFIWLPIGGAI	SDRIGRRPVL	-MGITLLALVTTLPVMNWLT
Chlorella		LLNTVVVGAVNVGSTLIAVMF	SDKFGRRFLL	IEGGIQCCLAMLTTGVVLAI
Arabidopsis		LMSAVVTGSVNVGATLVSIYG	VDRWGRRFLF	LEGGTQMLICQAVVAACIGA
Synechocystis		LLITVITGFINILTTLVAIAF	VDKFGRKPLL	LMGSIGMTITLGILSVVFGG
Leishmania		LVGNFVVMLWNFVTTLASIPL	SYVFTMRHVF	LFGSIFTFCMCLFMCGIPVY
Neurospora qa-y		LTTGIFGVVKMVLTIIWLLWL	VDLVGRRRIL	FIGAAGGSLCMWFIGAYIKI
Aspergillus qutD		LTTGIFGVVTAVITFVWLLYL	IDHFGRRNIL	LVGAAGSCVLWIVGGYIKI
E. coli	LacY	GYVTTMGELLNASIMFFAPLI	INRIGGKNAL	LLAGTIMSVRIIGSSFATSA
K. pneumoniae		GFVTTGGELLNALIMFCAPAI	INRIGAKNAL	LIAGLIMSVRILGSSFATSA
E. coli	MelB	FPYYLSYAGANLVTLVFFPRL	VKSLSRRILW	AGASILPVLSCGVLLLMALM

ficity, cation recognition, and other parameters of the molecular mechanism.

F. A Diffused Motif through Transmembrane Domains 4 and 5

The following pattern is conserved in the region including putative transmembrane domains 4 and 5 of the aligned monosaccharide transporters (Figs. 3–5):

$$R - - - G - - - G - - - - - - P - Y - - E - - - - - -$$
$$R G - - - - - - Q - - - - - G$$

No gaps are inserted to achieve this matching. This motif presumably has structural and/or functional significance, and it is conserved, albeit imperfectly, in the transport proteins for lactose (Chang and Dickson, 1988), citrate, and tetracycline (Henderson and Maiden, 1990; see Figs. 3 and 4 and Section VIII).

G. Residues That May Be Implicated in Sugar Binding from Comparisons with the Lactose Transporter

Many of the homologous sugar transport proteins have a conserved motif, PALL, in the predicted membrane-spanning region 6 (Fig. 7). Since mutation of the residue Ala-177 in the equivalent region of the lactose–H^+ transporter has been associated with changes in sugar specificity (Brooker and Wilson, 1985; see Section XII,B,C) it may be speculated that the PALL motif is also associated with sugar binding. By a similar, and admittedly tentative, argument the following conserved residues in the homologous proteins might be involved in sugar binding: a tyrosine residue at the end of transmembrane domain 7, a threonine residue in transmembrane domain 8, and an isoleucine residue in transmembrane domain 9. These are all altered in mutants of the LacY protein with altered binding of sugars (Markgraf *et al.*, 1985; Brooker, 1990). Site-directed mutagenesis experiments will be required to explore these possibilities.

FIG. 6 Conserved sequence motifs between putative helices 2 and 3 and 8 and 9 (see, e.g., Figs. 4 and 5) of transport proteins.

Helix 6

```
Human       GLUT1         IVLPF CPESPRFLLINRNEEN
Rabbit      GLUT1         IVLPL CPESPRFLLINRNEEN
Rat         GLUT1         ILLPF CPESPRFLLINRNEEN
Pig         GLUT1         VLLPF CPESPRFLLINRNEEN
Mouse       GLUT1         ILLPF CPESPRFLLINRNEEN
Human       GLUT2         LLLFF CPESPRYLYIKLDEEV
Rat         GLUT2         LLLLF CPESPRYLYLNLEEEV
Human       GLUT3         AALPF CPESPRFLLINRKEEE
Human       GLUT4         VLLPF CPESPRYLYIIQNLEG
Rat         GLUT4         LLLPF CPESPRYLYIIRNLEG
Mouse       GLUT4         ILLPF CPESPRYLYIIRNLEG
Human       GLUT5         LLLPF FPESPRYLLIQKKDEA
Yeast       SNF3          IGMFF LPESPRYYVLKDKLDE
Yeast       GAL2          GALTL VPESPRYLCEVNKVED
Yeast       MAL61         VGIFL APESPWWLVKKGRIDQ
Yeast       LAC12         IFGWL IPESPRWLVGVGREEE
E. coli     AraE          ILVVF LPNSPRWLAEKGRHIE
E. coli     XylE          MLLYT VPESPRWLMSRGKQEQ
E. coli     GalP          IGVFF LPDSPRWFAAKRRFVD
E. coli     TET           LGCFL MQESHKG-ERRPMPLR
E. coli     CIT           VLRRS LQETEAFLQRKHRPDT
Chlorella                 LGSLV LPESPNFLVEKGKTEK
Arabidopsis               IGSLV LPDTPNSMIERGQHEE
Synechocystis             VCAFL IPESPRYLVAQGQGEK
Leishmania                VLGIV TRESRAKFDGGEEGRA
Neurospora qa-y           LGSFW IPESPRWLYANGKREE
Aspergillus qutD          IGALL IRESPRWLFLRGNREK
E. coli     LacY          LFFAK -TDAPSSATVANAVGA
K. pneumoniae             LWVSK -PESSNSAEVIDALGA
E. coli     MelB          LRNVH EVFSSDNQPSAEGSHL
```

Helix 12

```
Human       GLUT1         FTYF KVPETKGRTFDEIASGF
Rabbit      GLUT1         FTYF KVPETKGRTFDEIASGF
Rat         GLUT1         FTYF KVPETKGRTFDEIASGF
Pig         GLUT1         FTYF KVPETKGRTFDEIASGF
Mouse       GLUT1         FTYF KVPETKGRTFDEIAFGF
Human       GLUT2         FTFF KVPETKGKSFEEIAAEF
Rat         GLUT2         FTFF KVPETKGKSFDEIAAEF
Human       GLUT3         FTFF KVPETRGRTFEDITRAF
Human       GLUT4         FTFL RVPETRGRTFDQISAAF
Rat         GLUT4         FTFL RVPETRGRTFDQISATF
Mouse       GLUT4         FTFL KVPETRGRTFDQISAAF
Human       GLUT5         YIFL IVPETKAKTFIEINQIF
Yeast       SNF3          VVYL TVYETKGLTLEEIDELY
Yeast       GAL2          YVFF FVPETKGLSLEEIQELW
Yeast       MAL61         WAVV DLPETAGRTFIEINELF
Yeast       LAC12         VIYF FFVETKGRSLEELEVVF
E. coli     AraE          ITFW LIPETKNVTLEHIERKL
E. coli     XylE          FMWK FVPETKGKTLEELEALW
E. coli     GalP          LTLW LVPETKHVSLEHIERNL
E. coli     TET           ALRR GAWSRATST
E. coli     CIT           TMLF ARLSSGYQTVENKL
Chlorella                 CAIF LLPETKGVPIERVQALY
Arabidopsis               FVYI FLPETKGIPIEEMGQVW
Synechocystis             FIWF FVKETKGKTLEQM
Leishmania                IQVF FLHPWDEERDGKKVVAP
Neurospora qa-y           FIYF FLPVTKSIPLEAMDRLF
Aspergillus qutD          FVFF LIPETKGVPLESMETLF
E. coli     LacY          VFTL SGPGPLSLLRRQVNEVA
K. pneumoniae             LFTL KGSKTLLPATA
E. coli     MelB          LYFR FYRLNGDTLRRIQIHIL
```

H. A Diffused Motif through Transmembrane Domain 7

Most of the aligned sugar transport proteins of Table I contain the following motif in putative transmembrane domain 7:

$$L - - - Q Q - - G I N - - F Y Y$$

Elements of this can also be recognized in the corresponding regions of the transport proteins for lactose, melibiose, citrate, and tetracycline (see Section VIII). The presence of the glutamine and asparagine residues generates the suspicion that the region may contribute to recognition of the sugar, but it may equally be involved in the translocation mechanism. Perhaps the presence of the IN pair here is a repetition of its presence at the end of putative transmembrane domain 1 in both the aligned glucose transporters (Fig. 3) and the LacY protein (Kaback, 1989).

I. A Diffused Motif through Transmembrane Domains 10 and 11

In this region there is a pattern of conserved residues as follows:

$$F - - - - G - - - W - - - - E - - - - - - R - - - - - - - - - - N W - - N F$$

The position and spacing of the glutamate and arginine reflect their occurrence in the corresponding region between transmembrane domains 4 and 5 in the N-terminal half of the protein. Photolabeling of a sugar transporter with cytochalasin B probably results from photoactivation of an aromatic amino acid residue in the protein rather than from activation of the inhibitor (Deziel et al., 1984; see IV,B). Peptide mapping experiments indicated that the covalently bound cytochalasin B was located between Phe-389 and Trp-412 in the mammalian D-glucose transporter (Cairns et al., 1984; Karim et al., 1987; Holman and Rees, 1987). The only aromatic residue conserved in this region of the bacterial transporters is the second tryptophan in transmembrane domain 11 of the above motif. Experiments are in progress to determine whether this is the actual site of labeling, and to discover why the XylE protein does not bind cytochalasin B. It may be relevant that this region of the MelB protein is suggested to be part of a sugar recognition site (Botfield and Wilson, 1988; Botfield et al., 1990).

FIG. 7 Conserved sequence motifs at the ends of putative helices 6 and 12 (see, e.g., Figs. 4 and 5) of transport proteins.

VIII. Similarities between Sugar, Carboxylate, and Antibiotic Transport Proteins

A. The Citrate—H$^+$ and α-Ketoglutarate Transporters

A plasmid-encoded citrate transport activity is thought to be proton linked (Reynolds and Silver, 1983). The *E. coli* citrate transporter was sequenced, via its gene, by two groups independently (Sasatsu *et al.*, 1985; Ishiguro and Sato, 1985). It is 63% identical to the *K. pneumoniae* citrate transporter (van der Rest *et al.*, 1990), and 33% identical to the *E. coli* α-ketoglutarate—H$^+$ transporter (Seol and Shatkin, 1990). At the primary sequence level these carboxylate transporters have only a slight similarity to the sugar transporters (Maiden *et al.*, 1987), but when analyzed by the algorithm of Eisenberg *et al.* (1984) they are predicted, like the sugar transporters, to have two groups of six transmembrane domains separated by a central hydrophilic domain (Maiden, 1987; van der Rest *et al.*, 1990; Seol and Shatkin, 1990). Furthermore, between the predicted transmembrane domains 2 and 3 and between domains 8 and 9 were the DRXGRR motifs, and through the predicted transmembrane domains 4 and 5 was the diffused motif described in Section VII,F (Fig. 5). There were other similarities between the carboxylate transporters and individual sugar transport proteins (Fig. 5), especially in their amino-terminal halves, but only a few more identities that occurred in every protein (Fig. 5). The most noticeable differences were the relative shortness of the central and C-terminal hydrophilic regions in the carboxylate transporters. These largely account for their shorter length (431–444 residues) compared to sugar transporters (472–522 residues).

B. The Tetracycline—H$^+$(K$^+$) and Other Antibiotic Transporters

One class of tetracycline genes encodes proteins that effect the active efflux of the antibiotic from the cell (McMurry *et al.*, 1980; Levy, 1984, 1988; Yamaguchi *et al.*, 1990a,b). The mechanism may be an obligatory coupling to the movement of protons (presumably antiport) or potassium ions (Dosch *et al.*, 1984). The sequences of several such tetracycline transporters from gram-negative organisms have been determined (Peden, 1983; Waters *et al.*, 1983; Hillen and Schollmeier, 1983). They are extensively homologous to each other (Hillen and Schollmeier, 1983; Sheridan and Chopra, 1991), but not obviously so to the sugar transporters. Nevertheless, each one is predicted, by the same algorithm used to analyze the sugar transporters, to have 12 transmembrane domains (Maiden, 1987; Eckert and Beck, 1989; Sheridan and Chopra, 1991; Fig. 5). Between the putative transmembrane domains 2 and 3 there is a perfect DRXGRR motif and between the putative transmembrane domains 8 and 9 is RXGEK, similar to RXGRR (Fig. 5). Through the putative transmembrane domains 4 and 5 is most of the diffused motif (Fig. 5). There are other similarities in primary sequence between the tetracycline, citrate, and individual sugar transporters (Fig. 5), but few of them are conserved in all of the transporters. Superficially, the structures of tetracycline and cytochalasin B appear similar (Henderson and Maiden, 1990), or at least they resemble each other more than they resemble a monosaccharide (but see Griffin et al., 1982). Perhaps the cytochalasin binding site on a sugar transport protein resembles a tetracycline binding site?

These observations were strengthened by the publication of the sequence of the QacA protein from *Staphylococcus aureus*, which catalyzes the energy-dependent efflux of intercalating dyes (ethidium, acriflavine) and quaternary ammonium antiseptic compounds (Rouch et al., 1990). It is predicted to have 514 amino acids and the hydropathic profile suggests the occurrence of 14 potentially membrane-spanning regions, rather than the usual 12. Five of these regions, 2, 5, 7, 10, and 13, demonstrate significant α-helical amphipathicity, "suggesting that these could facilitate transport through participation of their hydrophilic faces in the creation of suitable relatively hydrophilic regions across the membrane" (Rouch et al., 1990). The sequence can be readily aligned with those of the methylenomycin efflux protein from *Streptomyces coelicolor,* the aminotriazole efflux protein from *S. cerevisiae,* and tetracycline efflux proteins from gram-negative and gram-positive bacteria (Rouch et al., 1990). It is most closely related to the methylenomycin-resistance protein (Mmr) of *S. coelicolor* (Neal and Chater, 1987), and both of these are more closely related to the tetracycline efflux proteins found in gram-positive organisms than the latter group is to the tetracycline efflux proteins found in gram-negative organisms (Rouch et al., 1990; Sheridan and Chopra, 1991). Furthermore, comparisons with all the sugar transporter sequences at the same level of stringency demonstrate that the Mmr transporter is significantly related (5.4 SD) to the AraE sugar transporter of *E. coli,* and that the significance is actually greater when only the N-terminal halves of the proteins are compared (7.8 SD; M. E. Baker, J. K. Griffith, and D. A. Rouch, unpublished data). The regions of similarity often include individual residues that are very highly conserved in the aligned sequences of all the group I sugar transporters. They include an Asp residue in transmembrane domain 1, the Asp/Asn Arg X Gly Arg Arg/Lys motif between transmembrane domain 2 and transmembrane domain 3, a Leu – – – Arg – – – Gly motif at the beginning of transmembrane domain 4, an Arg residue at the beginning of transmembrane domain 5, and a Phe Leu–Pro Glu Ser motif at the end of transmembrane domain 6. Yoshida *et al.* (1990) have also remarked on the similarity of a quinolone efflux protein, which is predicted to have 12 transmembrane domains, to the XylE sugar transporter.

The corollary of these observations, that some antibiotics may interact with sugar transporters, was confirmed in Sections IV,B and C.

Interestingly, the tetracycline transporter has now been found to function as an Mg^{2+} tetracycline–H^+ antiporter (Yamaguchi *et al.*, 1990a; and this series). Uptake of tetracycline by passive diffusion as a relatively hydrophobic metal chelate probably explains the long-standing difficulty of identifying an uptake protein. The consequent increased efficiency of transport assays in vesicles now that the divalent metal requirement is recognized is also facilitating the characterization of mutants impaired in activity (Yamaguchi *et al.*, 1990b, 1991). Thus, mutagenesis of Asp-66 to Asn or Glu or of His-257 to Asp or Glu severely impaired activity, whereas changing Ser-65 to Ala or Cys was relatively inconsequential. The rationale for choosing these residues for mutagenesis was their high degree of conservation in the aligned sequences of tetracycline transporters (cf. Sheridan and Chopra, 1991). Previous structural studies showed that the C terminus of the tetracycline transporter is located on the cytoplasmic side of the membrane (Eckert and Beck, 1989), and that the N- and C-terminal halves appear to comprise separate domains that are functionally complementary (Rubin and Levy, 1990).

These similarities permit the speculation that the tetracycline transporters, which are representatives of a wider class of antibiotic-resistance factors (Neal and Chater, 1987; Rouch *et al.*, 1990; Sheridan and Chopra, 1991), have three-dimensional structures in the membrane that are fundamentally similar to those of the series of homologous sugar transporters.

IX. Structural Models of the Transport Proteins

Many membrane proteins are believed to be folded so that hydrophobic α-helices inserted through the membrane are joined by hydrophilic regions located in the head groups of the lipid and/or the aqueous environment (Kyte and Doolittle, 1982; von Heijne, 1988; Lodish, 1988). There are very few proteins where this model is supported by detailed structural information determined by electron diffraction or X-ray crystallographic measurements, the exceptions being bacteriorhodopsin from *Halobacterium halobium,* components of the photosynthetic reaction center of *Rhodopseudomonas viridis,* and the light-harvesting complex of plants (Henderson and Unwin, 1975; Henderson and Schertler, 1990; Diesenhofer *et al.*, 1985; Kuhlbrandt and Wang, 1991). Nevertheless, the burgeoning mass of primary sequence information shows that numerous other membrane proteins, including those involved in transport, also contain a series of hydrophobic regions, each sufficiently long to span the membrane as an α-helix, separated by hydrophilic regions (Kyte and Doolittle, 1982; von Heijne, 1988). Computerized algorithms have been derived to predict the occurrence of secondary structural features from primary sequence information (Garnier *et al.*, 1978; Eisenberg *et al.*, 1984; Eisenberg, 1985), and incorporated into models of the lactose (Foster *et al.*, 1983) and glucose (Mueckler *et al.*, 1985) transporters. Models of membrane protein structure based on such sequence data alone should be viewed with caution for several reasons (Lodish, 1988). However, they do generate testable hypotheses of structural features—for example, that the N terminus and the C terminus are on the same side of the membrane—and they set criteria for comparisons between different proteins, as illustrated above. When evidence is obtained that, for example, the C-terminal end of the mammalian glucose transporter is inside the cytoplasmic membrane (Davis *et al.*, 1987; Haspel *et al.*, 1988), or that the central hydrophilic region is also inside the membrane (Cairns *et al.*, 1987), the model is confirmed and the extrapolation to the related transporters becomes more tenable (see also Baldwin and Henderson, 1989).

Four examples of such models are given in Figs. 4 and 5, the xylose–H^+ and arabinose–H^+ transporters of *E. coli*, the citrate–H^+ transporter, and the tetracycline–$H^+(K^+)$ transporter, the last two of which are encoded on plasmids (see also Eckert and Beck, 1989; Henderson and Maiden, 1990). Importantly, they include the positions of residues conserved in all the monosaccharide transporters and some sequence motifs (see Section VII) apparently repeated in similar positions in both halves of the citrate and tetracycline transporters. For the latter, experiments using labeling reagents and proteolytic cleavage indicated that the N terminus and the C terminus are located inside the cytoplasmic membrane (Eckert and Beck, 1989), which is consistent with the results obtained with the glucose transporter and with the predictions in the models.

X. Identification of the Transport Proteins

The biochemical identification of a transport protein is an essential step in the characterization of any transport system. However, in bacteria such proteins usually constitute only 0.1–0.5% of the membrane proteins and are impossible to identify by conventional means. By combining genetic and biochemical techniques several strategies have evolved to overcome this problem, and they are illustrated briefly here. The specific covalent labeling of a transport protein is also a most important aid in its purification (see, e.g., Newman *et al.*, 1981).

A. Amplification of Gene Expression

In bacteria gene expression can be increased by factors of two to five by isolating mutants in which accumulation of the natural or gratuitous inducers is enhanced. Examples include galactokinase-negative, arabinose isomerase-negative, and xylose isomerase-negative strains of *E. coli*, in all of which expression of the corresponding sugar tranporter was increased (Henderson and Macpherson, 1986). Some mutants may express the transport (and other metabolism) genes constitutively, and this level is usually higher than is achieved in "fully induced" wild-type strains. Even these modest increases in expression depress the efficiency of cell growth for reasons that are not understood. More substantial levels of amplification require transfer of the transport gene to an appropriate vector, where its expression is suppressed during growth of the cells, and then activated.

Teacher *et al.* (1978) were the first to achieve amplified expression of a bacterial sugar transport protein, LacY, on a multicopy plasmid to the level where it could be identified as a Coomassie blue-stained protein of M_r ~30,000 by SDS-polyacrylamide gel electrophoresis (SDS-PAGE). The genes *araE, xylE,* and *fucP* have been transferred to vectors in which their amplified expression was achieved under the control of the efficient λ P_L promoter (Maiden *et al.*, 1988; McKeown, 1988; F. Gunn, C. G. Tate, and P. J. F. Henderson, unpublished data). The gene products, AraE and XylE, were then identified as Coomassie blue-stained proteins of apparent M_r 35,000–38,000 (Table I), constituting 15–20 and 5–7% of the inner membrane protein, respectively, while the apparent molecular weight of FucP was 33,000 and it constituted >20% of the membrane protein. The *galP* gene, which is negatively regulated, was overexpressed when cloned into plasmid pBR322; levels of GalP protein as high as 55% of the membrane protein were expressed constitutively (P. E. Roberts, unpublished data). Since the N-terminal sequences of the proteins were essentially as predicted from their DNA sequences, it was unlikely that they had undergone posttranslational modification (cf. Section VI). The melibiose transporter MelB has proved more difficult to overexpress, but exploitation of the T7 bacteriophage RNA polymerase expression system developed by Tabor and Richardson (1985) and protein labeling with [35S]methionine revealed a single protein of M_r 39,000 (Leblanc *et al.*, 1989).

B. Expression in Maxicells or Minicells

Overexpression of membrane proteins is often unsuccessful, presumably due to the toxic nature of these proteins. The protein products of plasmid-encoded genes expressed at low levels from a λ promoter can be identified by expression in maxicells or minicells (Stoker *et al.*, 1984). Expression in these systems results in only plasmid-encoded genes being expressed because the genomic DNA in the host strain is either degraded (maxicells) or absent (minicells). Inclusion of [^{35}S]Met when the λ promoter is turned on results in a radiolabeled gene product that can be detected by autoradiography of the proteins separated by SDS-PAGE. Thus, the RhaT protein was shown to migrate with M_r 27,000 (Tate *et al.*, 1992). An analogous system exists for detection of low levels of protein expressed from the T7 promoter (Studier and Moffatt, 1986; Tabor and Richardson, 1985).

C. Labeling with N-Ethylmaleimide

The proton-linked transporters for arabinose, xylose, or galactose from *E. coli* were all inhibited by *N*-ethylmaleimide (Table I). The protection by substrates was exploited to label specifically the susceptible transport proteins, AraE, XylE, and GalP, with radioactive *N*-ethylmaleimide. They were then identified as proteins that migrated with an apparent molecular weight of 35,000–40,000 in SDS-PAGE (Macpherson *et al.*, 1981, 1983; Henderson *et al.*, 1983b; Henderson and Macpherson, 1986; Davis, 1986; Maiden *et al.*, 1988), entirely consistent with the values observed in the amplified expression systems (above). By similar procedures the lactose–H^+ transport protein was the first to be identified using *N*-ethylmaleimide, with an apparent molecular weight of about 30,000 (Jones and Kennedy, 1969). Similarly, the proline–Na^+ transporter was identified as a protein of apparent molecular weight of about 35,000 (Hanada *et al.*, 1985). Such labeling experiments have been very important for the identification of hydrophobic transport proteins that are of low abundance (0.1–0.5% of the membrane proteins), although they are being superseded by the expression systems.

D. Labeling with Photoactivatable Substrate Analogs, Cytochalasin B, and Forskolin

It has already been mentioned how Kaczorowski *et al.* (1980) found that 4-nitrophenyl-α-D-galactoside (NPG) reacted covalently, and specifically, with the LacY protein when the two were exposed to UV light. Similarly, when activated by UV light the AraE and GalP transport proteins react covalently with radioactive cytochalasin B, and they can again be identified by SDS/PAGE as labeled proteins of apparent M_r 36,000–38,000. The presence of appropriate sugar substrates in the incubation mixture protected the transporters against this labeling (G. Smith, M. T. Cairns, K. R. Petro, and V. A. Lucas, unpublished data; M. T. Cairns *et al.*, 1989). Cytochalasin B will be a particularly convenient tool in purification studies on these proteins. The GalP protein was also labeled by a high-affinity photolabel/inhibitor derivative of forskolin, 3-[^{125}I]iodo-4-azido-phenethylamino-7-*O*-succinyldeacetylforskolin, in a substrate-protectable manner (Wadzinski *et al.*, 1990; G. E. Martin, M. F. Shanahan, P. E. Roberts, K. B. Seamon, and P. J. F. Henderson, unpublished data).

E. Anomalous Migration in SDS-polyacrylamide Gel Electrophoresis

An important general point, which has been made elsewhere (Leblanc *et al.*, 1989), is that all these sugar transport proteins, and the citrate, tetracycline, and proline transporters (Reynolds and Silver, 1983; Eckert and Beck, 1989; Hanada *et al.*, 1985), migrate in SDS-PAGE with an apparent molecular weight less than that predicted from the DNA sequence. The values for the sugar transporters are summarized in Table I. In the case of several of them there is evidence that no posttranslational modification has occurred that might have accounted for the phenomenon. It may be due to the binding of higher proportions of SDS than occurs with less hydrophobic proteins or to the retention of some secondary/tertiary structure.

XI. Reconstitution

A general observation is that membrane transport proteins of a variety of primary sequences can be solubilized under very similar conditions (usually with added lipid in 50 mM phosphate buffer, pH 7.5, 1 mM dithiothreitol), in the same detergent, octylglucoside (1.25% concentration; dodecylmaltoside is often a good substitute), and then successfully reconstituted by the convenient detergent dilution technique into proteoliposomes without much loss of activity. The pertinent examples are the mammalian glucose transporter (Kasahara and Hinkle, 1977), and the *E. coli* transporters for lactose–H$^+$ (Newman and Wilson, 1980), melibiose–Na$^+$ (Tsuchiya *et al.*, 1982), galactose–H$^+$ (Henderson *et al.*, 1983a), arabinose–H^+ (Henderson and Macpherson, 1986), and proline–Na^+ (Hanada *et al.*, 1988a,b).

The availability of a reconstitution assay is particularly important for purification of membrane transport proteins. Purification has been successful for the mammalian glucose transporter (Kasahara and Hinkle, 1977; Baldwin *et al.*, 1982), the *E. coli* lactose–H^+ transporter (Newman *et al.*, 1981; Wright *et al.*, 1983; Viitanen *et al.*, 1986; Page *et al.*, 1988), and the *E. coli* proline–Na^+ transporter (Hanada *et al.*, 1988b). It seems reasonable to expect that methods based on those used for these proteins will be successfully applied to the other sugar transport proteins described in this article when their expression is amplified sufficiently.

XII. Conclusions

A. Evolutionary and Structural Relationships

The family of sugar transport proteins listed in Table III and described in this article has been found in such diverse organisms so far that it seems likely to occur in most forms of life. This implies that investigations of the molecular mechanism of their transport activities in microorganisms, with the attendant conveniences of experimental manipulation, will uncover features equally valid for their operation in higher organisms, including humans. It seems that insight into the mechanism of transport of substrates other than sugars will also accrue, if the apparent similarities to the citrate, α-ketoglutarate, and antibiotic (and other?) transporters described here are real. Also, the homologies described imply a fundamental similarity between proteins that catalyze facilitated diffusion ("passive transport") and substrate–H^+ symport and antiport ("secondary active transport")— a perceived difference in mechanism seems not to require a profound difference in structure.

The purpose of cataloguing the substrate and inhibitor specificities of the transport proteins (Table II) is to gain insight into the nature of the ligand binding sites and to define the relationship between the proteins. The availability of a variety of sugars and their structural analogs, and the known sequence homologies of the arabinose, xylose, and galactose, and glucose transporters (Table III), make this a particularly promising series of membrane proteins in which to establish the characteristics of sugar recognition and their evolution. If the relatively unspecific glucose/galactose hexose transporters arose first, one can speculate that they then evolved to discriminate against the 6-OH residue and recognize the precise orientation of the 4-OH residue in order to become more specific for either of the pentoses, arabinose or xylose. When the substrate specificities of the homologous lactose and maltose transporters of yeast (Chang and Dickson, 1988; Yao *et al.*, 1989; Szkutnicka *et al.*, 1989) are known in more detail, then it may be possible to learn about molecular discrimination between mono- and disaccharides.

There also exist transport proteins with the same function of substrate–cation symport, but without any overall similarity in primary sequence, i.e., the lactose–H^+, melibiose–Na^+, fucose–H^+, rhamnose–H^+, and proline–Na^+ transporters of enterobacteria and the glucose–Na^+ transporter of mammalian intestine. Perhaps these arose by convergent evolution, and their tertiary structures and molecular mechanisms are actually similar to those of the homologous series. Among soluble proteins there are several examples of those with similar primary sequences that nevertheless have similar tertiary structures and molecular mechanisms, e.g., the periplasmic binding proteins (Quiocho, 1986, 1990) and the serine proteases (Fersht, 1985).

B. The Roles of Individual Amino Acids in the Transport Proteins

The structural and functional roles of individual amino acids may be evaluated by modification with reagents (e.g., *N*-ethylmaleimide for cysteine and diethylpyrocarbonate for histidine residues) and *in vivo* and *in vitro* mutagenesis (Botfield and Wilson, 1988; Kaback, 1987; Leblanc *et al.*, 1989; Kaback *et al.*, 1990; Brooker, 1990; Botfield *et al.*, 1990; Yamato and Anraku, 1987; Hinkle *et al.*, 1990; Wilson *et al.*, 1990; King and Wilson, 1990b). The difficulty of identifying the role of an *individual* amino acid acid is best overcome by exploiting the *in vivo* selection of mutants impaired in an identifiable function, as exemplified by experiments on the lactose–H^+ and melibiose–Na^+ symporters described above. The comparisons of aligned amino acid sequences (Figs. 3–7) serve to identify how essential *each and every one* of the residues is, but without determining their function. Directed mutagenesis studies, which are expensive and time consuming, are probably best undertaken after other techniques have indicated the significance of a particular residue or sequence motif.

Such genetic strategies are the most practicable for relating structure and function of membrane proteins at the present time, but they are unlikely to yield a molecular model of sufficient detail to understand entirely the different substrate specificities, the process of cation recognition, the mechanism of inhibitor action, the translocation mechanism, and the evolution of the transport proteins. This will also require knowledge of their three-dimensional structures.

C. Kinetic Studies

Since we can now manipulate the structure of transport proteins it is profoundly important to characterize properly the resultant changes in individual partial reactions, substrate specificity, inhibitor specificity, transport parameters, reaction with covalent modifiers, etc. Kinetically rigorous analyses and statistically valid comparisons between activities in mutants and wild-type proteins will enable real insights into molecular mechanism to be made.

Steady state kinetic studies of the transport activities described in this article are generally based on models involving the ordered combination of the transport protein with first the cation and then the sugar substrate, followed by the translocation step and an ordered dissociation of substrates on the other side of the membrane (see, e.g., Leblanc et al., 1989; Kaback, 1989). For the lactose–H^+ symporter the unloaded protein is perceived to carry a net negative charge and the loaded carrier is uncharged (Kaback, 1989), while for the melibiose–Na^+ carrier the unloaded protein is perceived to be uncharged and the loaded carrier has a net positive charge (Leblanc et al., 1989). In both cases the stoichiometry appears to be 1 mol of substrate transported with 1 mol of cation (West and Mitchell, 1973; Page et al., 1988; Pourcher et al., 1990a). More general statements of kinetic mechanisms for transport proteins may be accessed through articles by Lombardi (1981), Stein (1986), Walmsley (1988), Eddy (1989), and Severin et al. (1989).

It is very important that the kinetic parameters of novel mutations in transport proteins be determined rigorously and compared with features of the defined kinetic mechanism in the wild-type protein. Otherwise conclusions about the effect of the mutation on sugar specificity, or cation recognition, or translocation steps, or energization mechanism may be at the least naive and at the worst incorrect. For example, in a steady state ordered mechanism for a two-substrate reaction without any steps at thermodynamic equilibrium the apparent K_m measured for one substrate depends on the concentration of the other (Cleland, 1963; Fromm, 1979). Only when one substrate is at saturating concentrations (a situation that is very difficult to achieve in practice, but very easily extrapolated from measurements in a practicable range of substrate concentrations) is the K_m determined for the second substrate the true value (Cleland, 1963; Fromm, 1979. If a mutation alters the true (and apparent!) K_m for the sugar substrate then the apparent K_m for the cation will almost certainly be changed for the steady state mechanisms outlined above. This does not necessarily imply that there is a profound molecular link between the binding of the two ligands. Indeed, the proper measurement of the *true* K_m for the cation may well find that the value is unchanged. Furthermore, it is dangerous to compare only K_m values, when a much more rigorous criterion for the affinity of a protein for its substrates is the ratio of k_{cat}/K_m (Fersht, 1985).

Such basic considerations have usually been applied in kinetic studies on wild-type bacterial transport proteins, but the ability to generate interesting mutants seems sometimes to outpace the resources or will to characterize them quantitatively, with a consequent diminution in the validity of the conclusions drawn about them. The following rules should perhaps be implemented in future studies.

1. The steady state mechanism of the wild-type transport protein should be defined.

2. True K_m and k_{cat} values should be used in comparative studies, not apparent K_m and V_{max} values.

3. The numerical values of kinetic parameters should be determined by statistically rigorous methods quoting standard deviations (Cleland, 1979; Cornish-Bowden, 1979; Henderson, 1985).

4. Rigorous analytical methods should also be applied in determining dissociation constants and concentrations of binding sites in equilibrium binding measurements.

D. Determination of the Three-Dimensional Structures of the Proteins

A real understanding of the mechanism of transport is critically dependent on determination of the three-dimensional structure of transport proteins at the level of atomic resolution. Tentative models of transport protein structure such as those in Figs. 4 and 5 are the basis of experiments in which "topological" protein reagents and antibodies, and gene fusions (e.g., Cairns *et al.*, 1984; Carrasco *et al.*, 1986; Manoil and Beckwith, 1986; Broome-Smith and Spratt, 1986; Broome-Smith *et al.*, 1990; Page and Rosenbusch, 1988; Eckert and Beck, 1989; Calamia and Manoil, 1990; Manoil, 1990; Oka *et al.*, 1990; McKenna *et al.*, 1991), and other techniques, elucidate the actual folding of the protein in the membrane. However, the determination of detailed three-dimensional structure probably requires crystallization and analyses by X-ray diffraction or electron diffraction (Pattus, 1990; Henderson *et al.*, 1990; Kuhlbrandt and Wang, 1991). The first step is the purification of sufficient stable, undenatured protein, which has been achieved only for the lactose–H^+ transporter

(Page *et al.*, 1988; Kaback, 1989) and the erythrocyte glucose transporter (Walmsley, 1988).

The problems of determining structural information for membrane proteins have been admirably reviewed by Pattus (1990). Models of transport proteins such as the ones in Figs. 4 and 5 are legitimately based on our existing knowledge of hydropathy, sidedness of reaction with topological reagents, activities of fusion proteins, inhibitor-labeling sites, protease susceptibility, etc., but they serve best as a basis on which to design experiments rather than as a serious attempt to portray reality. Nevertheless, there have been advances that raise hopes of determining the complete three-dimensional structure of a membrane transport protein by physical methods.

The first is the increasingly successful amplification of membrane protein expression, particularly in bacteria, to levels sufficient for purification, structural studies, and crystallization trials. Bacterial sugar transporters can now be expressed in amounts equivalent to 5–50% of the membrane protein (Section X). This represents an amplification of up to 200 times the wild-type level, and facilitates the production of 10–100 mg of transport protein from a 15-liter culture. It seems likely that the mammalian glucose transporters will also be produced in large quantities by cloning the gene into baculovirus for expression in insect cell cultures. There is also the possibility of their expression in yeast, for which very large-scale fermentation facilities are already available. The second advance is the refined application of electron diffraction techniques and data analysis to tilted two-dimensional crystals of bacteriorhodopsin (Henderson *et al.*, 1990) and membrane proteins of the light-harvesting complex (Kuhlbrandt and Wang, 1991) to achieve three-dimensional models at the 6-Å (and better) level of resolution. These, at least, show the arrangement of α-helices in the membrane. The techniques should be capable of refinement to higher levels of resolution (Henderson *et al.*, 1990). Third, there is the improvement of crystals of the bacterial porin proteins, enabling X-ray crystallography to be used to obtain models of the structure at the 2Å level of resolution (Weiss *et al.*, 1990). Finally, there is the application of nuclear magnetic resonance (NMR) technology to determine the structure of small membrane-spanning peptides (e.g., Barsukov *et al.*, 1990). If individual, or a small number of combined, transmembrane domains of the lactose–H^+ (or any other) transporter can be expressed, purified, and reconstituted in the native form, as already achieved by Kaback *et al.* (1990) and Wrubel *et al.* (1990), it may be possible to determine the structure of separate parts of the protein by NMR and build up an overlapping picture of the whole.

Purification and crystallization may be difficult to accomplish for all these membrane transport proteins, but perhaps no more so than the apparent impossibility of establishing their primary sequences that we faced a decade ago.

Acknowledgments

Research in the authors' laboratory is supported by the SERC, the MRC, the SmithKline Foundation, the Wellcome Trust and SmithKline Beecham plc. We are indebted to Dr. M. E. Baker, Dr. E. O. Davis, Dr. J. K. Griffith, Prof. H. R. Kaback, Dr. G. Leblanc, Dr. M. C. J. Maiden, Dr. D. C. M. Moore, Dr. K. B. Seamon, Dr. M. Shanahan, Dr. G. Schmetterer, Prof. W. Tanner, and Prof. T. H. Wilson for unpublished information and helpful discussions. Mrs. Rowena Baxter provided skilled and invaluable assistance in the preparation of the manuscript.

References

Aslanidis, C., Schmid, K., and Schmitt, R. (1990). *J. Bacteriol.* **171,** 6753–6763.

Bachmann, B. J. (1987). *In* "*Escherichia coli* and *Salmonella typhimurium:* Cellular and Molecular Biology" (F. C. Neidhardt, J. L. Ingraham, K. B. Low, B. Magasanik, M. Schaechter, and H. E. Umbarger, eds.), pp. 807–876. Am. Soc. Microbiol., Washington, D.C.

Badia, J., Baldoma, L., Aguilar, J., and Boronat, A. (1989). *FEMS Microbiol. Lett.* **65,** 253–258.

Baldwin, J. M., Lienhard, G. E., and Baldwin, S. A. (1980). *Biochim. Biophys. Acta* **599,** 699–714.

Baldwin, S. A. (1990). *Curr. Opin. Cell Biol.* **2,** 714–721.

Baldwin, S. A., and Henderson, P. J. F. (1989). *Annu. Rev. Physiol.* **51,** 459–471.

Baldwin, S. A., Baldwin, J. M., and Lienhard, G. E. (1982). *Biochemistry* **21,** 3836–3842.

Baly, D. L., and Horuk, R. (1988). *Biochim. Biophys. Acta* **947,** 571–590.

Barnell, W. O., Yi, K. C., and Conway, T. (1991). *J. Bacteriol.* **172,** 7227–7240.

Barnett, J. E. G., Holman, G. D., and Munday, K. A. (1973). *Biochem. J.* **131,** 211–231.

Barsukov, I. L., Abdulaeva, G. V., Arseniev, A. S., and Bystrov, V. F. (1990). *Eur. J. Biochem.* **192,** 321–327.

Bassilana, M., Damiano-Ferano, E., and Leblanc, G. (1985). *Biochem. Biophys. Res. Commun.* **129,** 626–631.

Beauclerk, A. D. D., and Smith, A. J. (1978). *Eur. J. Biochem.* **82,** 187–197.

Beckwith, J. R., and Zipser, D. (1970). "The Lactose Operon." Cold Spring Harbor Lab., Cold Spring Harbor, New York.

Beyreuther, K., Bieseler, B., Ehring, R., and Muller-Hill, B. (1981). *In* "Methods in Protein Sequence Analysis" (M. Elzina, ed.), pp. 139–148. Humana Press, Clifton, New Jersey.

Birnbaum, M. J. (1989). *Cell (Cambridge, Mass.)* **57,** 305–315.

Birnbaum, M. J., Haspel, H. C., and Rosen, O. M. (1986). *Proc. Natl. Acad Sci. U.S.A.* **83,** 5784–5788.

Bloch, R. (1974). *J. Biol. Chem.* **249,** 1814–1822.

Botfield, M. C., and Wilson, T. H. (1988). *J. Biol. Chem.* **263,** 12909–12915.

Botfield, M. C., and Wilson, T. H. (1989a). *J. Biol. Chem.* **264,** 11643–11648.

Botfield, M. C., and Wilson, T. H. (1989b). *J. Biol. Chem.* **264,** 11649–11652.

Botfield, M. C., Wilson, D. M., and Wilson, T. H. (1990). *Res. Microbiol.* **141**, 328–331.
Bradley, S. A., Tinsley, C. R., Muiry, J. A. R., and Henderson, P. J. F. (1987). *Biochem. J.* **248**, 495–500.
Bremer, E., Silhavy, T. J., Weisemann, J. M., and Weinstock, G. M. (1984). *J. Bacteriol.* **158**, 1084–1093.
Brooker, R. J. (1990). *Res. Microbiol.* **141**, 309–315.
Brooker, R. J., and Wilson, T. H. (1985). *Proc. Natl. Acad. Sci. U.S.A.* **82**, 3959–3963.
Broome-Smith, J. K., and Spratt, B. G. (1986). *Gene* **49**, 341–349.
Broome-Smith, J. K., Tadayyon, M., and Zhang, Y. (1990). *Mol. Microbiol.* **4**, 1637–1644.
Buchel, D. E., Gronenborn, B., and Muller-Hill, B. (1980). *Nature (London)* **283**, 541–545.
Bush, D. R. (1990). *Plant Physiol.* **93**, 1590–1596.
Buttin, G. (1968). *Adv. Enzymol. Relat. Areas Mol. Biol.* **30**, 81–137.
Buvinger, W. E., and Riley, M. (1985). *J. Bacteriol.* **163**, 850–857.
Cairns, B. R., Collard, M. W., and Landfear, S. M. (1989). *Proc. Natl. Acad. Sci. U.S.A.* **86**, 7682–7686.
Cairns, M. T., Elliot, D. A., Scudder, P. R., and Baldwin, S. A. (1984). *Biochem. J.* **221**, 179–188.
Cairns, M. T., Alvarez, J., Panico, M., Gibbs, A. F., Morris, H. R., Chapman, D., and Baldwin, S. A. (1987). *Biochim. Biophys. Acta* **905**, 295–310.
Cairns, M. T., Smith, G., Henderson, P. J. F., and Baldwin, S. A. (1989). *Biochem Soc. Trans.* **17**, 552–553.
Cairns, M. T., McDonald, T. P., Horne, P., Henderson, P. J. F., and Baldwin, S. A. (1991). *J. Biol. Chem.* **266**, 8176–8183.
Calamia, J., and Manoil, C. (1990). *Proc. Natl. Acad. Sci. U.S.A.* **87**, 4937–4941.
Carrasco, N., Herzlinger, D., Danho, W., and Kaback, H. R. (1986). *In* "Methods in Enzymology" (S. Fleischer and B. Fleischer, eds.), vol. 125, pp. 453–467. Academic Press, Orlando, Florida.
Carruthers, A. (1990). *Physiol. Rev.* **70**, 1135–1176.
Carter, J. R., Fox, C. F., and Kennedy, E. P. (1968). *Proc. Natl. Acad. Sci. U.S.A.* **60**, 725–732.
Carter-Su, C., Pessin, J. E., Mora, R., Gitomer, W., and Czech, M. P. (1982). *J. Biol. Chem.* **257**, 5419–5425.
Celenza, J. L., Marshall-Carlson, L., and Carlson, M. (1988). *Proc. Natl. Acad. Sci. U.S.A.* **85**, 2130–2134.
Chang, Y.-D., and Dickson, R. C. (1988). *J. Biol. Chem.* **263**, 16696–16703.
Charalambous, B., Maiden, M. C. J., McDonald, T. P., Cunningham, I. J., and Henderson, P. J. F. (1989). *Biochem. Soc. Trans.* **17**, 441–444.
Cheng, Q., and Michels, C. (1989). *Genetics* **123**, 477–484.
Ciaraldi, T. P., Horuk, R., and Matthaei, S. (1986). *Biochem. J.* **240**, 115–123.
Cleland, W. W. (1963). *Biochim. Biophys. Acta* **67**, 104–137.
Cleland, W. W. (1979). *In* "Methods in Enzymology" (D. L. Purich, ed.), vol. 63, pp. 103–138. Academic Press, New York.
Cooper, R. A. (1986). *In* "Carbohydrate Metabolism in Cultured Cells" (M. J. Morgan, ed.), pp. 461–491. Plenum, London.
Cornish-Bowden, A. (1979). "Fundamentals of Enzyme Kinetics." Butterworth, London.
Daie, J. (1989). *Plant Mol. Biol. Rep.* **7**, 106–115.
Damiano-Forano, E., Bassilana, M., and Leblanc, G. (1986). *J. Biol. Chem.* **261**, 6893–6899.
Daruwalla, K. R., Paxton, A. T., and Henderson, P. J. F. (1981). *Biochem. J.* **200**, 611–627.
Davies, A., Meeran, K., Cairns, M. T., and Baldwin, S. A. (1987). *J. Biol. Chem.* **262**, 9347–9352.
Davis, E. O. (1986). Ph.D. Thesis, University of Cambridge.

Davis, E. O., and Henderson, P. J. F. (1987). *J. Biol. Chem.* **262,** 13928–13932.
Davis, E. O., Jones-Mortimer, M. C., and Henderson, P. J. F. (1984). *J. Biol. Chem.* **259,** 1520–1525.
Deves, R., and Krupka, R. M. (1978). *Biochim. Biophys. Acta* **510,** 339–348.
Deziel, M., Pegg, W., Mack, E., Rothstein, A., and Klip, A. (1984). *Biochim. Biophys. Acta* **772,** 403–406.
Diesenhofer, J., Epp, O., Miki, K., Huber, R., and Michel, H. (1985). *Nature (London)* **318,** 618–624.
Doolittle, R. F. (1981). *Science* **214,** 149–159.
Doolittle, R. F. (1990). *In* "Methods in Enzymology" (R. F. Doolittle, ed.), vol. 183, pp. 99–110. Academic Press, San Diego, California.
Dosch, D., Salvacion, F., and Epstein, W. (1984). *J. Bacteriol.* **160,** 1188–1190.
Eckert, B., and Beck, C. F. (1989). *J. Biol. Chem.* **264,** 11663–11670.
Eddy, A. A. (1982). *Adv. Microb. Physiol.* **23,** 1–78.
Eddy, A. A. (1990). *In* "Monovalent Cations in Biological Systems," (C. A. Pasternak, ed.) pp. 211–236. CRC Press, Inc., Baton Rouge, Florida.
Eiglmeier, K., Boos, W., and Cole, S. T. (1987). *Mol. Microbiol.* **1,** 251–258.
Eisenberg, D. (1985). *Annu. Rev. Biochem.* **53,** 595–623.
Eisenberg, D., Schwartz, E., Komaromy, M., and Wall, R. (1984). *J. Mol. Biol.* **179,** 125–142.
Fersht, A. R. (1985). "Enzyme Structure and Mechanism," pp. 317–331. Wiley, New York.
Flores, E., and Schmetterer, G. (1986). *J. Bacteriol.* **166,** 693–696.
Foster, D. L., Boublik, M., and Kaback, H. R. (1983). *J. Biol. Chem.* **258,** 31–34.
Fox, C. F., and Kennedy, E. P. (1965). *Proc. Natl. Acad. Sci. U.S.A.* **54,** 891–899.
Friedrich, M. J., and Kadner, R. J. (1987). *J. Bacteriol.* **169,** 3556–3563.
Fromm, H. J. (1979). *In* "Methods in Enzymology" (D. L. Purich, ed.), vol. 63, pp. 42–53. Academic Press, New York.
Fukumoto, H., Seino, S., Imura, H., Seino, Y., Eddly, R. L., Fukushima, Y., Byers, M. G., Shows, T. B., and Bell, G. I. (1988). *Proc. Natl. Acad. Sci. U.S.A.* **85,** 5434–5438.
Fukumoto, H., Kayano, T., Buse, J. B., Edwards, Y., Pilch, P. F., Bell, G. I., and Seino, S. (1989). *J. Biol. Chem.* **264,** 7776–7779.
Furlong, C. E. (1987). *In* "*Escherichia coli* and *Salmonella typhimurium:* Cellular and Molecular Biology" (F. C. Neidhardt, J. L. Ingraham, K. B. Low, B. Magasanik, M. Schaechter, and H. E. Umbarger, eds.), pp. 768–796. Am. Soc. Microbiol., Washington, D.C.
Garnier, J., Osguthorpe, D. J., and Rosen, B. (1978). *J. Mol. Biol.* **120,** 97–120.
Geever, R. F., Huiet, L., Baum, J. A., Tyler, B. M., Patel, U. B., Rutledge, B. J., Case, M. E., and Giles, N. H. (1989). *J. Mol. Biol.* **207,** 15–34.
Gogarten, P. G., and Bentrup, F.-W. (1989). *Planta* **178,** 52–60.
Gould, G. W., and Bell, G. I. (1990). *Trends Biochem. Sci.* **15,** 18–23.
Griffin, J. F., Rampal, A. L., and Jung, C. Y. (1982). *Proc. Natl. Acad. Sci. U.S.A.* **79,** 3759–3763.
Hanada, K., Yamato, I., and Anraku, Y. (1985). *FEBS Lett.* **191,** 278–282.
Hanada, K., Yamato, I., and Anraku, Y. (1988a). *Biochim. Biophys. Acta* **939,** 282–288.
Hanada, K., Yamato, I., and Anraku, Y. (1988b). *J. Biol. Chem.* **263,** 7181–7185.
Haspel, H. C., Rosenfeld, M. G., and Rosen, O. M. (1988). *J. Biol. Chem.* **263,** 398–403.
Hediger, M. A., Coady, M. J., Ikeda, T. S., and Wright, E. M. (1987). *Nature (London)* **330,** 379–381.
Hediger, M. A., Turk, E., and Wright, E. M. (1989). *Proc. Natl. Acad. Sci. U.S.A.* **86,** 5748–5752.

Henderson, P. J. F. (1985). In "Techniques in Protein and Enzyme Biochemistry—Part II Supplement" (K. F. Tipton, ed.), pp. 1–48. Elsevier, Dublin.

Henderson, P. J. F. (1986). In "Carbohydrate Metabolism in Cultured Cells" (M. J. Morgan, ed.), pp. 409–460. Plenum, London.

Henderson, P. J. F. (1990a). J. Bioenerg. Biomembr. 22, 525–569.

Henderson, P. J. F. (1990b). Res. Microbiol. 141, 316–328.

Henderson, P. J. F. (1991). Curr. Opin. Struct. Biol. 1, 590–601.

Henderson, P. J. F., and Kornberg, H. L. (1975). Ciba Found. Symp. 31, 243–269.

Henderson, P. J. F., and Macpherson, A. J. S. (1986). In "Methods in Enzymology" (S. Fleischer and B. Fleischer, eds.), vol. 125, pp. 387–429. Academic Press, Orlando, Florida.

Henderson, P. J. F., and Maiden, M. C. J. (1990). Philos. Trans. R. Soc. London, Ser. B 326, 391–410.

Henderson, P. J. F., Giddens, R. A., and Jones-Mortimer, M. C. (1977). Biochem. J. 162, 309–320.

Henderson, P. J. F., Hirata, H., and Kagawa, Y. (1983a). Biochim. Biophys. Acta 732, 204–209.

Henderson, P. J. F., Hirata, H., Horne, P., Jones-Mortimer, M. C., Kaethner, T. E., and Macpherson, A. J. S. (1983b). In "Biochemistry of Metabolic Processes" (D. L. F. Lennon, F. W. Stratman, and R. N. Zahlten, eds.), pp. 339–352. Elsevier, New York.

Henderson, R., and Schertler, G. (1990). Philos. Trans. R. Soc. London, Ser. B 326, 379–390.

Henderson, R., and Unwin, P. N. T. (1975). Nature (London) 257, 28–32.

Henderson, R., Baldwin, J. M., and Ceska, T. A. (1990). J. Mol. Biol. 213, 899–929.

Hickey, M. W., Hillier, A. J., and Jago, G. R. (1986). Appl. Environ. Microbiol. 51, 825–831.

Higgins, C. F., Gallagher, M. P., Hyde, S. C., Mimmack, M. L., and Pearce, S. R. (1990a). Philos. Trans. R. Soc. London, Ser. B 326, 353–366.

Higgins, C. F., Hyde, S. C., Mimmack, M. M., Gildeadi, U., Gill, D. R., and Gallagher, M. P. (1990b). J. Bioenerg. Biomembr. 22, 571–592.

Higgins, C. J. (1989). Nature (London) 341, 103.

Hillen, W., and Schollmeier, K. (1983). Nucleic Acids Res. 11, 525–539.

Hinkle, P. C., Hinkle, P. V., and Kaback, H. R. (1990). Biochemistry 29, 10989–10994.

Holman, G. D., and Rees, W. D. (1987). Biochim. Biophys. Acta 897, 395–405.

Horne, P. (1980). Ph.D. Thesis, University of Cambridge.

Horne, P., and Henderson, P. J. F. (1983). Biochem. J. 210, 699–705.

Hutchings, V. (1978a). Planta 138, 229–235.

Hutchings, V. (1978b). Planta 138, 237–241.

Ishiguro, N., and Sato, G. (1985). J. Bacteriol. 164, 977–982.

Jackowski, S., and Alix, J. H. (1990). J. Bacteriol. 172, 3842–3844.

James, D. E., Strube, M., and Mueckler, M. (1989). Nature (London) 338, 83–87.

Jones, T. D. H., and Kennedy, E. P. (1969). J. Biol. Chem. 244, 5981–5987.

Jones-Mortimer, M. C., and Henderson, P. J. F. (1986). In "Methods in Enzymology" (S. Fleischer and B. Fleischer, eds.), vol. 125, pp. 157–180. Academic Press, Orlando, Florida.

Joset, F., Buchou, T., Zhang, C.-C., and Jeanjean, R. (1988). Arch. Microbiol. 149, 417–421.

Jund, R., Weber, E., and Chevallier, M.-R. (1988). Eur. J. Biochem. 171, 417–424.

Jung, C. Y., and Rampal, A. L. (1977). J. Biol. Chem. 252, 5456–5463.

Kaback, H. R. (1972). Biochim. Biophys. Acta 265, 367–416.

Kaback, H. R. (1974). Science 186, 882–892.

Kaback, H. R. (1986). Ann. N. Y. Acad. Sci. 456, 291–304.

Kaback, H. R. (1987). Biochemistry 26, 2071–2076.

204 PETER J. F. HENDERSON *ET AL.*

Kaback, H. R. (1989). *Harvey Lect.* **83,** 77–105.
Kaback, H. R. (1990). *Philos. Trans. R. Soc. London, Ser. B* **326,** 425–436.
Kaback, H. R., Bibi, E., and Roepe, P. D. (1990). *Trends Biochem. Sci.* **15,** 309–314.
Kaczorowski, G. J., Leblanc, G., and Kaback, H. R. (1980). *Proc. Natl. Acad. Sci. U.S.A.* **77,** 6319–6323.
Kaestner, K. M., Christy, R. J., McLenithan, J. C., Braiterman, L. T., Cornelius, P., Pekala, P. M., and Lane, M. D. (1989). *Proc. Natl. Acad. Sci. U.S.A.* **86,** 3150–3154.
Kane, S. E., Pastan, I., and Gottesman, M. M. (1990). *J. Bioenerg. Biomembr.* **22,** 593–618.
Karim, A. R., Rees, W. D., and Holman, G. D. (1987). *Biochim. Biophys. Acta* **902,** 402–405.
Kartner and Ling (1989). *Sci. Am.* March, pp. 26–33.
Kasahara, M., and Hinkle, P. (1977). *J. Biol. Chem.* **252,** 7384–7390.
Kasanicki, M. A., and Pilch, P. F. (1990). *Diabetes Care* **13,** 219–227.
Kayano, T., Fukumoto, H., Eddy, R. L., Fan, Y. S., Byers, M. G., Shows, T. B., and Bell, G. I. (1988). *J. Biol. Chem.* **263,** 15245–15248.
Kennedy, E. P. (1970). *In* "The Lactose Operon" (J. R. Beckwith and D. Zipser, eds.), pp. 49–92. Cold Spring Harbor Lab., Cold Spring Harbor, New York.
King, S. C., and Wilson, T. H. (1989a). *J. Biol. Chem.* **264,** 7390–7394.
King, S. C., and Wilson, T. H. (1989b). *Biochim. Biophys. Acta* **982,** 253–264.
King, S. C., and Wilson, T. H. (1990a). *Mol. Microbiol.* **4,** 1433–1438.
King, S. C., and Wilson, T. H. (1990b). *J. Biol. Chem.* **265,** 9638–9644.
Kohara, Y., Akiyama, K., and Isono, K. (1987). *Cell (Cambridge, Mass.)* **50,** 495–508.
Kolodrubetz, D., and Schleif, R. F. (1981). *J. Mol. Biol.* **151,** 215–227.
Komor, E., and Tanner, W. (1974). *Eur. J. Biochem.* **44,** 219–223.
Kornberg, H. L. (1990). *Philos. Trans. R. Soc. London, Ser. B* **326,** 505–514.
Kosiba, B. E., and Schleif, R. F. (1982). *J. Mol. Biol.* **156,** 53–56.
Kotyk, A. (1983). *J. Bioenerg. Biomembr.* **15,** 307–319.
Kruckeberg, A. L., and Bisson, L. F. (1990). *Mol. Cell. Biol.* **10,** 5903–5913.
Kuhlbrandt, W., and Wang, D. N. (1991). *Nature (London)* **350,** 130–134.
Kyte, J., and Doolittle, R. F. (1982). *J. Mol. Biol.* **157,** 105–132.
Lam, V. M. S., Daruwalla, K. D., Henderson, P. J. F., and Jones-Mortimer, M. C. J. (1980). *J. Bacteriol.* **143,** 396–402.
Laurenza, A., Sutkowski, E. M., and Seamon, K. B. (1989). *Trends Pharmacol. Sci.* **10,** 442–447.
Leblanc, G., Pourcher, T., and Bassilana, M. (1989). *Biochimie* **71,** 969–979.
Lefevre, P. G. (1961). *Pharmacol. Rev.* **13,** 39–70.
Lengeler, J. W., Titgemeyer, F., Vogler, A. P., and Wohrl, B. M. (1990). *Philos. Trans. R. Soc. London, Ser. B* **326,** 489–504.
Levy, S. (1984). *In* "Antimicrobial Drug Resistance" (L. E. Brian, ed.), pp. 191–240. Academic Press, New York.
Levy, S. (1988). *ASM News* **54,** 418–421.
Lewis, D. A., and Bisson, L. F. (1990). *J. Cell. Biochem.* **14E,** 26.
Lin, E. C. C. (1987). *In* "*Escherichia coli* and *Salmonella typhimurium:* Cellular and Molecular Biology" (F. C. Neidhardt, J. L. Ingraham, K. B. Low, B. Magasanik, M. Schaechter, and H. E. Umbarger, eds.), pp. 244–284. Am. Soc. Microbiol., Washington, D.C.
Lodish, H. F. (1988). *Trends Biochem. Sci.* **13,** 332–334.
Lombardi, F. J. (1981). *Biochim. Biophys. Acta* **649,** 661–679.
Lowe, A. G., and Walmsley, A. R. (1986). *Biochim. Biophys. Acta* **857,** 146–154.
Lu, Z., and Lin, E. C. C. (1989). *Nucleic Acids Res.* **17,** 4883–4884.
Macpherson, A. J. S., Jones-Mortimer, M. C., and Henderson, P. J. F. (1981). *Biochem. J.* **196,** 269–283.

Macpherson, A. J. S., Jones-Mortimer, M. C., Horne, P., and Henderson, P. J. F. (1983). *J. Biol. Chem.* **258**, 4390–4396.

Madore, M. A., and Lucas, W. J. (1987). *Planta* **171**, 197–204.

Maiden, M. C. J. (1987). Ph.D. Thesis, University of Cambridge.

Maiden, M. C. J., Davis, E. O., Baldwin, S. A., Moore, D. C. M., and Henderson, P. J. F. (1987). *Nature (London)* **325**, 641–643.

Maiden, M. C. J., Jones-Mortimer, M. C., and Henderson, P. J. F. (1988). *J. Biol. Chem.* **263**, 8003–8010.

Maloney, P. C. (1990a). *Philos. Trans. R. Soc. London, Ser. B* **326**, 437–454.

Maloney, P. C. (1990b). *Res. Microbiol.* **141**, 374–383.

Manoil, C. (1990). *Proc. Natl. Acad. Sci. U.S.A.* **87**, 4937–4941.

Manoil, C., and Beckwith, J. (1986). *Science* **233**, 1403–1408.

Markgraf, M., Bocklage, H., and Muller-Hill, B. (1985). *Mol. Gen. Genet.* **198**, 473–475.

Marshall-Carlson, L., Celenza, J. L., Laurent, B. C., and Carlson, M. (1990). *Mol. Cell. Biol.* **10**, 1105–1115.

McKenna, E., Hardy, D., Pastore, J. C., and Kaback, H. R. (1991). *Proc. Natl. Acad. Sci. U.S.A.* **88**, 2969–2973.

McKeown, B. J. (1988). CPGS Thesis, University of Cambridge.

McMorrow, I., Chin, D. T., Fiebig, K., Pierce, J. L., Wilson, D. M., Reeve, E. C. R., and Wilson, T. H. (1988). *Biochim. Biophys. Acta* **945**, 315–323.

McMurry, L. M., Petrucci, R. R., and Levy, S. B. (1980). *Proc. Natl. Acad. Sci. U.S.A.* **77**, 3974–3977.

Menick, D., Lee, J. A., Brooker, R. J., Wilson, T. H., and Kaback, H. R. (1987). *Biochemistry* **26**, 1132–1136.

Mitchell, P. (1961). *Nature (London)* **191**, 144–148.

Mitchell, P. (1963). *Biochem. Soc. Symp.* **22**, 142–169.

Mitchell, P. (1973). *J. Bioenerg.* **4**, 63–91.

Mueckler, M., Caruso, C., Baldwin, S. A., Panico, M., Blench, I., Morris, H. R., Allard, W. J., Lienhard, G. E., and Lodish, H. (1985). *Science* **229**, 941–945.

Muiry, J. A. R. (1989). Ph.D. Thesis, University of Cambridge.

Nakao, T., Yamato, I., and Anraku, Y. (1987). *Mol. Gen. Genet.* **208**, 70–75.

Neal, R. J., and Chater, K. F. (1987). *Gene* **58**, 229–241.

Newman, M. J., and Wilson, T. H. (1980). *J. Biol. Chem.* **255**, 10583–10586.

Newman, M. J., Foster, D. L., Wilson, T. H., and Kaback, H. R. (1981). *J. Biol. Chem.* **256**, 11804–11808.

Nordlie, R. C. (1985). *In* "Metabolic Regulation" (R. S. Ochs, R. W. Hanson, and J. Hall, eds.), pp. 60–69. Elsevier, Amsterdam.

Oka, Y., Asano, T., Shibasaki, Y., Lin, J. L., Tsukuda, A., Katagiri, H., Akanuma, Y., and Takaku, F. (1990). *Nature (London)* **345**, 550–553.

Overath, P., Weigel, U., Neuhaus, J.-M., Soppa, J., Seckler, R., Riede, I., Bocklage, H., Muller-Hill, B., Aichelle, G., and Wright, J. K. (1987). *Proc. Natl. Acad. Sci. U.S.A.* **84**, 5535–5539.

Pacholocyk, T., Blakely, R. D., and Amara, S. G. (1991). *Nature (London)* **350**, 350–354.

Page, M. G., and Rosenbusch, J. P. (1988). *J. Biol. Chem.* **263**, 15906–15914.

Page, M. G., Rosenbusch, J. P., and Yamato, I. (1988). *J. Biol. Chem.* **263**, 15897–15905.

Patel, L., Garcia, M. L., and Kaback, H. R. (1982). *Biochemistry* **21**, 5805–5810.

Pattus, F. (1990). *Curr. Opin. Cell Biol.* **2**, 681–685.

Peden, K. W. (1983). *Gene* **22**, 277–280.

Petro, K. (1988). M. Phil. Thesis, University of Cambridge.

Poolman, B., Royer, T. J., Mainzer, S. E., and Schmidt, B. F. (1989). *J. Bacteriol.* **171**, 244–253.

Postma, P. W., and Lengeler, J. W. (1985). *Microbiol. Rev.* **49**, 232–269.
Pourcher, T., Bassilana, M., Sarkar, H. K., Kaback, H. R., and Leblanc, G. (1990a). *Philos. Trans. R. Soc. London, Ser. B* **326**, 411–424.
Pourcher, T., Bassilana, M., Sarkar, H., Kaback, H. R., and Leblanc, G. (1990b). *Biochemistry* **29**, 690–696.
Pourcher, T., Sarkar, H., Bassilana, M., Kaback, H. R., and Leblanc, G. (1990c). *Proc. Natl. Acad. Sci. U.S.A.* **87**, 468–472.
Quiocho, F. A. (1986). *Annu. Rev. Biochem.* **55**, 287–315.
Quiocho, F. A. (1990). *Philos. Trans. R. Soc. London, Ser. B* **326**, 341–352.
Raboy, B., and Padan, E. (1978). *J. Biol. Chem.* **253**, 3287–3291.
Rampal, A. L., and Jung, C. Y. (1987). *Biochim. Biophys. Acta* **896**, 287–294.
Rausch, T., Raszeja-Specht, A., and Koepsell, H. (1989). *Biochim. Biophys. Acta* **985**, 133–138.
Rees, W. D., and Holman, G. D. (1981). *Biochim. Biophys. Acta* **646**, 251–260.
Reynolds, C. H., and Silver, S. (1983). *J. Bacteriol.* **156**, 1019–1024.
Riordan, C., and Kornberg, H. L. (1977). *Proc. R. Soc. London, Ser. B* **198**, 401–410.
Riordan, C. *et al.* (1989). *Science* **245**, 1066–1073.
Rippka, R., Deruelles, J., Waterbury, J. B., Herdman, M., and Stanier, R. Y. (1979). *J. Gen. Microbiol.* **111**, 1–61.
Robbins, J. D., Laurenza, A., Kosley, R. W., O'Malley, G. J., Spahl, B., and Seamon, K. B. (1991). In press.
Roepe, P. D., Consler, T. J., Menezes, M. E., and Kaback, H. R. (1990). *Res. Microbiol.* **141**, 290–308.
Romano, A. H. (1986). *In* "Carbohydrate Metabolism in Cultured Cells" (M. J. Morgan, ed.), pp. 225–244. Plenum, London.
Rosen, B. P. (1990). *Res. Microbiol.* **141**, 336–341.
Rotman *et al.* (1968). *J. Mol. Biol.* **36**, 247–260.
Rouch, D. A., Cram D. S., DiBerardino, D., Littlejohn, T., and Skurray, R. A. (1990). *Mol. Microbiol.* **4**, 2051–2062.
Rubin, R. A., and Levy, S. B. (1990). *J. Bacteriol.* **172**, 2303–2312.
Saier, M. H. (1985). "Mechanisms and Regulation of Carbohydrate Transport in Bacteria." Academic Press, New York.
Sasatsu, M., Misra, T. K., Chu, L., Ladagu, R., and Silver, S. (1985). *J. Bacteriol.* **164**, 983–993.
Sauer, N. (1986). *Planta* **168**, 139–144.
Sauer, N., and Tanner, W. (1989). *FEBS Lett.* **259**, 43–46.
Sauer, N., Caspari, T., Klebl, F., and Tanner, W. (1990a). *Proc. Natl. Acad. Sci. U.S.A.* **87**, 7949–7952.
Sauer, N., Friedlander, K., and Graml-Wicke, U. (1990b). *EMBO J.* **9**, 3045–3050.
Seol, W., and Shatkin, A. J. (1990). *Proc. Natl. Acad. Sci. U.S.A.* **88**, 3802–3806.
Severin, J., Langel, P., and Hofer, M. (1989). *J. Bioenerg. Biomembr.* **21**, 321–334.
Shanahan, M. F. (1982). *J. Biol. Chem.* **257**, 7290–7293.
Sheridan, R. P., and Chopra, I. (1991). *Mol. Microbiol.* **5**, 895–900.
Silhavy, T. J., Berman, M. L., and Enquist, L. W. (1984). "Experiments with Gene Fusions." Cold Spring Harbor Lab., Cold Spring Harbor, New York.
Stack, S. P., Stein, D. A., and Landfear, S. M. (1990). *Mol. Cell. Biol.* **10**, 6785–6790.
Stein, W. (1986). "Transport and Diffusion across Cell Membranes." Academic Press, Orlando, Florida.
Stoker, N. G., Pratt, J. M., and Holland, I. B. (1984). *In* "Transcription and Translation: A Practical Approach" (B. D. Hames and S. J. Higgins, eds.), pp. 153–177. IRL Press, Oxford.

Stoner, C., and Schleif, R. F. (1983). *J. Mol. Biol.* **171**, 369–381.

Studier, F. W., and Moffatt, R. A. (1986). *J. Mol. Biol.* **189**, 113–130.

Sumiya, M. (1989). Ph.D. Thesis, University of Cambridge.

Szkutnicka, K., Tschopp, J. F., Andrews, L., and Cirillo, V. P. (1989). *J. Bacteriol.* **171**, 4486–4493.

Tabor, S., and Richardson, C. C. (1985). *Proc. Natl. Acad. Sci. U.S.A.* **82**, 1074–1078.

Tate, C. G., Muiry, J. A. R., and Henderson, P. J. F. (1992). *J. Biol. Chem.* **267**.

Teather, R. M., Muller-Hill, B., Abrutsch, U., Aichele, G., and Overath, P. (1978). *Mol. Gen. Genet.* **159**, 239–248.

Thorens, B., Sarkar, H. K., Kaback, H. R., and Lodish, H. F. (1988). *Cell (Cambridge, Mass.)* **55**, 281–291.

Tobin, J. F., and Schleif, R. F. (1987). *J. Mol. Biol.* **196**, 789–799.

Tsuchiya, T., Ottina, K., Moriyama, Y., Newman, M. J., and Wilson, T. H. (1982). *J. Biol. Chem.* **257**, 5125–5128.

Turk, E., Zabel, B., Mundlos, S., Dyer, J., and Wright, E. M. (1991). *Nature (London)* **350**, 354–356.

van der Rest, M. E., Schwarz, E., Oesterhelt, D., and Konings, W. N. (1990). *Eur. J. Biochem.* **189**, 401–407.

van Dijken, P., and Scheffers, B. (1986). *FEMS Microbiol. Rev.* **32**, 199–224.

van Leeuwen, C. C. M., Postma, E., van den Brock, P. J. A., and van Steveninck, J. V. (1991). *J. Biol. Chem.* **266**, 12146–12151.

Viitanen, P., Newman, M. J., Foster, D. L., Wilson, T. H., and Kaback, H. R. (1986). *In* "Methods in Enzymology" (S. Fleischer and B. Fleischer, eds.), vol. 125, pp. 429–452. Academic Press, Orlando, Florida.

von Heijne, G. (1987). "Sequence Analysis in Molecular Biology—Treasure Trove or Trivial Pursuit." Academic Press, London.

von Heijne, G. (1988). *Biochim. Biophys. Acta* **947**, 307–333.

Wadzinski, B. E., Shanahan, M. F., Seamon, K. B., and Ruoho, A. E. (1990). *Biochem. J.* **272**, 151–158.

Walmsley, A. R. (1988). *Trends Biochem. Sci.* **13**, 226–231.

Waters, S. H., Rogowsky, J., Grinstead, J., Altenbuchner, J., and Schmitt, R. (1983). *Nucleic Acids Res.* **11**, 6089–6140.

Weiss, M. S., Wacker, T., Wecklesser, J., Welte, W., and Schulz, G. E. (1990). *FEBS Lett.* **267**, 268–272.

West, I. C. (1970). *Biochem. Biophys. Res. Commun.* **41**, 655–661.

West, I. C., and Mitchell, P. (1972). *J. Bioenerg.* **3**, 445–462.

West, I. C., and Mitchell, P. (1973). *Biochem. J.* **132**, 587–592.

Wheeler, T. J., and Hinkle, P. C. (1985). *Annu. Rev. Physiol.* **47**, 503–517.

Wilson, D. M., and Wilson, T. H. (1987). *Biochim. Biophys. Acta* **904**, 191–200.

Wilson, D. M., Tsuchiya, T., and Wilson, T. H. (1986). *In* "Methods in Enzymology" (S. Fleischer and B. Fleischer, eds.), vol. 125, pp. 377–387.

Wilson, T. H., Yunker, T. L., and Hansen, C. L. (1990). *Biochim. Biophys. Acta* **1029**, 113–116.

Wright, J. K., Teather, R. M., and Overath, P. (1983). *In* "Methods in Enzymology" (S. Fleischer and B. Fleischer, eds.), vol. 97, pp. 158–175. Academic Press, New York.

Wrubel, W., Stochaj, U., Sonnewald, U., Theres, C., and Ehring, R. (1990). *J. Bacteriol.* **172**, 5374–5381.

Yamaguchi, A., Udagawa, T., and Sawai, T. (1990a). *J. Biol. Chem.* **265**, 4809–4813.

Yamaguchi, A., Ono, N., Akasaka, T., Noumi, T., and Sawai, T. (1990b). *J. Biol. Chem.* **26**, 15525–15530.

Yamaguchi, A., Adachi, K., Akasaka, T., Ono, N., and Sani, T. (1991). *J. Biol. Chem.* **266**, 6045–6051.

Yamato, I., and Anraku, Y. (1989). *Biochem. J.* **258**, 389–396.

Yao, B., Sollitti, P., and Marmur, J. (1989). *Gene* **79**, 189–197.

Yazyu, H., Shiota-Niija, S., Shimamoto, T., Kanazawa, H., Futai, M., and Tsuchiya, T. (1984). *J. Biol. Chem.* **259**, 4320–4326.

Yazyu, H., Shiota-Niija, S., Futai, M., and Tsuchiya, T. (1985). *J. Bacteriol.* **162**, 933–937.

Yoshida, H., Bogaki, M., Nakamura, S., Ubukata, A., and Konno, M. (1990). *J. Bacteriol.* **172**, 6942–6949.

Zhang, C.-C., Durand, M.-C., Jeanjean, R., and Joset, F. (1989). *Mol. Microbiol.* **3**, 1221–1229.

Zilberstein, D., and Dwyer, D. M. (1985). *Proc. Natl. Acad. Sci. U.S.A.* **82**, 1716–1720.

Zilberstein, D., Dwyer, D. M., Matthei, S., and Horuk, R. (1986). *J. Biol. Chem.* **261**, 15053–15057.

Molecular and Cellular Physiology of GLUT-2, a High-K_m Facilitated Diffusion Glucose Transporter

Bernard Thorens

Institute of Pharmacology, University of Lausanne, CH-1005 Lausanne, Switzerland

I. Introduction

Glucose enters the organism through a transport system located in the epithelial cells lining the intestine and is reabsorbed by kidney nephron by a similar transepithelial transport process. Maintaining systemic glucose homeostasis further involves glucose uptake by liver, muscles, adipose tissue, and brain and release into the blood of glucose stored in the liver as glycogen or synthesized *de novo* by hepatic gluconeogenesis. A major aspect of the regulation of glucose homeostasis is the balanced secretion of insulin and glucagon by the pancreatic islet β and α cells, respectively (Unger and Orci, 1981a,b). Insulin stimulates the storage of glucose in the liver and the absorption of glucose by adipose cells and muscle by a mechanism stimulating the rate of glucose uptake. Glucagon stimulates the degradation of hepatic glycogen stores and the production of glucose by the gluconeogenic pathway. The consequent release of glucose prevents the development of hypoglycemia. These mechanisms of glucose uptake, release, and sensing by pancreatic α and β cells require the presence in the plasma membrane of these cells of integral membrane proteins known as facilitated diffusion glucose transporters, which catalyze the transport of D-glucose and closely related molecules down their chemical concentration gradients.

The human erythrocyte glucose transporter was the first glucose carrier to be characterized at the biochemical level (Wheeler and Hinkle, 1985). Antibodies raised against the erythrocyte transporter (Lienhard *et al.*, 1982) permitted the molecular cloning of its cDNA by screening an expression library constructed from the human hepatoma cell line HepG2

(Mueckler *et al.*, 1985). The availability of a cDNA probe for the erythrocyte glucose transporter (referred to as GLUT-1) allowed the study of its tissue localization. Although GLUT-1 was detected in a number of tissues (Mueckler *et al.*, 1985; Birnbaum *et al.*, 1986; Flier *et al.*, 1987), some organs important for the control of glucose homeostasis, such as liver and pancreatic islet β cells, were found to express almost undetectable levels of this transporter (Birnbaum *et al.*, 1986; Flier *et al.*, 1987; Thorens *et al.*, 1988). This led to the search for related isoforms of GLUT-1. At the present time five different functional glucose transporter isoforms have been characterized by molecular cloning and functional expression. These transporters are expressed in a tissue-specific manner, display specific kinetic properties, and are differentially regulated by insulin or glucose levels. Therefore each glucose carrier appears to perform specific functions in the general control of glucose homeostasis. A wealth of information has accumulated over the years on the physiology of GLUT-1 (the erythrocyte glucose transporter), GLUT-2 (the liver/β cell glucose transporter), and GLUT-4 (the muscle/adipose cell glucose transporter). Various reviews discuss these results (Gould and Bell, 1990; Bell *et al.*, 1990; Mueckler, 1990; Thorens *et al.*, 1990a).

In the present article we will discuss the data accumulated so far on the molecular and cellular physiology of GLUT-2, the liver/β cell glucose transporter. We will emphasize that the relatively high K_m for glucose of GLUT-2 is necessary for its specific transport functions in the tissues in which it is expressed: liver, intestine, kidney and pancreatic islet β cells. In particular, we will discuss the involvement of this high-K_m transporter in the control of glucose-induced insulin secretion by the pancreatic β cells and its role in the development of β cell dysfunction characteristic of insulinomas and diabetic rat islets.

II. Structural and Functional Characterization of GLUT-2

A. Molecular Cloning of GLUT-2

The cloning of the liver glucose transporter was achieved by screening rat (Thorens *et al.*, 1988) and human (Fukumoto *et al.*, 1988) liver cDNA libraries with a cDNA probe for GLUT-1, using reduced stringency conditions. This technique similarly yielded cDNA clones for all the other transporter isoforms cloned so far. The liver glucose transporter, GLUT-2, is 55% identical in amino acid sequence to GLUT-1. Its hydropathy plot is similar to GLUT-1, displaying the same alternation of hydrophobic and hydrophilic segments, predicting a protein with 12 putative transmembrane domains. Both GLUT-1 and GLUT-2 sequences are colinear with the exception of the exoplasmic, glycosylated loop connecting the first and second transmembrane domains: this loop is 64 amino acids long in GLUT-2 versus 32 amino acids long in GLUT-1 and the other GLUT molecules (Bell *et al.*, 1990). Figure 1 shows the comparison of the sequences of rat, mouse (Asano *et al.*, 1989; Suzue *et al.*, 1989), and human GLUT-2.

Comparison of the amino acid sequence of the human, rat, and mouse GLUT-2 proteins (Fig. 1) shows a sequence identity of 82% between the human and the rat transporter and 95% between mouse and rat forms. This level of conservation of the primary sequence between different species is lower than for GLUT-1. Indeed, the human GLUT-1 amino acid sequence is 97% identical to rat GLUT-1. In that context, it is interesting to note that, despite a relatively low degree of sequence conservation among species, a potential phosphorylation site for the cyclic AMP-dependent protein kinase has been conserved in the COOH-terminal cytoplasmic tail of GLUT-2 (Fig. 1). A similar phosphorylation site is present in GLUT-4 and its phosphorylation is stimulated following an increase in the intracellular concentration of cyclic AMP (Lawrence *et al.*, 1990). Whether GLUT-2 is phosphorylated and what role the phosphorylation may play in GLUT-2 function are not known.

The characterization of GLUT-2 as a glucose carrier was completed by expressing the transporter in bacteria. Mutant *Escherichia coli* cells lacking the phosphotransferase systems for glucose and mannose and one of the transport system for galactose (Bouma *et al.*, 1987) were used to express GLUT-2 utilizing the T7 promoter/T7 polymerase system (Tabor and Richardson, 1985). GLUT-2 was inserted into the bacterial membrane and catalyzed glucose uptake. Transport was stereoselective for D-glucose but not for L-glucose, and was inhibited by phloretin and mercury chloride, known inhibitors of the glucose transporter (Thorens *et al.*, 1988). Injection of *in vitro* synthesized mRNA for GLUT-2 into *Xenopus* oocytes and consequent measurement of glucose uptake have further confirmed that GLUT-2 is a glucose carrier (Vera and Rosen, 1989, 1990).

B. Kinetic Properties of GLUT-2

To understand the specific function of GLUT-2 in the control of glucose homeostasis, it is important to understand its kinetic properties, its tissue and cellular localization, as well as its regulated expression.

In this section we will review data pertaining to the kinetics of glucose uptake by isolated hepatocytes, by vesicles prepared from the basolateral membrane of intestinal epithelial cells, and by pancreatic islet cells. As will

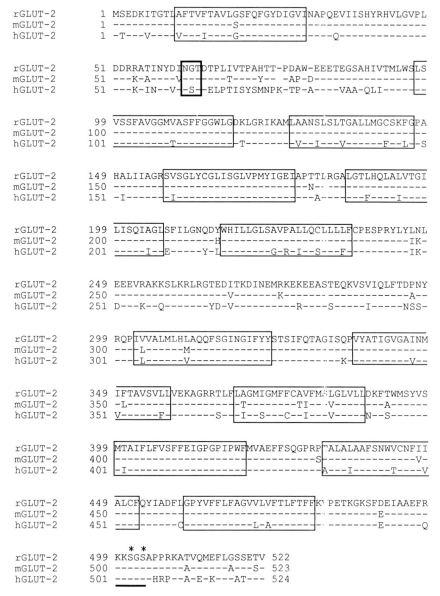

FIG. 1 Amino acid sequences of the rat, mouse, and human forms of GLUT-2. For the mouse and human sequences, only the amino acids different from the rat sequence are shown. The gaps in the rat and mouse first exoplasmic loop have been introduced to allow maximum homology between the three sequences. The 12 hydrophobic putative transmembrane segments are boxed. The N-glycosylation site present in the exoplasmic loop between the first and second transmembrane domain is delineated by a bold box. An evolutionary conserved consensus phosphorylation site for cyclic AMP-dependent protein kinase (Edelman *et al.*, 1987) is present in the COOH-terminal tail and is underlined; the possible phosphoserines are indicated by asterisks.

be discussed below, GLUT-2 is the predominant facilitated diffusion glucose transporter in these tissues. Therefore we will assume that the kinetic data for glucose uptake reflect the specific properties of GLUT-2. In the following section, the kinetic data together with the cellular and subcellular localization of GLUT-2 will form the basis for a discussion of the specific physiological role of this transporter isoform in the control of glucose homeostasis.

Studies of glucose uptake by intact liver showed the presence of a stereospecific system for D-glucose with a high K_m for glucose (Williams *et al.*, 1968). Further studies on isolated hepatocytes characterized the transport system in greater detail and showed that (1) the K_m for glucose was relatively high, about 15–20 mM, as compared to a K_m of 1–3 mM for the glucose transporter from the human erythrocyte (Craik and Elliott, 1978); (2) the transporter was symmetric, i.e., the K_m for glucose was the same (about 20 mM) for influx or efflux in zero trans conditions or for equilibrium exchange entry or exit (Craik and Elliott, 1978; Ciaraldi *et al.*, 1986). This contrasts to the transport of glucose by the human erythrocyte carrier, which is asymmetrical, with a K_m for glucose efflux (20–30 mM) at least one order of magnitude greater than that for glucose influx (1–3 mM)(Stein, 1986); (3) the concentration of cytochalasin B producing half-maximum inhibition of glucose uptake (K_i) by isolated hepatocytes is 1.9 μM, a value about 10-fold higher than the K_i for inhibition of the human erythrocyte glucose transporter (Axelrod and Pilch, 1983).

The presence of a high-K_m glucose transporter in the basolateral membrane of intestinal epithelial cells has been demonstrated in kinetic experiments measuring the rate of glucose uptake into vesicles prepared from the basolateral membrane of these cells (Maenz and Cheeseman, 1987). In these experiments, the K_m for glucose efflux was 23 and 48 mM for influx, suggesting a possible small degree of asymmetry. The K_i for cytochalasin B inhibition of glucose transport was 0.11 μM, similar to that for the erythrocyte transporter but lower than that of hepatocytes. Because GLUT-2 has been localized to the basolateral membrane of intestinal absorptive cells (see Section B,1 below), the measured glucose transport kinetics most probably reflect GLUT-2 activity.

β cells have been known for a long time to have a very efficient stereoselective uptake system for glucose, and initial kinetic measurements showed a K_m for glucose of about 50 mM (Hellman *et al.*, 1971). Using 3-O-methyl-glucose to measure the kinetics of uptake by preparations of dispersed islet cells, Johnson *et al.* (1990a) demonstrated the presence of two components for glucose uptake in islet cells (Fig. 2). One component has a high K_m (17 mM), high V_{max} (32 mmol/min/liter) for glucose and the other a low K_m (1.4 mM), low V_{max} (5.5 mmol/min/liter). The high K_m component is thought to represent the activity of GLUT-2. Indeed, in islet

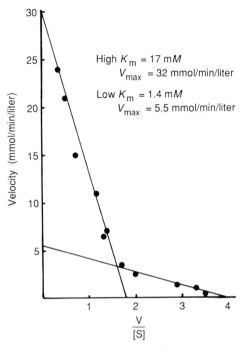

FIG. 2   Eadie–Hofstee plot of the concentration dependence of 3-O-methyl D-glucose uptake by dispersed rat islet cells. This plot shows that the glucose uptake process has two components: a high-K_m component, which is accounted for by GLUT-2, and a low-K_m component. The molecular nature of the low-K_m component is not yet established. (Reprinted with permission from Johnson et al., 1990a.)

cells that no longer express GLUT-2 as a result of a hyperinsulinemic clamp, which maintains the glycemia at about 50 mg/dl, the high-K_m component for glucose uptake disappears while the low-K_m component is unchanged (Chen et al., 1990). The glucose transporter isoform responsible for the low-K_m uptake component is not yet known. The other already characterized transporter isoforms (GLUT-1, -3, -4, and -5) have a low K_m for glucose, but by Western or Northern blotting they cannot be detected in protein or RNA preparations made from freshly isolated islets. By screening at reduced stringency a rat islet cDNA library made in λgt11, using a GLUT-1 cDNA probe, only GLUT-2 and GLUT-1 could be detected. GLUT-1 cDNAs clones were, however, 40 times less abundant than GLUT-2 clones (10 versus 400 clones in 1 million plaques screened) (B. Thorens, unpublished observations). Therefore it is not clear whether the low level of GLUT-1 expression can account for the low-K_m glucose uptake component of islet cells or whether an as yet uncharacterized transporter is expressed by pancreatic islets. This additional transporter would, however, be of low sequence similarity to GLUT-1 because with the screening conditions utilized the probe hybridized to cDNAs having nucleotide sequence identities with GLUT-1 as low as 40%.

The above data, together with the cellular localization and regulated expression of GLUT-2 discussed below, strongly suggest that GLUT-2 is the high-K_m glucose transporter. The kinetics of glucose uptake by GLUT-2 have been measured directly by expression of the transporter in *Xenopus* oocytes injected with GLUT-2 mRNA (Vera and Rosen, 1989). These experiments showed a K_m for glucose of 7 m*M*, a value lower than that measured in intact tissues (15 to 20 m*M*). The discrepancy between these values may result from the use, in the uptake measurements in oocytes, of a glucose analog (2-deoxy-D-glucose) that can be phosphorylated by hexo-kinase instead of the nonmetabolizable analog 3-*O*-methyl-glucose.

The presence in the plasma membrane of a glucose transporter with a K_m of about 15–20 m*M* allows the rate of glucose uptake to increase in proportion to the rise in blood glucose concentration from the normal value of about 5 m*M* to values as high as 20 m*M*. This occurs, for instance, after a carbohydrate-rich meal in the portal circulation and in the arterial blood traversing the pancreatic islets. A low-K_m glucose transporter with a K_m of about 1–3 m*M* would be almost saturated at the normal blood glucose concentration and will permit only a slight increase in the rate of glucose uptake as the blood glucose rises to higher concentrations. In contrast, the presence of the high-K_m GLUT-2 in certain organs allows the rate of glucose uptake to increase in proportion to the variation of blood glucose concentration above the normal value of 5 m*M*. As will be discussed below, this property is required for the proper control of glucose homeostasis.

A high K_m for glucose is a property unique to the GLUT-2 isoform. The other glucose transporters have K_m valves for glucose of about 1 to 5 m*M* (Bell *et al.*, 1990). GLUT-1, the erythrocyte glucose transporter, has been extensively studied and its K_m for glucose uptake is between 1 and 3 m*M* (Wheeler and Hinkle, 1985; Stein, 1986). The K_m for glucose of GLUT-4, which is the major transporter isoform expressed in muscle and adipo-cytes, has a very similar value. The K_m for glucose of GLUT-3 may be lower than that of GLUT-1 (Bell *et al.*, 1990).

The structural basis for the high K_m and low affinity for glucose of GLUT-2 is not known. However, a unique structural feature of GLUT-2 is the length of the glycosylated exoplasmic loop present between the first and second transmembrane domain (Fig. 1; Bell *et al.*, 1990). Is this long loop important for the high affinity for glucose? The ability to express glucose transporters in *Xenopus* oocytes (Gould and Lienhard, 1989; Vera and Rosen, 1989, 1990; Keller *et al.*, 1990) and to perform glucose uptake kinetic measurements may permit an answer to this question by expressing chimeric or mutated transporters in these heterologous systems.

III. Tissue and Cellular Localization of GLUT-2

GLUT-2 is expressed in a restricted set of organs that are of critical importance for the control of glucose homeostasis: hepatocytes, the absorptive epithelial cells of intestine and kidney, and the pancreatic islet β cells (Thorens *et al.*, 1988; Fukumoto *et al.*, 1988). In this section we will discuss the involvement of GLUT-2 in glucose uptake by hepatocytes and in the release into the blood of glucose stored in the liver; we will discuss its role in the mechanism of glucose absorption and reabsorption by intestinal and kidney epithelial cells. Finally, we will discuss why its expression by pancreatic islet β cells is required for normal glucose sensing and insulin secretion. The polarized expression of GLUT-2 will be compared in the different cell types in which it is expressed, and its tissue-specific regulated expression will be correlated with the physiological function of the different tissues in the normal and pathological states.

A. GLUT-2 in Hepatocytes

The blood glucose in the portal circulation can raise rapidly over the normal concentration of about 5 mM after a carbohydrate-rich meal and most of this glucose is cleared from the circulation into hepatocytes, where it is either catalyzed by glycolysis or stored in the form of glycogen. In contrast, during a fasting period, the liver releases glucose into the blood to keep its concentration at the normal level of about 5 mM (Unger and Orci, 1981a,b). The glucose released into the blood comes from the breakdown of glycogen and from the formation of glucose by the gluconeogenic pathway. Insulin stimulates the storage of glucose by liver mostly by increasing the level of glucokinase, which traps glucose by converting it to glucose 6-phosphate, by stimulating glycogen synthesis, and by inhibiting the transcription of the gene for phosphoenolpyruvate carboxykinase (PEPCK), a key enzyme in the gluconeogenic pathway (Granner and Pilkis, 1990). In contrast, during a fasting episode or in a period of stress, hepatic glucose output is stimulated by counterregulatory hormones (glucagon, epinephrine, glucocorticoid) (Unger and Orci, 1981a,b). These hormones increase the intracellular concentration of cyclic AMP, which stimulates the rate of transcription of the PEPCK gene. Glucagon also has a inhibitory effect on the expression of glucokinase, thereby preventing phosphorylation of the newly formed glucose, which then exits the cell by the glucose transporter (Granner and Pilkis, 1990). Thus, the flux of glucose into or out of hepatocytes is mostly regulated by controlling the activity of enzymes involved in key steps of glycolysis, glycogen synthesis, or gluconeogenesis and very little or no control of the glucose transport system has been noted in hepatocytes (Williams *et al.*, 1968; Ciaraldi *et al.*, 1986).

1. Immunolocalization of GLUT-2 in Hepatocytes

The characterization of GLUT-2 by molecular cloning from human and rat liver cDNA libraries and the deduction of the primary amino acid sequence allowed the preparation of transporter-specific antibodies. In particular, the COOH-terminal sequence of GLUT-2, and of all the glucose transporters characterized so far, is unique. Antibodies to this region of GLUT-2 were extremely useful for the characterization of the protein by Western blotting and for its localization by immunohistochemical techniques both at the light and electron microscope level. By Western blotting, the protein migrated as a polypeptide of 51 kDa in liver, 57 kDa in intestine, 55 kDa in kidney, and 53 kDa in pancreatic islets (Thorens *et al.*, 1988). The differences in the electrophoretic mobility of GLUT-2 present in the different organs are due to differences in posttranslational modifications, as cDNAs for GLUT-2 have been isolated from libraries made from each tissue and have been shown by restriction mapping and nucleotide sequencing to encode proteins of identical amino acid sequence (Thorens *et al.*, 1990b; Johnson *et al.*, 1990a).

GLUT-2 was localized by immunofluorescence microscopy using antibodies to the COOH-terminal sequence applied on frozen sections of rat liver (Thorens *et al.*, 1990b). GLUT-2 is expressed only in hepatocytes (Fig. 3) and is not detected in Kuppfer cells, expression is restricted to the sinusoidal membrane and is absent from the apical, bile canalicular membrane. No staining for GLUT-2 is detected in internal membranes, an observation in agreement with the report that cytochalasin B binding sites are measurable only in plasma membrane fractions prepared from total liver (Ciaraldi *et al.*, 1986).

The absence of a sizeable pool of GLUT-2 in intracellular vesicles—which should, however, be confirmed by electron microscopy—is in contrast to expression of other transporter isoforms in other tissues. For instance, GLUT-1 and GLUT-4 present in adipocytes and muscles, tissues in which the rate of glucose uptake is regulated by insulin, are localized mostly in intracellular vesicles. After insulin stimulation, glucose transporter-containing vesicles translocate to the plasma membrane, thereby increasing the number of cell surface-expressed transporters and, conse-

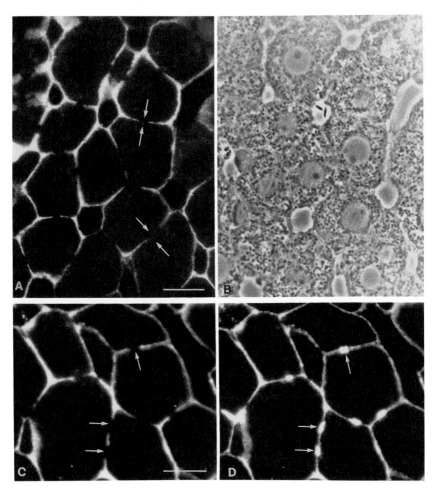

FIG. 3   Immunolocalization of GLUT-2 in rat liver. (A) Immunofluorescence staining; staining is restricted to the sinusoidal membrane of hepatocytes and is absent from bile canaliculi
membranes (arrows). Kupffer cells (thick arrows) are not labeled. (B) Phase contrast.
(C) Immunodetection of GLUT-2. (D) Immunodetection of HA4 antigen [which labels the
bile canalicular membrane (Hubbard *et al.*, 1985); arrows] on the same section as in (C), with
a rhodamine-coupled secondary antibody. Bar = 15 μm. (From Thorens *et al.*, 1990c.)

quently, the rate of glucose uptake (Suzuki and Kono, 1980; Cushman and
Wardzala, 1980; Kasanicki and Pilch, 1990). The absence of an intracellular pool of GLUT-2 in hepatocytes is therefore consistent with the unre-
sponsiveness of the glucose transport rate to the action of insulin. Interestingly, however, when expressed in *Xenopus* oocytes, GLUT-2 is present
to some extent in intracellular vesicles and its translocation to the cell
surface can be induced by insulin addition to the oocytes (Vera and Rosen,

1990). Therefore, it seems that the translocation properties of a given transporter may be determined by the cellular context in which it is expressed and that the primary structure of the transporter is not the unique determinant of its intracellular localization.

2. Regulation of GLUT-2 Expression in States of Imbalanced Glucose Homeostasis

The level of expression of GLUT-2 in hepatocytes shows minimal regulation. Streptozocin-induced diabetes has been reported to have no effect (Thorens *et al.*, 1990c) or to raise slightly (Oka *et al.*, 1990) the levels of GLUT-2 protein in rat liver. In this latter study, insulin treatment returned the level of GLUT-2 to control values. In Zucker diabetic rats (Johnson *et al.*, 1990b) and in neonatal streptozocin rats (Thorens *et al.*, 1990d), two models of non-insulin-dependent type II diabetes (see Section C,3,*c* below), no significant changes in the amount of GLUT-2 protein or mRNA have been noticed as a result of the hyperglycemia. Fasting rats for 2 days decreased the level of GLUT-2 mRNA by 50%, and its level was increased to about fivefold the control value of 24 hr after refeeding. However, GLUT-2 protein levels are minimally changed, showing a significant increase of about 70% 6 hr after refeeding and returning to control levels after 18 hr (Thorens *et al.*, 1990c). The discrepancy between the mRNA and protein levels is surprising but may reflect complex regulation of GLUT-2 expression by translational or posttranslational mechanisms. A similar discrepancy between mRNA and protein levels during fasting and refeeding has been described for glucokinase (Iynedjian *et al.*, 1987).

The localization of GLUT-2 only on the plasma membrane of hepatocyte and not in intracellular vesicles, and its lack of regulation in situations of imbalanced glucose homeostasis, are consistent with the dual role of this transporter. It is responsible not only for the uptake of glucose from the arterial circulation when glucose is above the normal concentration of about 5 mM, but also for the release of glucose into the blood to prevent hypoglycemia. This role is also consistent with the observed symmetry of glucose transport in hepatocytes. The high K_m (15–20 mM) for glucose of GLUT-2 ensures that the rate of glucose uptake will increase in proportion to the variations in extracellular glucose concentrations over the range to which hepatocytes may be exposed, i.e., up to about 10–15 mM.

3. GLUT-2: The Major Glucose Transporter Isoform in Liver

Early work using the cDNA probe for GLUT-1 showed that this transporter isoform was not expressed or present at a very low level in the liver (Mueckler *et al.*, 1985; Birnbaum *et al.*, 1986; Flier *et al.*, 1987). Increased levels of GLUT-1 could, however, be detected in the liver of fasted rats or in isolated hepatocytes kept in culture (Rhoads *et al.*, 1988). GLUT-1 is indeed expressed in normal liver and its mRNA is present at a level corresponding to 1–3% of the level of GLUT-2 mRNA. However, after fasting, the levels of GLUT-1 mRNA and protein are increased several-fold, in contrast to GLUT-2 protein, which is unchanged (Thorens *et al.*, 1990c). By immunofluorescence microscopy GLUT-1 was shown to be expressed by only a subset of hepatocytes, those forming the first row around each central vein (Tal *et al.*, 1990). The increase in GLUT-1 expression seen in fasted rats corresponds to the expression of GLUT-1 by additional hepatocytes around the central veins. These perivenous hepatocytes also express GLUT-2. The physiological role of GLUT-1 in this subset of hepatocytes, which are known to be more glycolytic that the periportal hepatocytes, is not presently understood, nor is it known why fasting induces expression of GLUT-1 in more perivenous hepatocytes. These data indicate, however, that GLUT-2 is the major glucose transporter isoform in the liver and that GLUT-1 is a minor species. Also, the regulation of the expression of these two transporters is different in the same tissues, and the increased expression of GLUT-1 in hepatocytes after fasting is in contrast to its expression in adipocytes. In these cells GLUT-1 levels are slightly reduced, rather than increased, after a fasting period (Kahn *et al.*, 1989). The differential regulated expression of these two transporter isoforms in liver is unexpected from the available physiological data on glucose uptake by hepatocytes and suggest that, at least in the perivenous hepatocytes, some control of glucose uptake by the high-affinity GLUT-1 may be important for the physiological function of these hepatocytes. The perivenous hepatocytes are known to be more glycolytic than the other hepatocytes. Thus, there is a correlation between the glycolytic activity of these cells and expression of GLUT-1. A similar situation is observed in the kidney nephron: different cell types express GLUT-1 at different levels, and the levels of GLUT-1 expression correlate with the known glycolytic activity of these cells (Thorens *et al.*, 1990e). Therefore GLUT-1 may provide the cells with a glucose uptake system important for the control of glycolysis.

B. GLUT-2 in Intestinal and Kidney Absorptive Epithelial Cells

In mammals, transepithelial glucose transport systems permit the absorption of glucose into the organism through the epithelial cells lining the small intestine and the reabsorption of glucose through the kidney proximal tubules. These two processes are phenomenologically very similar but may actually be carried out by different sets of transporter proteins, as we will discuss below.

The simultaneous presence of three different membrane proteins, located at the different poles of the epithelial cells, are required for the transepithelial transport process to take place (also see M. J. Birnbaum, this volume). The first step is the transport of glucose into the cells against its concentration gradient: this step is carried out by an Na^+-dependent glucose symporter located in the apical brush border of the epithelial cells (Semenza et al., 1984; Hediger et al., 1987). This symporter transport Na^+ down its electrochemical gradient, a mechanism providing the driving force to cotransport and concentrate glucose into the cells. A low intracellular concentration of Na^+ is maintained by the Na^+/K^+-ATPase localized in the basolateral membrane. The glucose accumulated into the cells is then released into the extracellular space adjacent to blood capillaries by a mechanism of facilitated diffusion also located on the basolateral membrane.

The molecular structure of the apical Na^+-dependent glucose transporter has been elucidated by analysis of cDNA clones obtained by expression cloning (Hediger et al., 1987). This symporter is a very hydrophobic membrane protein with 11 putative transmembrane domains. It shows a high level of structural similarity to the bacterial Na^+–proline (Hediger et al., 1989) and Na^+–glutamate cotransporters (Deguchi et al., 1990) and is probably the prototype for a new family of Na^+-dependent symporters. Importantly, this symporter shows no sequence similarity to any of the facilitated diffusion glucose transporters.

1. Immunolocalization of GLUT-2 in Absorptive Epithelial Cells

In intestine, GLUT-2 is expressed at high level in the basolateral membrane of the absorptive epithelial cells (Fig. 4) (Thorens et al., 1990c,e), a location similar to that of the Na^+/K^+-ATPase (Amerongen et al., 1989), and is not present in the brush border. The expression of GLUT-2 in the basolateral membrane of enterocytes suggests that it is responsible for the release of glucose close to the blood capillaries. Interestingly, GLUT-2 is expressed only in enterocytes present well above the base of the villi. Intestinal epithelial cells are produced from dividing stem cells present in the crypts of Lieberkuhn and differentiate into mature absorptive cells as they migrate toward the tips of the villi. Therefore, GLUT-2 expession in the intestine is a marker of the differentiated state of epithelial cells. A new glucose transporter, GLUT-5, has been cloned from a human intestine cDNA library (Kayano et al., 1990). The mRNA for this transporter isoform is detected at high levels in the intestine and kidney. So far there are no data concerning its cellular and subcellular localization in the intestine or the kidney. It is therefore not known if it colocalizes with

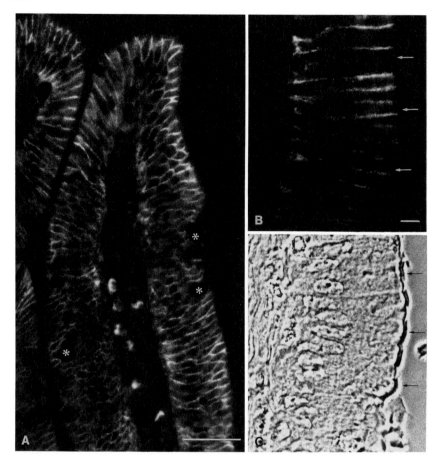

FIG. 4   Immunolocalization of GLUT-2 in rat duodenum. (A) Low magnification showing immunostaining over the top part of a villus. Staining is restricted to the basolateral membrane of absorptive epithelial cells. The brush border is not labeled nor are the goblet cells (*). Intensely stained cells in the lamina propria are most probably mast cells, which often absorb immunoreagents nonspecifically. (B and C) Immunofluorescence staining and phase-contrast images of a segment of a villus cut parallel to the plane of the epithelial cells, showing the basolateral staining of absorptive cells. Brush border (arrows) is not labeled. Bar in (A) = 40 μm; bar in (B) = 10 μm. (From Thorens et al., 1990d.)

GLUT-2 or whether it is expressed in a different membrane of the epithelial cells.

In Kidney, GLUT-2 is present only in the basolateral membrane of the cells forming the first (S_1) part of the proximal convoluted Tubule (Thorens et al., 1990c,e) (Fig. 5). This segment of the kidney nephron is where the

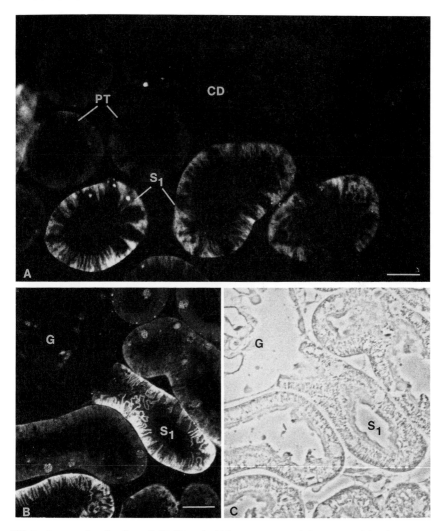

FIG. 5   Immunolocalization of GLUT-2 in kidney cortex. (A) Immunostaining is on the basolateral membrane of cells in the proximal convoluted tubule (S_1); no staining is detectable on the brush border of these cells. Some segments of the proximal tubule (PT) are not labeled, nor is the collecting duct (CD). (B) This section shows that the segment of the proximal tubule connected to a glomerulus (G) is very intensely stained, indicating that GLUT-2 is expressed in the S_1 part of the proximal tubule. (C) Phase-contrast picture of the field shown in (B). Bar = 25 μm. (From Thorens *et al.*, 1990e.)

bulk of glucose reabsorption takes place and glucose absorption is probably mediated by the same Na^+–glucose symporter present in intestine (Coady et al., 1990). In the proximal straight tubule (S_3), glucose reabsorption takes place from a low luminal glucose concentration. In this segment, the uptake of glucose through the brush border is performed by a high-energy symporter, which transports two Na^+ to move one molecule of glucose (Turner and Moran, 1982; Burg, 1986), and the cells forming this segment express only GLUT-1 and not GLUT-2 in their basolateral membranes (Thorens et al., 1990e). Therefore, in kidney, the reabsorption of glucose in different segments of the nephron requires the presence of two different pairs of apical, Na^+-dependent, and basolateral, Na^+-independent, glucose transporters. An Na^+-dependent glucose symporter, probably identical to the intestine one, and the low-affinity GLUT-2, are responsible for the bulk of glucose reabsorption in the S_1 part of the proximal convoluted tubule while reabsorption of the remaining low concentrations of glucose present in the S_3 segment of the proximal tubule is effected by a high-energy, apical, Na^+-dependent glucose symporter coupled with the high-affinity, basolateral GLUT-1. The kidney transepithelial reabsorption process is therefore apparently more complex than that of the intestine.

2. Regulated Expression of GLUT-2 in Intestine and Kidney

The rate of glucose transport across the intestinal mucosa is increased by hyperglycemia, consequent either to streptozocin-induced diabetes or to perfusion with a glucose solution. In guinea pig, the increased transepithelial glucose absorption results from an increase in the rate of transport across the brush border and across the basolateral membrane (Fischer and Lauterbach, 1984). In rat, hyperglycemia induced by glucose perfusion leads within 2 hr to a 3.5-fold increase in the V_{max} for glucose uptake by vesicles prepared from the basolateral membrane of jejunal epithelial cells (Maenz and Cheeseman, 1987). Interestingly, this increased rate of glucose uptake is not paralleled by an increase in the number of glucose-inhibitable cytochalasin B binding sites—a measure of the number of glucose transporters. Furthermore, cycloheximide injection prior to glucose perfusion decreased glucose uptake rates by 80% with no change in the number of cytochalasin B binding sites (Cheeseman and Maenz, 1989). These observations suggest that the rate of glucose transport through the basolateral membrane may result from changes in the intrinsic activity of the transporter. Whether the measured glucose transport is the result of GLUT-2 activity or a combination of GLUT-2 and GLUT-5 is still unresolved. The availability of antibody and cDNA probes for these glucose transporter isoforms will permit one to address these questions. In kidney,

GLUT-2 mRNA and proteins levels are increased in streptozocin diabetic rats, an effect completely prevented by insulin treatment (Garvey *et al.*, 1990). Therefore, the present data suggest that GLUT-2 expression may be regulated in the absorptive epithelial cells as a result of imbalanced glucose homeostasis. This is in contrast to its mostly unregulated expression in hepatocytes, but similar to its expression in β cells, as we will now discuss.

C. GLUT-2 in Pancreatic Islet β Cells

GLUT-2 expressed in rat pancreatic islets has an electrophoretic mobility slightly slower than the liver form of this transporter (Thorens *et al.*, 1988). However, the molecular cloning and nucleotide sequencing of a cDNA for the rat islet form of GLUT-2 showed that its sequence was identical to the liver cDNA (Johnson *et al.*, 1990a). The identity between the liver and pancreatic islet form of human GLUT-2 has also been established (Permutt *et al.*, 1989). Therefore, the difference in apparent size may be due to islet-specific posttranslational modifications of the transporter such as glycosylation or phosphorylation.

1. Immunolocalization of GLUT-2 in Pancreatic Islets

By immunofluorescence microscopy GLUT-2 was shown to be expressed by pancreatic islet β cells (Thorens *et al.*, 1988), and to be absent from glucagon-, somatostatin-, or pancreatic polypeptide-containing cells (Orci *et al.*, 1989). Electron microscopic localization of GLUT-2 by the protein A–gold technique revealed a very interesting distribution of the transporter at the surface of the β cells (Fig. 6) (Orci *et al.*, 1989). The transporter is preferentially associated with microvilli present on "lateral" segments of the plasma membrane that face adjacent β cells, and is present at a relatively low level on flat segments of the plasma membrane. The lateral segments of the plasma membrane, which contains the microvilli, form between β cells a structure similar to a canaliculus (Weir and Bonner-Weir, 1990). The restricted localization of GLUT-2 to this domain reveals three interesting aspects of the structure and function of β cells. First, although β cells do not possess tight junctions as in intestine or kidney epithelial cells, they still maintain expression of a plasma membrane protein (GLUT-2) on a defined segment of the cell surface, thereby assuming a polarized structure. Second, association of GLUT-2 with microvilli in β cells is surprising compared to its expression on intestinal or kidney epithelial cells, where GLUT-2 is localized to the basolateral membrane. We do not know what determines the polarized expression of GLUT-2 in these different cell types. The localization of the Na^+/K^+-

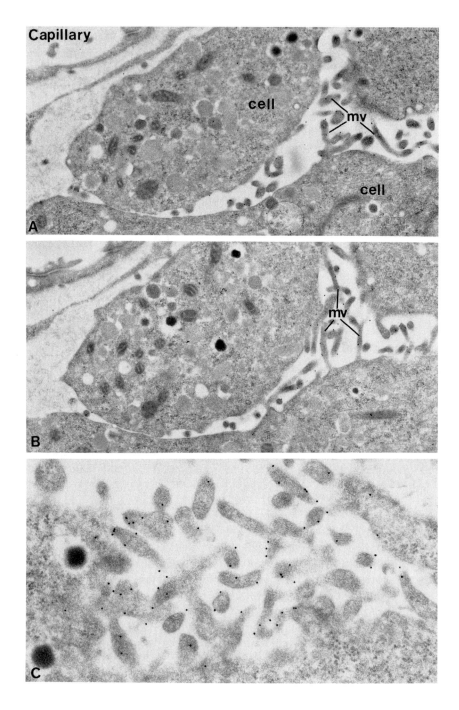

ATPase to the basolateral membrane of kidney epithelial cells is probably maintained through an interaction of the ATPase with ankyrin (Nelson and Veshnock, 1987), which serves as a link between the membrane proteins and the underlying cytoskeleton. A similar interaction with a subtype of ankyrin is thought to restrict the expression of the Na^+ channel to the node of Ranvier in myelinated neurons (Srinivasan et al., 1988; Baines, 1990). Therefore, it could be hypothesized that an interaction of GLUT-2 with ankyrin may account for its polarized expression in absorptive epithelial cells. This, however, needs to be demonstrated. The localization of GLUT-2 to microvilli in β cells is surprising. Indeed, in intestine, microvilli are at the apical pole of the cell and have a core of bundles of microfilaments. Specialized proteins, such as myosin I, mediate the interaction between the plasma membrane and the actin filaments (Mooseker and Coleman, 1989; Mooseker, 1989). The microvilli present on the lateral membrane of β cells also contain microfilaments (Orci et al., 1972). What maintains GLUT-2 associated with the microvilli? Ankyrin is not thought to be part of these structure; but a myosin I-like protein could be responsible for this interaction. Or there could be a β cell-specific protein interacting with GLUT-2 and the microfilaments. These questions are still unresolved but are of considerable interest because these observations suggest the existence of a cell-specific targeting of GLUT-2 to a particular domain of the plasma membrane. Also, because GLUT-2 expression in β cells is a key element for the normal process of glucose-induced insulin secretion, the polarized expression of GLUT-2 on β cells may be important for the normal physiology of these cells.

Third, the localization of GLUT-2 to microvilli present on lateral membranes implies that the glucose taken up by the β cells has first filtrated through the fenestrae of the epithelial cells and percolated through the canaliculi present between β cells. Because glucose sensing by β cells first requires glucose to enter the cell, the lateral membrane domain containing GLUT-2 can be considered as the glucose-sensing domain of the β cells. Why the glucose transporter is not localized on the membranes which are closest to the blood capillaries, and what the possible physiological advantage of this particular arrangement might be, is not known. Vasoconstrictors (such as neuropeptide Y or endothelin) may modify the rate of fluid

---

FIG. 6   Ultrastructural localization of GLUT-2 on β cells by the protein A–gold method. (A and B) Serial thin sections showing numerous gold particles labeling the microvilli (mv) projecting into the intercellular space. The labeling on the microvilli is about sixfold higher than on the flat membranes. (C) Higher magnification showing the association of GLUT-2-immunoreactive sites with microvilli. [(A and B) ×16,262; (C) ×31,020). (From Orci et al., 1989, copyright © 1989 by the AAAS.)

outflow through the fenestrae and through the canaliculi, therefore, influencing glucose-induced insulin secretion.

2. Requirement for GLUT-2 for Normal Glucose Sensing by β Cells

The major physiological function of islet β cells is secretion of insulin, a phenomenon mostly controlled by variations in blood glucose concentrations. Insulin secretion proceeds at a low basal rate at the normal glucose concentration of about 5 mM. The secretion rate increases with the rise in extracellular glucose concentration according to a sigmoidal dose–response curve and reaches a maximum at about 20 mM glucose (Meglasson and Matschinski, 1986). For glucose signaling to occur, both glucose uptake and metabolism are required. The catabolism of glucose through the glycolytic pathway generates a second messenger, which may be ATP (Cook *et al.*, 1988), and which induces the closure of an ATP-dependent K^+ channel (Henquin, 1978; Ashcroft *et al.*, 1984; Cook and Hales, 1984). The decreased K^+ efflux depolarizes the plasma membrane, which leads to opening of a voltage-gated Ca^+ channel. The influx of calcium into the β cells then initiates a cascade of events that culminates in the exocytosis of insulin-containing secretory granules (MacDonald, 1990). The proximal part of this signaling system (up to the closure of the ATP-dependent K^+ channel) requires that the rate of production of the glycolytic intermediate(s) (ATP?) reflects the variations in extracellular glucose concentration. The rate-limiting step of glycolysis in pancreatic β cells is the phosphorylation of glucose by glucokinase, an isoenzyme of hexokinase with a relatively high K_m for glucose (6 mM) (Vischer *et al.*, 1987). Because this K_m value is in the range of the normal blood glucose concentration, glucokinase controls the rate of glycolysis when extracellular glucose concentrations vary over the stimulatory range of 5 to 20 mM. However, for glucokinase to be the rate-limiting step, glucose uptake must not be limiting. Several reports have shown that the equilibration of glucose across the cell membrane is extremely fast and that the intracellular concentration of glucose equals that of the extracellular medium (Meglasson and Matschinski, 1986). Therefore, glucose uptake is not limiting for metabolism in normal β cells.

The demonstration of a high level of expression of GLUT-2 in the β cell plasma membrane together with its high K_m (17 mM) for glucose explains the observed high rate of glucose equilibration even at high extracellular glucose concentrations. Therefore, the concomitant expression of a high-K_m glucose transporter and a high-K_m glucokinase is required for the normal functioning of the β cell glucose sensor. GLUT-2 plays mostly a permissive role in glucose metabolism in normal islets and is the most proximal component of the glucose sensor.

The availability of molecular probes for GLUT-2 has permitted the investigation of some aspects of its regulated expression and its role in β cell dysfunctions characteristic of diabetic states. These studies, which are discussed below, lead to the observation that GLUT-2 expression in the plasma membranes of β cells or of insulinoma cells was greatly reduced in situations in which the cells no longer respond to an acute challenge by an otherwise maximally stimulatory glucose concentrations (16.7 mM). In these circumstances, a reduced expression of the high-K_m GLUT-2 decreases the rate of glucose uptake into the β cells when the glucose concentration rises over the stimulatory range (5 to 20 mM). It is therefore proposed that in glucose-unresponsive β cells, the uptake step may become rate limiting for glucose metabolism, consequently preventing normal glucose sensing (see also Unger, 1991).

3. Reduced GLUT-2 Expression and the Loss of Glucose Sensing

a. Insulinomas   Most of the insulinoma cell lines have either lost their responsiveness to glucose or respond to variations of glucose concentrations between 0 and 5 mM, a concentration range much lower than that stimulating insulin secretion by normal β cells (5 to 20 mM). For instance, insulinoma cell lines (βTC-1, βTC-3) have been obtained from transgenic β cells expressing the simian virus 40 (SV-40) large T antigen under the control of the rat insulin promoter (Hanahan, 1985; Efrat *et al.*, 1988). These cell lines are still able to secrete insulin after glucose stimulation, but the maximum rate of insulin secretion is achieved when the extracellular glucose concentration is increased from 0 to 1.5 mM (Efrat *et al.*, 1988). Interestingly, these cells express a greatly reduced amount of GLUT-2 but a very high level of GLUT-1 (Tal *et al.*, 1992), a glucose transporter isoform with a low K_m for glucose (1–3 mM). The same observation— maximum rate of insulin secretion at 2.8 mM glucose and high expression of GLUT-1 with very low expression of GLUT-2—has been made for the insulinoma cell line RIN5F (Praz *et al.*, 1983; Thorens *et al.*, 1988). Whether the reduced expression of GLUT-2 and/or the high expression of GLUT-1 are the direct causes for the abnormal glucose-induced insulin secretion in these insulinoma cell lines is not known. However, a publication describes the establishment of two insulinoma cell lines, MIN6 and MIN7, derived from transgenic mice expressing the SV-40 large T antigen in their β cells by a protocol very similar to the one used for the generation of the βTC-1 and βTC-3 cell lines (Miyazaki *et al.*, 1990). The

MIN6 cells have a sensitivity to glucose similar to that of normal β cells, while the MIN7 cells show the abnormal sensitivity characteristic of the other insulinoma lines. Importantly, the MIN6 cells express a high level of GLUT-2 and little GLUT-1, while the glucose-unresponsive MIN7 cells express low levels of GLUT-2 and high levels of GLUT-1. These data are compatible with the notion that a normal expression of GLUT-2 is required for normal glucose responsiveness of β cells. The persistent expression of GLUT-2 in glucose-responsive MIN6 cells is probably necessary for the normal glucose sensing but may also indicate that these cells have remained more differentiated than the MIN7 cells and are more closely related to nontransformed β cells, therefore exhibiting an apparently normal glucose sensitivity.

A study by Tal *et al.*, (1992) has addressed more directly the relationship between expression of GLUT-1 and GLUT-2 and glucose responsiveness in a transgenic mouse model in which expression of the SV-40 large T antigen in β cells triggers the development of insulinomas in a predictable manner (Hanahan, 1985). In these mice, the SV-40 large T antigen is expressed in every β cell under the control of the rat insulin promoter. Islets have a normal appearance until about the fifth week of life, at which point some islets become hyperplastic. At 12 weeks of age, some of the hyperplastic islets have generated β cell tumors that cause the death of the animals, probably by provoking hyperinsulinemia. GLUT-2 expression in β cells is normal until the development of the first hyperplastic islets, then GLUT-2 expression starts to decrease, reaching very low levels in β cells from tumor-bearing mice; no GLUT-1 is detectable in β cells *in vivo* at any stage of tumor development. Cultured β cell lines (βTC cells) derived from the tumors have almost completely lost GLUT-2 expression and express high levels of GLUT-1. GLUT-1 expression by these insulinoma cells is due to the *in vitro* culture of these cells and not to the oncogenic transformation per se, and is completely reversed in secondary tumors formed from cultured cells injected into syngeneic mice. Interestingly, the first defect in glucose-induced insulin secretion—no secretion at 0 mM glucose and maximal secretion at 5 mM—is observed in islets prior to any notable reduction in GLUT-2 expression and without expression of GLUT-1. A second phase of deregulated insulin secretion—high insulin secretion in the absence (0 mM) glucose and very little stimulation by increases in extracellular glucose concentration—is reached when β cells are in the hyperplastic stage, at which point they expressed a reduced amount of GLUT-2. Therefore, in this model abnormal insulin secretion in transgenic β cells may occur without any change in glucose transporter isoform expression, but a complete loss of glucose sensitivity still correlates with a decreased expression of GLUT-2 and does not need the expression of GLUT-1. The observation that the MIN6 cells, which are derived from the same type of transgenic mice as are βTC cells, retain a normal glucose responsiveness and a high level of GLUT-2 expression suggests that different sites of integration of the transgene may affect to various extents the transformation and state of differentiation of β cells, a phenomenon ultimately reflected in different glucose-sensing capabilities.

b. Reduced Expression of GLUT-2 at the Onset of Diabetes in the BB/W Rat Before the onset of type I diabetes, a decreased responsiveness of the β cells to glucose but not to nonglucose secretagogues has been observed in humans (Srikanta *et al.*, 1983) and in the BB/W rat (Tominaga *et al.*, 1986), a rodent model of type I diabetes (Rossini *et al.*, 1985). The autoantigens that serve as targets for the autoimmune destruction of the β cells are as yet poorly characterized (Castano and Eisenbarth, 1990). A 64-kDa β cell autoantigen (Baekkeskov *et al.*, 1982) has been identified as the enzyme glutamic acid decarboxylase, which catalyzes the formation of γ-aminobutyric acid (GABA) from glutamic acid (Baekkeskov *et al.*, 1990). However, it has been observed that the sera from diabetic patients reduce glucose-induced insulin secretion by perifused rat islet (Kanatsuma *et al.*, 1983). Also, preincubation of islet cells with immunoglobulins purified from the sera of diabetic patients decreases the rate of glucose uptake by the high-K_m transporter Johnson *et al.*, 1990c). In the BB/W rat, Unger and collaborators (Tominaga *et al.*, 1986), have shown that on the day of onset of diabetes, the pancreatic β cells fail to respond to increases in glucose concentrations while they maintain normal sensitivity to arginine. Simultaneous to the loss of glucose responsiveness, these authors measured a 90% reduction in the rate of glucose uptake by pancreatic islets. The above data suggest that (1) autoantibodies can decrease the glucose-induced insulin secretion by normal β cells; (2) the activity of the high-K_m β cell glucose transporter is reduced by interaction of the autoantibodies either with the transporter itself or with an antigen associated to it; (3) reduced expression or a decreased activity of the β cell glucose transporter is associated with, and may be the cause of, the loss of glucose responsiveness of the β cell at the onset of diabetes.

To test more directly the expression of the high-K_m β cell glucose transporter in islets of prediabetic and diabetic BB/W rats, specific antibodies against GLUT-2 were used to determine its level of expression by immunofluorescence and immunoelectron microscopic analysis (Orci *et al.*, 1990a). This study indicated that GLUT-2 expression was reduced by 27% in the nondiabetic diabetes-prone rats and by about 50% in the diabetic rats on the first day of diabetes, as compared to nondiabetic diabetes-resistant rats. Therefore, the reduced expression of GLUT-2, together with the presence in the serum of diabetic rats of autoantibodies capable of reducing the rate of glucose uptake by β cells, may well account for the observed reduced rate of glucose uptake. The reduced rate of glucose uptake may in turn prevent a free access of glucose to glucokinase, thereby causing the impaired glucose sensing characteristic of these islet β cells (Tominaga et al., 1986).

c. Reduction of GLUT-2 Expression in Glucose-Unresponsive β Cells from Type II Diabetic Rats   Non-insulin-dependent type II diabetes is characterized by a loss of sensitivity of peripheral tissues to the action of insulin and by a dysfunction of β cells, in particular, a complete loss of glucose-induced insulin secretion while retaining the capability to respond to other secretagogues such as arginine. In humans, the disease has a strong genetic component (Harris et al., 1987) but the molecular defects underlying the development of the disease are for the most part unknown. A few exceptions are the mutations of the insulin receptors, which cripple its function, and which have been associated with the severe insulin resistance characteristic of acanthosis nigricans and leprechaunism (Yoshimasa et al., 1988; Kadowaki et al., 1988; Taira et al., 1989; Odawara et al., 1989).

The expression of GLUT-2 has been studied in a pharmacological and a genetic model of type II diabetes. In the pharmacological model, newborn rats are injected at 2 days of age with a single dose of streptozocin. This induces a transient hyperglycemia over the first week of life but thereafter the rats are normoglycemic for several weeks. At about 6 weeks of age, they start to develop a hyperglycemia (Bonner-Weir et al., 1981; Weir et al., 1981; Portha et al., 1974; Giroix et al., 1983). When tested by the pancreas perfusion technique, β cells from these rats no longer respond to glucose but are still able to secrete insulin after a challenge with arginine (Weir et al., 1981; Giroix et al., 1983). The expression of GLUT-2 in these islets was assessed by immunofluorescence microscopy and by Western blot analysis (Thorens et al., 1990d). By both techniques, GLUT-2 expression was shown to be decreased and the extent of reduction was proportional to the hyperglycemia of the rats. Therefore in this model of type II diabetes, there is a direct correlation between the reduced expression of GLUT-2 and the loss of glucose sensitivity of the β cells.

The male Zucker diabetic fatty rat (ZDF/DRT-*fa*) (Clark et al., 1983) represents a genetic model of type II diabetes. In this colony only male rats develop diabetes while females, which are obese and insulin resistant, do not progress to overt diabetes. Johnson and collaborators (1990b; Orci et al., 1990b) have shown that the initial rate of high-K_m glucose uptake by islets from the diabetic rats was reduced to 19% of the value of control rat islets and that islets from females had a near normal rate of glucose uptake. They further showed that the decreased glucose uptake activity correlated with a decreased expression of GLUT-2 on the β cell surface. Interestingly, the loss of glucose-responsiveness by β cells appeared whenever the number of GLUT-2 positive β cells was reduced by 40%, a situation corresponding to a blood glucose concentration of over 200 mg/dl. Prevention of hyperglycemia by acarbose treatment corrected most of the hyperglycemia but failed to prevent a decreased expression of GLUT-2. Therefore, in this genetic model of type II diabetes, the loss of GLUT-2 expression is not the result of hyperglycemia. Instead, a reduced expression of GLUT-2 may be at the origin of the β cell dysfunction.

The decreased expression of GLUT-2 in the β cells from diabetic rats is specific to these cells, as the level of expression of GLUT-2 in liver from the same animals is only minimally altered by the diabetes.

What controls GLUT-2 expression in islets is not yet known. However, it has been shown that perfusion of rats with insulin to maintain glycemia at about 50 mg/dl for up to 12 days induces a complete suppression of GLUT-2 mRNA expression (Chen et al., 1990). In contrast, maintaining a high blood glucose concentration (about 200 mg/dl) by perfusing rats with a 50% glucose solution for several days increased GLUT-2 mRNA by about 50%. However, a simple effect of glucose or insulin is not likely to explain the regulated expression of GLUT-2. Indeed, in Zucker diabetic rats (Johnson et al., 1990b; Orci et al., 1990b) and in db/db mice (Shafrir, 1990; B. Thorens, et al., 1992), hyperinsulinemia and hyperglycemia are present at the same time, yet GLUT-2 expression is decreased. In the neonatal streptozocin rats, a reduced GLUT-2 expression exists in the presence of a slight hypoinsulinemia and hyperglycemia. Finally, obese female Zucker rats have a high basal insulin level, are normoglycemic, and express normal levels of GLUT-2 (Johnson et al., 1990b; Orci et al., 1990b).

The decreased expression of GLUT-2 in β cells correlates with their inability to secrete insulin in response to an acute glucose challenge. A decreased expression of GLUT-2 may reduce the rate of glucose uptake to such an extent that the access of glucose to glucokinase may become the rate-limiting step in glucose metabolism, thereby making the glycolytic flux insensitive to changes in extracellular glucose. The decreased expression of GLUT-2 in rodent models of type II diabetes suggests that the primary defect in β cells from these diabetic animals may be at the level of expression of GLUT-2. It cannot be excluded, however, that other β cell defects, besides a decrease in GLUT-2 expression, may also be responsible for the observed loss of glucose sensing. In any circumstances, a consistent decrease in GLUT-2 expression in glucose-unresponsive β cells has been observed. Since expression of this transporter is normal in liver of the same diabetic animals, there must be a β cell-specific control of GLUT-2 expression. At what level the control of GLUT-2 expression occurs is still unknown. The understanding of the regulated expression of GLUT-2 in β cells is a major goal of future research, as the possibility to revert its downregulation may be important in restoration of the normal glucose sensitivity of diabetic islets.

IV. Conclusion

The molecular cloning of a cDNA for GLUT-1 has permitted the identification of four additional glucose transporter isoforms. The five facilitated diffusion glucose transporters so far characterized appear to be each involved in different aspects of the control of glucose homeostasis. GLUT-2 is unique in being present in hepatocytes, absorptive epithelial cells from intestine and kidney, and in pancreatic islet β cells. Functionally, it is characterized by a relatively high K_m for glucose (15–20 mM). This property confers to the cells in which it is expressed the ability to take up glucose at rates proportional to blood glucose concentrations even when these concentrations are well above the normal value of about 5 mM. This kinetic property of GLUT-2 is necessary for the rapid entry of glucose into hepatocytes, and for its consequent metabolism, and for the efficient transport of glucose through intestine and kidney absorptive epithelial cells.

In β cells, the uptake of glucose is the first step in a cascade of events that lead to the secretion of insulin. In normal β cells, the high-K_m GLUT-2 allows unrestricted access of glucose to glucokinase, even at glucose concentrations stimulating maximal rates of insulin secretion (about 20 mM). This property is required for proper glucose sensing and insulin secretion. Importantly, however, at the onset of type I diabetes or in type II diabetic rats a reduced level of GLUT-2 expression by pancreatic β cells correlates with the loss of glucose-induced insulin secretion. In these circumstances, the uptake step by the high-K_m GLUT-2 may restrict access of glucose to glucokinase, therefore preventing normal glucose sensing. The reduced expression of GLUT-2 may therefore be an initial step in the development of the β cell dysfunction characteristic of islets from diabetic rats. The understanding of the factors controlling the β cell-specific expression of GLUT-2 may provide means to revert its low expression in diabetes and possibly ameliorate the glucose-unresponsiveness of diabetic β cells.

Acknowledgments

I wish to thank Dr. Harvey F. Lodish, in whose laboratory all the work I have done on GLUT-2 was carried out, for his continuous support and for his critical reading of this article. I also thank Dr. Giulia Baldini for many helpful comments on this manuscript. This work has been supported by National Institute of Health Grants GM 40916 and HL 41484 to H.F.L. B.T. was the recipient of postdoctoral fellowships from the European Molecular Biology Organization and the Swiss National Science Foundation.

References

Amerongen, H. M., Mack, J. A., Wilson, J. M., and Neutra, M. R. (1989). *J. Cell Biol.* **109,** 2129–2138.

Asano, T., Shibasaki, Y., Lin, J.-L., Akanuma, Y., Takaku, F., and Oka, Y. (1989). *Nucleic Acids Res.* **17,** 6386.

Ashcroft, F. M., Harrison, D. E., and Ashcroft, S. J. H. (1984). *Nature (London)* **312,** 446–448.

Axelrod, J. D., and Pilch, P. F. (1983). *Biochemistry* **22,** 2222–2227.

Baekkeskov, S., Nielsen, J. H., Marner, B., Bilde, T., Ludvigsson, J., and Lernmark, A. (1982). *Nature (London)* **298,** 167–169.

Baekkeskov, S., Aanstoot, H.-J., Christgau, S., Reetz, A., Solimena, M., Cascalho, M., Folli, F., Richter-Olesen, H., and Camilli, P.-D. (1990). *Nature (London)* **347,** 151–156.

Baines, A. J. (1990). *Trends Neurosci.* **13,** 119–121.

Bell, G. I., Kayano, T., Buse, J. B., Burant, C. F., Takeda, J., Lin, D., Fukumoto, H., and Seino, S. (1990). *Diabetes Care* **13,** 198–208.

Birnbaum, M. J., Haspel, H. C., and Rosen, O. M. (1986). *Proc. Natl. Acad. Sci. U.S.A.* **83,** 5784–5788.

Bonner-Weir, S., Trent, D. F., Honey, R. N., and Weir, G. C. (1981). *Diabetes* **30,** 64–69.

Bouma, C. L., Meadow, N. D., Stover, E. W., and Roseman, S. (1987). *Proc. Natl. Acad. Sci. U.S.A.* **84,** 930–934.

Burg, M. B. (1986). *In* "The Kidney" (B. M. Brenner and F. C. Rector, eds.), 3rd ed., pp. 147–175. Saunders, Philadelphia, Pennsylvania.

Castano, L., and Eisenbarth, G. S. (1990). *Annu. Rev. Immunol.* **8,** 647–679.

Cheeseman, C. I., and Maenz, D. D. (1989). *Am. J. Physiol.* **256,** G878–G883.

Chen, L., Alam, T., Johnson, J. H., Newgard, C. B., and Unger, R. H. (1990). *Proc. Natl. Acad. Sci. U.S.A.* **87,** 4088–4092.

Ciaraldi, T. P., Horuk, R., and Matthei, S. (1986). *Biochem. J.* **240,** 115–123.

Clark, J. B., Palmer, C. J., and Shaw, W. N. (1983). *Proc. Soc. Exp. Biol. Med.* **173,** 68.

Coady, M. J., Pajor, A. M., and Wright, E. M. (1990). *Am. J. Physiol.* **259,** C605–C610.

Cook, D. L., and Hales, N. (1984). *Nature (London)* **311,** 271–273.

Cook, D. L., Satin, L. S., Ashford, M. L. J., and Hales, C. N. (1988). *Diabetes* **37,** 495–498.

Craik, J. D., and Elliott, K. R. F. (1978). *Biochem. J.* **182,** 503–508.

Cushman, S. W., and Wardzala, L. J. (1980). *J. Biol. Chem.* **255,** 4758–4762.

Deguchi, Y., Yamato, I., and Anraku, Y. (1990). *J. Biol. Chem.* **265,** 21704–21708.

Edelman, A. M., Blumenthal, D. K., and Krebs, E. G. (1987). *Annu. Rev. Biochem.* **56,** 567–613.

Efrat, S., Linde, S., Hofod, H., Spector, D., Delannoy, M. Grant, S., Hanahan, D., and Baekkeskov, S. (1988). *Proc. Natl. Acad. Sci. U.S.A.* **85,** 9037–9041.

Fischer, E., and Lauterbach, F. (1984). *J. Physiol. (London)* **335**, 567–586.

Flier, J. S., Mueckler, M., McCall, A. L., and Lodish, H. F. (1987). *J. Clin. Invest.* **97**, 657–661.

Fukumoto, H., Seino, S., Imura, H., Seino, Y., Eddy, R. L., Fukushima, Y., Byers, M. G., Shows, T. B., and Bell, G. I. (1988). *Proc. Natl. Acad. Sci. U.S.A.* **85**, 5434–5438.

Garvey, W. T., Maianu, L., Camp, K., and Dominguez, J. H. (1990). *J. Am. Soc. Nephrol.* **1**, 718.

Giroix, M.-H., Portha, B., Kergoat, M., Bailbe, D., and Picon, L. (1983). *Diabetes* **32**, 445–451.

Gould, G. W., and Bell, G. I. (1990). *Trends Biochem. Sci.* **15**, 18–23.

Gould, G. W., and Lienhard, G. E. (1989). *Biochemistry* **28**, 9447–9452.

Granner, D., and Pilkis, S. (1990). *J. Biol. Chem.* **265**, 10173–10176.

Hanahan, D. (1985). *Nature (London)* **315**, 115–122.

Harris, M. I., Hadden, W. C., Knowler, W. C., and Bennett, P. H. (1987). *Diabetes* **36**, 523–534.

Hediger, M. A., Coady, M. J., Ikeda, T. S., and Wright, E. M. (1987). *Nature (London)* **330**, 379–381.

Hediger, M. A., Turk, E., and Wright, E. M. (1989). *Proc. Natl. Acad. Sci. U.S.A.* **86**, 5748–5752.

Hellman, B., Sehlin, J., and Taljedal, I. (1971). *Biochim. Biophys. Acta* **241**, 147–154.

Henquin, J. C. (1978). *Nature (London)* **271**, 271–273.

Hubbard, A. L., Bartles, R., and Breiterman, L. T. (1985). *J. Biol. Chem.* **100**, 1115–1125.

Iynedjian, P. B., Ucla, C., and Mach, B. (1987). *J. Biol. Chem.* **262**, 6032–6038.

Johnson, J. H., Newgard, C. B., Milburn, J. L., Lodish, H. F., and Thorens, B. (1990a). *J. Biol. Chem.* **265**, 6548–6551.

Johnson, J. H., Crider, B. P., McCorkle, K., Orci, L., and Unger, R. H. (1990b). *N. Engl. J. Med.* **322**, 653–659.

Johnson, J. H., Ogawa, A., Chen, L., Orci, L., Newgard, C. B., Alam, T., and Unger, R. H. (1990c). *Science* **250**, 546–549.

Kadowaki, T., Bevins, C. L., Cama, A., Ojamaa, K., Marcus-Samuel, B., Kadowaki, H., Beitz, L., McKeon, C., and Taylor, S. I. (1988). *Science* **240**, 787–790.

Kahn, B. B., Cushman, S. W., and Flier, J. S. (1989). *J. Clin. Invest.* **82**, 199–204.

Kanatsuma, T., Baekkeskov, S., Lernmark, A., and Ludvigsson, J. (1983). *Diabetes* **32**, 520–524.

Kasanicki, M. A., and Pilch, P. F. (1990). *Diabetes Care* **13**, 219–227.

Kayano, T., Burant, C. F., Fukumoto, H., Gould, G. W., Fan, Y.-S., Eddy, R. L., Byers, M. G., Shows, T. B., Seino, S., and Bell, G. I. (1990). *J. Biol. Chem.* **265**, 13276–13282.

Keller, K., Strube, M., and Mueckler, M. (1990). *J. Biol. Chem.* **264**, 18884–18889.

Lawrence, J. C., Hiken, J. K., and James, D. E. (1990). *J. Biol. Chem.* **265**, 2324–2332.

Lienhard, G. E., Kim, H. H., Ransome, K. J., and Gorga, J. C. (1982). *Biochem. Biophys. Res. Commun.* **105**, 1150–1156.

MacDonald, M. J. (1990). *Diabetes* **39**, 1461–1466.

Maenz, D. D., and Cheeseman, C. I. (1987). *J. Membr. Biol.* **97**, 259–266.

Meglasson, M. D., and Matschinski, F. M. (1986). *Diabetes/Metab. Rev.* **2**, 163–214.

Miyazaki, K., Araki, K., Yamato, E., Ikegami, H., Asano, T., Shibasaki, Y., Oka, Y., and Yamamura, K. (1990). *Endocrinology (Baltimore)* **127**, 127–132.

Mooseker, M. S. (1989). *Nature (London)* **340**, 505.

Moosekar, M. S., and Coleman, T. R. (1989). *J. Cell Biol.* **108**, 2395–2400.

Mueckler, M. (1990). *Diabetes* **39**, 6–11.

Mueckler, M., Caruso, C., Baldwin, S. A., Panico, M., Blench, I., Morris, H. R., Allard, J. W., Lienhard, G. E., and Lodish, H. F. (1985). *Science* **229**, 941–945.

Nelson, W. J., and Veshnock, P. J. (1987). *Nature (London)* **328**, 533–536.

Odawara, M., Kadowaki, T., Yamamoto, R., Shibbasaki, Y., Tobe, K., Accili, D., Bevins, C., Mikami, Y., Matsuura, N., Akanuma, Y., Takaku, F., Taylor, S. I., and Kasuga, M. (1989). *Science* **245**, 66–68.

Oka, Y., Asano, T., Shibsaki, Y., Lin, J. L., Tsikida, K., Akanuma, Y., and Takaku, F. (1990). *Diabetes* **39**, 441–446.

Orci, L., Gabay, K. H., and Malaisse, W. J. (1972). *Science* **175**, 1128–1130.

Orci, L., Thorens, B., Ravazzola, M., and Lodish, H. F. (1989). *Science* **245**, 295–297.

Orci, L., Unger, R. H., Ravazzola, M., Ogawa, A., Komiya, I., Baetens, D., Lodish, H. F., and Thorens, B. (1990a). *J. Clin. Invest.* **86**, 1615–1622.

Orci, L., Ravazzola, M., Baetens, D., Inman, L., Amherdt, M., Peterson, R. G., Newgard, C. B., Johnson, J. H., and Unger, R. H. (1990b). *Proc. Natl. Acad. Sci. U.S.A.* **87**, 9953–9957.

Permutt, M. A., Koranyi, L., Keller, K., Lacy, P. E., Scharp, D. W., and Mueckler, M. (1989). *Proc. Natl. Acad. Sci. U.S.A.* **86**, 8688–8692.

Portha, B., Levacher, C., Picon, L., and Rosselin, G. (1974). *Diabetes* **23**, 889–895.

Praz, G. A., Halban, P. A., Wollheim, C. B., Strauss, A. J., and Renold, A. E., (1983). *Biochem. J.* **210**, 345–352.

Rhoads, D. B., Takano, M., Gattoni-Celli, S., Chen, C.-C., and Isselbacher, K. J. (1988). *Proc. Natl. Acad. Sci. U.S.A.* **85**, 9042–9046.

Rossini, A. A., Mordes, J. P., and Like, A. A. (1985). *Annu. Rev. Immunol.* **3**, 289–320.

Semenza, G., Kessler, M., Hosang, M., Weber, J., and Schmidt, U. (1984). *Biochim. Biophys. Acta* **779**, 343–379.

Shafrir, E. (1990). In "Diabetes Mellitus, Theory and Practice" (H. Rifkin and D. Porte, Jr., eds.), 4th ed., pp. 299–340. Elsevier, New York.

Srikanta, S., Ganda, O. P., Eisenbarth, G. S., and Soeldner, J. S. (1983). *N. Engl. J. Med.* **308**, 322–325.

Srinivasan, Y., Elmer, L., Davis, J., Bennett, V., and Angelides, K. (1988). *Nature (London)* **333**, 177–179.

Stein, W. D. (1986). "Transport and Diffusion across Cell Membranes." Academic Press, Orlando, Florida.

Suzue, K., Lodish, H. F., and Thorens, B. (1989). *Nucleic Acids Res.* **17**, 10099.

Suzuki, K., and Kono, T. (1980). *Proc. Natl. Acad. Sci. U.S.A.* **77**, 2542–2545.

Tabor, S., and Richardson, C. C., (1985). *Proc. Natl. Acad. Sci. U.S.A.* **82**, 1074–1078.

Taira, M., Taira, M., Hashimoto, N., Shimada, F., Suzuki, Y., Kanatsuka, A., Nakamura, F., Ebina, Y., Tatibana, M., Makino, H., and Yoshida, S. (1989). *Science* **245**, 63–66.

Tal, M., Schneider, D. L., Thorens, B., and Lodish, H. F. (1990). *J. Clin. Invest.* **86**, 986–992.

Tal, M., Thorens, B., Surana, M., Fleischer, N., Lodish, H. F., Hanahan, D., and Efrat, S. (1992). *Mol. Cell. Biol.*, **E12**, 422–432.

Thorens, B., Sarkar, H. S., Kaback, H. R., and Lodish, H. F. (1988). *Cell (Cambridge, Mass.)* **55**, 281–290.

Thorens, B., Charron, M. C., and Lodish, H. F. (1990a). *Diabetes Care* **13**, 209–218.

Thorens, B., Flier, J. S., Lodish, H. F., and Kahn, B. B. (1990b). *Diabetes* **39**, 712–719.

Thorens, B., Cheng, Z.-Q., Brown, D., and Lodish, H. F. (1990c). *Am. J. Physiol.* **259**, C279–C285.

Thorens, B., Lodish, H. F., and Brown, D. (1990d). *Am. J. Physiol.* **259**, C287–C294.

Thorens, B., Weir, G. C., Leahy, J. L., Lodish, H. F., and Bonner-Weir, S. (1990e). *Proc. Natl. Acad. Sci. U.S.A.* **87**, 6492–6496.

Thorens, B., Wu, Y.-J., Leahy, J. L., and Weir, G. C. (1992). *J. Clin. Invest.* In press.

Tominaga, M., Komiya, I., Johnson, J. H., Inmanm L., Alam, T., Moltz, J., Crider, B.,

Stefan, Y., Baetens, D., McKorkle, K., Orci, L., and Unger, R. H. (1986). *Proc. Natl. Acad. Sci. U.S.A.* **83,** 9749–9753.

Turner, R. J., and Moran, A. (1982). *Am. J. Physiol.* **242,** F406–F414.

Unger, R. H. (1991). *Science* **251,** 1200–1205.

Unger, R. H., and Orci, L. (1981a). *N. Engl. J. Med.* **304,** 1518–1504.

Unger, R. H., and Orci, L. (1981b). *N. Engl. J. Med.* **304,** 1575–1580.

Vera, J. C., and Rosen, O. M. (1989). *Mol. Cell. Biol.* **9,** 4287–4195.

Vera, J. C., and Rosen, O. M. (1990). *Mol. Cell. Biol.* **10,** 743–751.

Vischer, U., Blondel, B., Wollheim, C. B., Hoppner, W., Seitz, H. J., and Iynedjian, P. B. (1987). *Biochem. J.* **210,** 345–352.

Weir, G. C., and Bonner-Weir, S. (1990). *J. Clin. Invest.* **85,** 983–987.

Weir, G. C., Clore, E. T., Zmachinski, C. J., and Bonner-Weir, S. (1981). *Diabetes* **30,** 590–595.

Wheeler, T. J., and Hinkle, P. C. (1985). *Annu. Rev. Physiol.* **47,** 503–517.

Williams, T. F., Exton, J. H., Park, C. R., and Regen, D. M. (1968). *Am. J. Physiol.* **215,** 1200–1209.

Yoshimasa, Y., Seino, S., Whitaker, J., Kakehi, T., Kosaki, A., Kuzuya, H., Imura, H., Bell, G. I., and Steiner, D. F. (1988). *Science* **240,** 784–787.

The Insulin-Sensitive
Glucose Transporter

Morris J. Birnbaum
Department of Cellular and Molecular Physiology, Harvard Medical School,
Boston, Massachusetts 02115

I. Introduction

The limiting membranes of virtually all cells are essentially impermeable to small polar solutes, such as sugars. Since many organisms are at some time dependent on the metabolism of sugars to provide cellular energy, they have evolved integral membrane proteins that serve to transport solutes across the hydrophobic barrier. All transporters must be designed in such a way as to solve problems inherent in the need for specificity in the recognition of substrate as well as in overcoming the unfavorable energetics of carrying hydrophilic molecules across a lipophilic membrane. However, glucose transporters in higher eukaryotes have the additional task of regulating the redistribution after a meal of potential energy from simple sugars into more efficient long-term storage forms: triglyceride, glycogen, and protein. Moreover, the process must be accomplished with exquisite tissue specificity in such a way that hexose is redistributed primarily to muscle and adipose tissue during the absorptive period, but the glucose supply of the central nervous system is protected during fasting. Thus, glucose transport presents an unusual regulatory problem in which a virtually ubiquitous function is modulated differentially in a limited number of cell types. It has been recognized for some time that the polypeptide hormone insulin plays a pivotal role in redirecting glucose into muscle and fat. The challenge that has occupied investigators in this field for the past 50 years has been to define in detail the succession of biochemical events initiated by binding of insulin to its cell surface receptor and leading to an increase in glucose flux. This article concentrates on the terminal steps of this process, the regulation of the glucose transport proteins. Several excellent recent reviews have discussed this and related topics (Bell et al., 1990; Carruthers, 1990; Klip and Paquet, 1990; Mueckler, 1990).

II. The Translocation Hypothesis

A. Kinetic Experiments

Experiments performed during the first half of this century led to the concept that insulin promotes the disposal of glucose via two major routes: storage and oxidation. These studies, however, did not address the cellular compartment or biochemical step at which insulin exerts its major effect. In fact, the prevailing belief was that an augmentation of intracellular metabolism passively "pulled" glucose out of the circulation into the cell. It was the work of Levine and colleagues that provided the first coherent formulation of a model in which the predominant action of insulin is to promote the uptake of sugars at the cell membrane (Levine and Goldstein, 1955). Levine's ideas were based on a number of earlier experiments in which, for example, insulin was able to stimulate muscle glycogen deposition in the absence of changes in hexose oxidation. In addition, it had been shown that the intracellular concentration of glucose is essentially zero except in the presence of extraordinarily high circulating sugar. The first direct demonstration of an effect of insulin on transport was a series of experiments in eviscerated dogs in which insulin promoted the rate of uptake and volume of distribution of non- or poorly metabolizable hexoses (Levine and Goldstein, 1955). In spite of a rudimentary appreciation of the biophysical nature of the limiting cell membrane, Levine correctly concluded that insulin must be increasing permeability to a defined class of sugars, and noted the similarity in structural specificity between the insulin-dependent transfer of sugars into peripheral tissues and the constitutive uptake into human erythrocytes, a process for which a discrete carrier had already been postulated. Independent studies from Cori's laboratory on the uptake of sugars into frog sartorius muscle also led to the conclusion that flux was saturable, carrier mediated, and the primary site of action of insulin (Narahara et al., 1960). Initially, glucose was used as a substrate; at 19°C there was little hexose oxidation in the absence of insulin, and thus some kinetic analyses could be attempted. A significant advance was the introduction of the truly nonmetabolizable sugar, 3-O-methyl-glucose (3-O-MG), whose uptake nonetheless was regulated by insulin in eviscerated animals in a manner similar to glucose (Csary and Wilson, 1956; Landau et al., 1958). Uptake and efflux of 3-O-MG into frog sartorii showed saturation kinetics; importantly, insulin increased the flux of hexose by altering the V_{max} without appreciable effect on the K_m (Narahara and Özand, 1963). The original report describing these experiments contained the insightful observation that "insulin may increase the number of sites that participate in the transport of sugar" (Narahara and Özand, 1963).

The first experiments aimed at unraveling the biochemical pathways that lead from insulin to increased glucose transport were designed to assess more accurately the kinetic parameters, in the hope of extrapolating these data to a mechanism of action. Thus, technical considerations were of major import, both in defining an experimentally tractable system and a reliable protocol for flux measurements. Adipose tissue, recognized as an important insulin-responsive organ for the conversion of carbohydrate to lipids in the absorptive phase, provided a sensitive *in vitro* system. In rat epididymal fat pads, the uptake of glucose was rate limiting for utilization and served as the primary site for control by insulin (Crofford and Renold, 1965a,b). A major limitation of adipose tissue as an experimental system was the relatively high ratio of extracellular to intracellular water space, making diffusion rate limiting at low concentrations of glucose. This problem was largely overcome by the development of methods to prepare isolated rat adipocytes (Rodbell, 1964). The other barrier to the accurate measurement of hexose transport was the subsequent metabolism of glucose. 3-*O*-Methyl-glucose, which is not phosphorylated within the cell, provided a solution, but the rapid time of equilibration required the development of techniques to halt virtually instantaneously the uptake process. A number of studies were performed directed at assessing kinetic parameters of hexose transport in isolated adipocytes in the presence and absence of insulin, with the prevailing conclusion that insulin primarily augmented the maximal velocity with little change in the apparent affinity (Table I). In a careful series of experiments, the flux of 3-*O*-MG across the adipocyte plasma membrane was measured by a protocol that involved arresting uptake with the transport inhibitor phloretin, then rapidly separating the cells by oil flotation (Vinten *et al.*, 1976; Whitesell and Gliemann, 1979). In measurements of efflux under equilibrium exchange conditions, insulin was found to increase the V_{max} without substantially altering the apparent K_m. Taylor and Holman (1981) performed a more detailed study of 3-*O*-MG transport in isolated rat adipocytes, utilizing rate equations for the calculation of kinetic parameters. They also concluded that the effect of insulin was on maximal velocity and not apparent affinity. In addition, they noted symmetrical kinetics for the transporter, suggesting that the affinities for substrate were the same on both sides of the membrane, and that the occupied and unoccupied tranporters "move" with equal efficiency (Taylor and Holman, 1981; May and Mikulecky, 1982). This provided strong evidence against models in which an alteration in the intrinsic asymmetry of the carrier provided a site for insulin action. The uniform conclusion of these studies was that the most likely mechanism of altered transport in response to insulin was an increase in the number of functional plasma membrane hexose carriers. In addition, the similarity in the Arrhenius activation energies for basal and insulin-stimulated transport sup-

TABLE I

Kinetic Parameters for Insulin-Stimulated Glucose Transport[a]

Cell type	Substrate	Measurement	K_m (mM)	V_{max}	K_m (mM)	V_{max}	Ref.
			− Insulin		+ Insulin		
Adipose tissue[b]	Glucose	zt,[c] influx	90	1.5 μmol/3 hr/mg	7	1.5 μmol/3 hr/mg	Crofford and Renold (1965a)
Adipocytes	Glucose	zt, influx	~6[d]	0.77 mg/hr/g	~6	1.94 mg/hr/g	Denton et al. (1966)
	2-DOG	zt, influx	2	0.6 nmol/min/10⁶ cells	2	1.8 nmol/min/10⁶ cells	Olefsky (1975)
	3-O-MG	ee,[e] efflux	5	0.2 mmol/sec/liter	5	1.7 mmol/sec/liter	Vinten et al. (1976)
	2-DOG	zt, influx	2.4	7.3 nmol/min/10⁶ cells	2.4	23 nmol/min/10⁶ cells	Olefsky (1978)
	3-O-MG	ee, efflux	3.5	0.13 mmol/sec/liter	3.1	0.8 mmol/sec/liter	Whitesell and Gliemann (1979)
	3-O-MG	zt, influx	5.4	0.034 mmol/sec/liter	6.1	1.2 mmol/sec/liter	Taylor and Holman (1981)
		zt, efflux	4.1	0.15 mmol/sec/liter	2.7	1.2 mmol/sec/liter	Taylor and Holman (1981)
		ee, influx	4.2	0.06 mmol/sec/liter	4.5	0.84 mmol/sec/liter	Taylor and Holman (1981)
	3-O-MG	zt, influx	10	0.1 mmol/sec/liter	9.4	0.57 mmol/sec/liter	May and Mikulecky (1982)
	3-O-MG	zt, influx	35	0.15 mmol/sec/liter	3	0.75 mmol/sec/liter	Whitesell and Abumrad (1985)
	3-O-MG	ee, influx	6.4	0.05 mmol/sec/liter	6.3	1.06 mmol/sec/liter	Martz et al. (1986)

Tissue	Sugar	Method	K_m	V_{max}	K_m	V_{max}	Reference
	3-O-MG	ee, influx	17.4	0.24 nmol/sec/10⁶ cells	7.1	2.0 nmol/sec/10⁶ cells	Suzuki (1988)
Brown adipocytes	3-O-MG	zt, influx	8.0	0.1 nmol/sec/10⁶ cells	6.3	3.9 nmol/sec/10⁶ cells	Joost et al. (1988)
	3-O-MG	zt, influx	4[f]	18 pmol/min/10⁶ cells			Czech et al. (1974)
Frog sartorius	Glucose	zt, influx	6	1.4 μmol/hr/ml	3.2	3.3 μmol/hr/ml	Narahara et al. (1960)
	3-O-MG	zt, influx[g]	3.3				Narahara and Özand (1963)
Gastrocnemius Red	3-O-MG	zt, influx	70	54 nmol/min/g	5.4	420 nmol/min/g	Ploug et al. (1987)
White	3-O-MG	zt, influx	69	34 nmol/min/g	7.3	100 nmol/min/g	Ploug et al. (1987)
Soleus			12.8	26 nmol/min/g	6.7	600 nmol/min/g	Ploug et al. (1987)
Skeletal muscle[h]	Glucose	zt, influx	5.6	80 nmol/min/g	5.6	275 nmol/min/g	Nesher et al. (1985)
Heart	Glucose	ee, influx	21.6	1.4 nmol/sec/mg	19.4	4.4 nmol/sec/mg	Sternlicht et al. (1988)
	3-O-MG	zt, influx	7	0.8 μmol/min/g	6	13.0 μmol/min/g	Cheung et al. (1978)
	3-O-MG	ee, influx	7	0.8 μmol/min/g	3	6.4 μmol/min/g	Cheung et al. (1978)
	3-O-MG	zt, efflux	6.9	17.6 μmol/min/g	2.7	34.1 μmol/min/g	Zaninetti et al. (1988)

[a] Representative examples of measurements of apparent K_m and V_{max} are included; this is not a comprehensive list.

[b] Species is rat unless otherwise indicated.

[c] zt, Zero trans.

[d] Double reciprocal plots were nonlinear, with estimated K_m values ranging from 5.2 to 7.9 mM.

[e] ee, Equilibrium exchange.

[f] Insulin increased the V_{max} without altering the K_m, although the actual values were not reported.

[g] Summer frogs.

[h] Rat epitrochlearis muscle, in vitro.

ported a model in which the absolute number of transporters increased, as opposed to a conformational change in a fixed number of transporters (Olefsky, 1978; Vinten, 1978; Ludvigsen and Jarett, 1980).

On the other hand, there have been a number of convincing reports of alterations in apparent affinity for hexose accompanying the increased transport rates in response to insulin. Whitesell and colleagues have argued that cell agitation, substrate starvation, and reduction of temperature below 37°C all artifactually reduce the basal K_m, obscuring an effect of insulin on affinity. They have reported a 10-fold reduction in the apparent K_m for influx of 3-O-MG accompanied by a moderate increase in V_{max} (Whitesell and Abumrad, 1985; 1986). Although it was later suggested that the higher apparent K_m in these experiments was due to incomplete mixing of the radioactive hexose (Martz et al., 1986), the blockade of the effects of agitation by depletion of cellular ATP argues for a more physiological basis for the discrepancies in K_m (Whitesell and Abumrad, 1986). Recently, other investigators have corroborated a decrease in apparent K_m of a lesser magnitude associated with hormonal stimulation (Suzuki, 1988). When initally observed, the insulin-dependent changes in affinity were interpreted as evidence for a hormone-stimulated increase in the intrinsic activity of resident plasma membrane transporter proteins (Whitesell and Abumrad, 1985). However, a reappraisal of the kinetics of basal 3-O-MG uptake has revealed that the observed alterations in apparent K_m are equally compatible with a model in which a low-affinity transporter is constitutively present on the surface of basal cells and a high-affinity transporter is added in response to insulin (Suzuki, 1988; Whitesell et al., 1989). As described below, this idea is congruent with the current understanding of translocation of glucose transporter proteins.

B. Formulation of the Model

Despite the preponderance of kinetic data consistent with an insulin-stimulated increase in the number of surface transporters, a formal test of this model required a way of ascertaining carrier number independent of transport activity in situ. Two methods served this purpose: cytochalasin B binding and reconstitution of transport in liposomes. Cytochalasin B, a metabolite of the mold Helminthosporum dermatoideum, has been long recognized to inhibit glucose utilization in a number of mammalian cell types by specifically interfering with hexose transport (Mizel and Wilson, 1972; Kletzien and Perdue, 1973). The block to uptake is rapid and reversible, with a K_i in the submicromolar range. Studies in human erythrocytes identified a single, high-affinity binding site for cytochalasin B located on the intracellular surface of the glucose carrier (Jones and Nickson, 1981).

Transported hexoses competitively inhibited the binding of cytochalasin B to membranes isolated from human red blood cells, whereas cytochalasin B noncompetitively inhibited the influx of hexose into intact erythrocytes. In white and brown adipose cells, cytochalasin B inhibited both basal and insulin-stimulated hexose transport (Czech *et al.*, 1973, 1974; Loten and Jeanrenaud, 1974). Wardzala *et al.* (1978) took advantage of the binding of cytochalasin B to the fat cell glucose transporter to measure plasma membrane transporter number (Czech, 1976). Equilibrium [^{3}H]cytochalasin B binding was performed in the absence and present of 500 mM D-glucose, and the difference taken as a measure of the specific binding to glucose carrier. Cytochalasin E, which does not inhibit hexose flux, was included in the assay to further reduce nonspecific binding. The dissociation constant for D-glucose-inhibitable cytochalasin B binding was found to be 120 nM, which was comparable to the K_i for the inhibition of adipose cell glucose uptake. Plasma membranes prepared from adipocytes treated with insulin demonstrated about fourfold greater specific cytochalasin B binding than membranes from untreated cells, with no change in the K_d. The conclusion was that insulin increased the number of functional glucose transport systems in the plasma membrane of the rat adipocyte. The weakness of these experiments was that they did not distinguish between the uncovering of latent cell surface carriers and the recruitment of functional carriers from an alternative site of storage. This shortcoming was addressed in a subsequent study in which adipocytes were fractionated by the method of McKeel and Jarett and the localization of glucose transporter tracked by cytochalasin B equilibrium binding (McKeel and Jarett, 1970; Cushman and Wardzala, 1980). The increased number of glucose transporters present in the plasma membrane was confirmed, and it was also noted that this was paralleled by a precisely reciprocal decrement in hexose carriers in an intracellular fraction, called "low-density microsomes." The total cellular complement of glucose transporters remained unchanged in response to insulin, as did the recovery of marker enzymes (Cushman and Wardzala, 1980). At about the same time, Suzuki and Kono (1980) independently assessed the subcellular distribution of glucose transport proteins in adipocytes exposed to insulin using reconstitution of glucose uptake activity into liposomes as an assay. Nonetheless, the conclusion was quite similar; insulin increased the number of glucose transporters residing on the cell surface, while simultaneously inducing a decrease in transporters at another subcellular site. Both groups interpreted the experiments as showing that, in the basal state, glucose transporters reside in a latent intracellular compartment. Whether these intracellular transporters possess the same catalytic activity as those on the cell surface was and remains unclear, although no suppression of low-density microsome transporters is required, since their exclusion from contact with extracellular sugars is sufficient to render them physiologically insignificant. When the cell is exposed to insulin, the latent transporters redistribute to the cell surface such that there is an increase in the number of catalytically active carriers exofacially disposed. On removal of hormone, these transporters return to a storage site in the cell interior. The initial experiments indicated that the intracellular transporters had the same biochemical characteristics as those in the plasma membrane, as determined either by cytochalasin B binding or kinetic analysis after reconstitution into liposomes (Simpson *et al.*, 1983; Smith *et al.*, 1984). This provided additional evidence that insulin was effecting the movement of a single class of transporter molecules, as opposed to activating and simultaneously inactivating two different types of carriers.

   The first studies characterizing the translocation process, whether assaying transporter by cytochalasin B binding or reconstitution into liposomes, yielded surprisingly consistent results. In essentially all cases, the changes in plasma membrane glucose transporter represented precisely the inverse of alterations in low-density microsome transporter. Moreover, the level of cell surface carrier usually reflected in general terms the rate of hexose flux. For example, translocation of transporters and rates of 3-*O*-MG uptake displayed identical dose–response profiles to insulin, with half-maximal stimulation occuring at about 0.11 n*M* hormone (Karnieli *et al.*, 1981b). A number of insulin-mimetic agents, including H_2O_2, vanadate, and trypsin, stimulated glucose transporter translocation with potencies roughly approximating their abilities to activate hexose flux (Kono *et al.*, 1982). The time course of redistribution of transporter in response to insulin was rapid, with a half-time of about 2.5 min, and deactivation after removal of hormone was about four times slower (Karnieli *et al.*, 1981b; Kono *et al.*, 1982). Interestingly, the latter process was virtually superimposable with the decay in hexose transport, although there appeared to be about a 1.5-min delay between the appearance of carrier on the plasma membrane and augmentation in hexose flux (Karnieli *et al.*, 1981b). This discrepancy may reflect a need for transporters to be activated following recruitment, or may be an artifact related to the imprecision inherent in "instantaneously" halting translocation prior to cell fractionation.

   When insulin acts on a cell to stimulate redistribution of glucose transporters, no new protein synthesis is required (Karnieli *et al.*, 1981b). However, cellular energy is necessary for both the translocation of transporters onto the cell surface, and their redistribution to intracellular sites after the removal of hormone. Agents such as azide, dinitrophenol, and cyanide, all of which deplete cellular ATP, effectively blocked both processes (Kono *et al.*, 1982). Interestingly, manipulations such as the lowering of the temperature of incubation, increasing the extracellular pH, or hyperosmolarity, which activated hexose transport in adipocytes in the absence of insulin, also stimulated redistribution of carriers in a manner dependent on cellular energy (Ezaki and Kono, 1982; Toyoda *et al.*, 1986). The precise subcellular compartment in which the transporter resides was unclear in these early studies. The low-density microsomal fraction was defined rather empirically, as those membranes sedimenting at 212,000 g but not 48,000 g, or the peak of transporter protein sedimenting at about 15% sucrose on a continuous density gradient, partially overlapping galactosyl transferase (Cushman and Wardzala, 1980; Suzuki and Kono, 1980; Ezaki and Kono, 1982; Simpson *et al.*, 1983). In fact, the terminology is misleading; the glucose transporter-containing vesicles in the low-density microsomes are of the same density as the plasma membrane fraction, as determined by equilibrium sucrose gradient centrifugation (Shibata *et al.*, 1987). Although some enzyme markers for the Golgi complex co-purified with low-density microsomes, it was quite clear that these did not represent the same cellular compartment. There was and still remains no known marker for transporter-containing vesicles other than the transporter itself. Thus, since there was no way of assessing contamination of the plasma membrane fraction by intracellular vesicles, it was impossible to accurately quantitate translocation. Nonetheless, most investigators were impressed by the closeness with which the rough estimates of hormone-dependent redistribution approximated the degree of simulation of hexose transport by insulin (Karnieli *et al.*, 1981b; Kono *et al.*, 1982).

One issue not addressed by the initial experiments was whether low-density microsomes could have represented a unique domain of the plasma membrane as opposd to a legitimate intracellular organelle. Oka and Czech (1984) approached this problem by selectively antagonizing cell surface cytochalasin B binding using an impermeant sugar analog, ethylidene glucose. In this study, binding was not assayed directly but inferred from the covalent labeling of the transporter. Exposure of membranes derived from a number of cell types to ultraviolet light in the presence of [^{3}H]cytochalasin B leads to the stable attachment of radionucleotide to the glucose transporter protein (Carter *et al.*, 1982; Pessin *et al.*, 1982; Shanahan *et al.*, 1982). When such a reaction was performed in intact rat adipocytes, essentially all pools of transporter were equally labeled (Oka and Czech, 1984). On the other hand, the inclusion of ethylidene glucose in the extracellular medium selectively antagonized photolabeling of a pool of glucose transporter whose abundance increased in the presence of insulin and which cofractionated with plasma membranes. Moreover, the photolabeling of glucose transporters that fractionated as low-density microsomes, whether in basal or hormone-treated cells, was not inhibited by hexose restricted to the extracellular space. These data were interpreted as showing that low-density microsomes were truly derived from intracellular membranes. Additionally, the inhibition of labeling of all transporters by 3-O-MG indicated that the internal carrier was potentially active, at least as inferred from its ability to bind substrate (Oka and Czech, 1984).

C. The Discovery of the "Insulin-Sensitive" Glucose Transporter

The years following the development of the translocation hypothesis produced a number of studies examining several problems with the model. One discrepancy was between the the magnitude of the stimulation of hexose uptake in intact adipocytes and the redistribution of transporter as determined by either cytochalasin B binding or reconstitution of transport activity. 3-O-methyl-glucose uptake increased from 10- to 30-fold, whereas the abundance of transporter in the plasma membrane fraction rose only 2- to 4-fold after exposure of cells to insulin. The likely explanation for this was the inevitable and unquantifiable contamination of the plasma membrane fraction with internal transporter. In addition, the translocation assay was quite difficult, with any number of minor perturbations leading to an artifactual increase in basal plasma membrane transporter. In a careful study in which many of these effects were scrupulously controlled, the activation of transport came much closer to approximating the measured translocation of transporter (Joost et al., 1988). Thus, it was believed quite plausible that much of the lack of correlation between increases in hexose flux and recruitment, as measured by cytochalasin B binding, stemmed from problems inherent in subcellular fractionation.

Much more difficult to reconcile were discrepancies between the number of transporters in a subcellular fraction as determined by cytochalasin B binding and that measured by immunological methods. The first antibody reagents available were polyclonal antisera raised against the purified human erythrocyte transporter. Lienhard et al. (1982) reported that insulin increased the quantity of transporter in the plasma membrane of rat adipocytes about threefold, as determined by either cytochalasin B binding or reactivity with anti-erythrocyte glucose transporter antisera. The ratio of microsomal to plasma membrane transporter in the basal state was 8.1 as determined by cytochalasin B binding, but only 3.2 as assayed by Western immunoblot. A number of plausible interpretations are compatible with these data, including an intracellular pool of translocatable transporter that binds cytochalasin B but is not recognized by antibody. Wheeler et al. observed a a similar anomaly in the ratio of microsomal to plasma membrane transporter but, in addition, measured the increase in cell surface transporter in response to insulin was 4.2-fold by cytochalasin B binding but only 1.3-fold by antibody reactivity (Wheeler *et al.*, 1982). Analogous inconsistencies have been apparent when transporter was quantitated by reconstitution into liposomes instead of cytochalasin B binding (Ezaki *et al.*, 1986; Joost *et al.*, 1988). The problem was addressed more systematically by evaluating for different cell types the reactivity of anti-erythrocyte carrier antisera toward a constant number of transporters, as determined by cytochalasin B binding (Wang, 1987). The antisera appeared to react 5 to 10 times more effectively with transporter from either human or neonatal rat erythrocytes as compared to rat adipocytes. This can be interpreted as indicative either of posttranslational modifications that alter the immunoreactivity of some of the transporters, or of an immunologically silent pool of intracellular transporter that binds cytochalasin B.

There have been several other provocative studies revealing a lack of correlation between transporter number and other parameters of transporter function. For example, in an ontological study of insulin-stimulatable hexose transport in rat diaphragm, insulin responsiveness peaked at about 20 days of age, whereas glucose-inhibitable cytochalasin B binding steadily decreased postnatally (Wang, 1985). This was interpreted as indicating that a factor distinct from intracellular transporter number determines the magnitude of the response to insulin. Another intriguing observation was made in the cultured murine adipocyte cell line, 3T3-L1 (see below). Antisera against the human red blood cell (RBC) transporter reacted more efficiently per unit of cytochalasin B binding toward 3T3-L1 adipocyte than rat adipocyte transporter (Schroer *et al.*, 1986). Last, in an elegant morphological study of transporter redistribution using immunoelectron microscopy, Blok *et al.* (1988) observed only a 3-fold increase in plasma membrane transporter as detected by an affinity-purified antiserum against the human erythrocyte protein, in spite of a 15-fold stimulation in transport in response to insulin.

The cloning of cDNAs encoding a human hepatoma glucose transporter and subsequently a rat brain transporter, both homologs of genes also encoding the human erythrocyte transporter, provided important tools for addressing the issue of insulin-responsive glucose uptake (Mueckler *et al.*, 1985; Birnbaum *et al.*, 1986). Initially, the observation that the "erythrocyte" glucose transporter was expressed in adipocytes as well as brain and most other tissues was viewed as evidence in support of a single transporter providing insulin-stimulatable and constitutive glucose transport (Birnbaum *et al.*, 1986; Flier *et al.*, 1987). Nonetheless, knowledge of the primary amino acid sequence of a glucose transporter allowed experiments in which the discrepancies cited above could be reevaluated using anti-peptide antisera specific for the erythrocyte transporter. Using such a reagent, insulin-stimulated translocation again was measured to be about two- to threefold in rat adipocytes, in spite of a considerably greater stimulation of transport (Haspel *et al.*, 1988). In another study, cell surface glycoproteins on 3T3-L1 adipocytes were labeled by oxidation with galactose oxidase followed by reduction by sodium borohydride. Immunoprecipitation with an antiserum directed against a synthetic peptide corresponding to the carboxyl terminus of the erythrocyte transporter permitted measurement of plasma membrane transporter without potential artifacts of cellular fractionation. This technique revealed a 2- to 3-fold translocation although hexose transport in intact cells increased 10-fold after exposure to insulin (Calderhead and Lienhard, 1988). Moreover, the two processes were dissociated in time, translocation of immunodetectable transporter preceding the hormone-induced increases in hexose uptake (Gibbs *et al.*, 1988). One of two alternative models appeared to explain the latter data. Either insulin-stimulated glucose transport required translocation of hexose carriers to the cell surface followed by augmentation of their intrinsic activity, or there could be a latent population of intracellular transporters not recognized by the then available antisera. Support for the latter was provided in a series of experiments by Oka and colleagues (1988). Glucose transporters from human erythrocytes, rat brain, and rat adipocytes were photolabeled with [^{3}H]cytochalasin B. Anti-peptide antisera prepared against several regions of the erythrocyte transporter immunoprecipitated all of the labeled erythrocyte transporter, but only 30 and 3% of the photolabeled transporter from brain and fat cell, respectively (Oka *et al.*, 1988). In addition, one-dimensional maps of trypsin-digested erythrocyte and adipocyte [^{3}H]cytochalasin B-labeled transporter generated distinct patterns.

More definitive evidence in support of a latent, adipose tissue-specific hexose transporter derived from efforts to purify glucose transporter containing intracellular vesicles, starting with rat adipocyte low-density microsomes (James *et al.*, 1987). Monoclonal antibodies were raised against the 10-fold purified vesicles, and screened for reactivity with constituent proteins. One antibody, 1F8, recognized a 45-kDa protein present in low-density microsomes that translocated to the plasma membrane in response to insulin, increasing in the latter compartment about 10-fold (James *et al.*, 1988). Moreover, 1F8 detected a protein only in insulin-responsive tissues, i.e., adipose tissue, skeletal muscle, and heart. Last, the 45-kDa protein appeared to be a transporter, since it could be photolabeled by [^{3}H]cytochalasin B in a glucose-inhibitable manner. These data strongly supported the existence of a novel intracellular form of the glucose transporter in insulin-responsive tissues, but, as noted by the authors, did not distinguish between a new gene product or a posttranslational modification of the erythrocyte transporter.

Resolution to this question occurred in early 1989 when five research groups independently screened rat, human, or mouse muscle or adipose tissue libraries at reduced stringency with an erythrocyte glucose transporter probe and identified cDNAs encoding a putative insulin-responsive transporter (Birnbaum, 1989; Charron *et al.*, 1989; Fukumoto *et al.*, 1989; James *et al.*, 1989b; Kaestner *et al.*, 1989). All were clearly homologs of the same gene; for example, the rat and human proteins are identical at about 95% of the amino acid residues. The adipose/muscle transporter is encoded by a gene distinct from the erythrocyte, liver, fetal muscle, or jejunal glucose transporters. For clarity, the nomenclature adopted here will be that suggested by Bell, based on the order in which the first cDNA encoding each transporter family member was cloned. The muscle/adipose transporter and its gene will be referred to as GLUT4 (Gould and Bell, 1990). The other transporters are denoted as follows: erythrocyte/hepatoma/brain, GLUT1; liver/β cell, GLUT2; fetal muscle/brain, GLUT3; jejunal, GLUT5. In the literature of the last several years, GLUT4 has carried many names: IRGT (James *et al.*, 1989b), GT2 (Kaestner *et al.*, 1989), GTIII (Garcia de Herreros and Birnbaum, 1989b), GT5 (Carruthers, 1990) and the M-type glucose transporter (Kasanicki and Pilch, 1990). Predictably, this has generated enormous confusion, which hopefully will be resolved as more investigators adopt the "biologically neutral" nomenclature used here.

In spite of being 17 amino acids longer, GLUT4 is more similar to GLUT1 than any other known glucose transporter protein, sharing identity in about 65% of amino acids (Bell *et al.*, 1990). The hydropathy plot for GLUT4 is virtually superimposable on that of GLUT1, suggesting a common transmembrane topology, and conservation of the single site for asparagine-linked glycosylation. As originally proposed for GLUT1 (Mueckler *et al.*, 1985), GLUT4 probably consists of 12 α-helical membrane-spanning segments, with both amino and carboxyl termini located in the cytoplasm (Fig. 1). It should be noted that the dispositions of only the first extracellular domain, the major cytoplasmic loop, and the carboxyl terminus have been proven biochemically for GLUT1, from which the topology of GLUT4 is inferred (Mueckler *et al.*, 1985; Andersson and Lundahl, 1988; Haspel *et al.*, 1988). Therefore, this model must be regarded as tentative. Interestingly, the sequence similarity between GLUT1 and GLUT4 is not randomly distributed (Fig. 1). With the exception of the third membrane-spanning domain, the regions putatively traversing the membrane are the most highly conserved. The most disparate domains are the glycosylated extracellular loop, the termini of the protein, and the large intracellular region connecting membrane-spanning segments 6 and 7. One plausible interpretation of this distribution of similarity is that those parts of the transporters most involved in the flux of sugars, i.e., the membrane-spanning segments that presumably form the

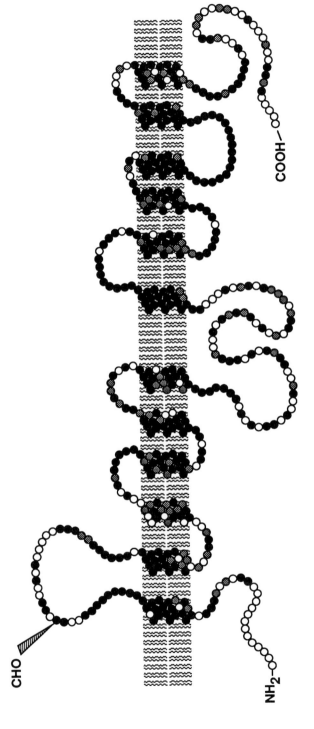

FIG. 1 Putative topology of the rat GLUT4 glucose transporter protein. The transporter is drawn with the cytoplasm surface of the plasma membrane facing downward. The probable single site of N-linked glycosylation is indicated (CHO), as are the amino and carboxy termini. Amino acids that are identical in the rat GLUT1 and GLUT4 glucose transporters are represented by black circles and conservative differences are shown as shaded circles. The precise locations of the borders of the transmembrane segments are arbitrary and drawn for convenience of representation. The overall sequence identity between the two transporters is 63.5%. However, the identity between the putative membrane-spanning domains is 73.0%, and the identity between the three large cytoplasmic domains of GLUT1 and GLUT4 is 40.5%. (From Birnbaum, 1989.)

glucose "channel," are most highly conserved, while the regions primarily involved in regulatory functions are most divergent. This model is appealing since it explains the remarkable similarity in the kinetics of hexose transport in insulin-responsive and -nonresponsive peripheral tissues. Also striking is the remarkable similarity in general two-dimensional topology, i.e., six transmembrane segments followed by a cytoplasmic domain followed by six transmembrane segments, shared with many prokaryotic and eukaryotic permeases, including the "ABC" family of active transport proteins (Baldwin and Henderson, 1989; Juranka et al., 1989; P. J. F. Henderson et al., this volume).

What is the evidence that GLUT4 encodes an insulin-responsive glucose transport protein? First, the protein encoded by the GLUT4 cDNA appears to be identical to that recognized by the 1F8 monoclonal antibody. RNA transcribed in vitro from a rat or human GLUT4 cDNA template and translated in a reticulocyte extract cell-free system directed the synthesis of a protein that was immunoprecipitated by 1F8 (Fukumoto et al., 1989; James et al., 1989b). Second, the GLUT4 cDNA encodes a catalytically active hexose carrier. Injection of in vitro synthesized GLUT4 mRNA into oocytes from the African frog Xenopus laevis resulted in a significant increase in cytochalasin B-inhibitable 2-deoxyglucose uptake (Birnbaum, 1989). Next, the tissue distribution of GLUT4, as assayed by Northern RNA hybridization blot, corresponded to what one might predict for an acutely insulin-regulated transporter, i.e., expressed in those tissues most hormone responsive in terms of glucose transport: brown and white adipose tissue, red and white skeletal muscle, heart, and to a lesser degree smooth muscle and possibly kidney (Birnbaum, 1989; Charron et al., 1989; Fukumoto et al., 1989; James et al., 1989b; Kaestner et al., 1989). It is not clear whether the latter represents contamination with perirenal or caliceal fat tissue or true expression in renal cells. An immunoelectron microscopy study of the distribution of GLUT4 claiming expression in blood vessel endothelial cells of adipose tissue and muscle has been refuted in some detail and is probably incorrect (Vilaro et al., 1989; Slot et al., 1990; Douen et al., 1991; Friedman et al., 1991). Antisera raised against either a synthetic peptide corresponding to the carboxyl terminus or a bacterial fusion protein that includes the last cytoplasmic domain of GLUT4 recognized a protein with an apparent mobility on sodium dodecyl sulfate (SDS)-polyacrylamide electrophoresis of about 45 kDa and a tissue distribution identical to that of the GLUT4 mRNA (Birnbaum, 1989; James et al., 1989a). Most importantly, when these antisera were used for quantitating the degree of translocation of GLUT4 in membrane fractions prepared from adipocytes treated with insulin, there was an approximately 50% decrease in GLUT4 protein present in the low-density microsomes and a concomitant increase in the plasma membrane fraction of about 7 to

10-fold (Birnbaum, 1989; James *et al.*, 1989b). As described above, this corresponds quite well to the redistribution of glucose-inhibitable cytochalasin B binding sites, and approaches the measured increase in hexose flux.

The mouse and human genes encoding GLUT4 have been cloned (Bell *et al.*, 1990; Kaestner *et al.*, 1990). Both span about 8 kb of genomic DNA and include 10 introns, located in the same positions as for GLUT1, but with an additional intron in the sequences encoding putative membrane-spanning domain 4 (Fukumoto *et al.*, 1988; Williams and Birnbaum, 1988). Thus, as for GLUT1, introns occur predominantly in sequences encoding extramembranous portions of the protein, suggesting that it may have been assembled from transmembrane segments. The human GLUT4 gene has been mapped to chromosome 17p13 (Bell *et al.*, 1989).

The data presented in the preceding paragraphs provide a compelling picture of an association between the expression of the GLUT4 glucose transporter protein and the presence of markedly insulin-responsive hexose uptake. However, these data do not establish whether GLUT4 is sufficient, or in fact even necessary, for the full hormone response. In other words, the expression of GLUT4 in the appropriate tissue distribution does not prove that its primary structure makes it intrinsically more capable of redistributing to the plasma membrane in adipose tissue in response to insulin. While these issues are still under study, a number of experiments have been performed that provide strong, if not definitive, support for this hypothesis.

One of the most useful systems for studying the mechanism of action of insulin as well as the acquisition of insulin-responsiveness is the murine adipocyte cell line 3T3-L1. These cells were originally established as a subclone of Swiss 3T3 fibroblasts and have the remarkable characteristic of, under appropriate culture conditions, differentiating into adipocyte-like cells (Green and Kehinde, 1973). Accompanying this fatty conversion, 3T3-L1 cells initiate expression of genes encoding the enzymes of fatty acid synthesis and other adipose-specific proteins, resulting in the accumulation of triglyceride droplets visible with the light microscope (Mackall *et al.*, 1976; Kuri-Harcuch and Green, 1977; Reed *et al.*, 1977). More importantly, the murine adipocyte cells develop the markedly acute insulin-responsive glucose transport characteristic of authentic adipose tissue (Rubin *et al.*, 1978; Resh, 1982). Thus, if expression of GLUT4 were integral to the hormone-stimulatable hexose flux, it should correlate with the increase in responsiveness in these cells.

In two of the original reports describing the cloning of GLUT4, it was reported that 3T3-L1 adipocytes have higher levels of GLUT4 mRNA than is present in preadipocytes (James *et al.*, 1989b; Kaestner *et al.*, 1989). Subsequently, this problem was studied in more detail. Insulin stimulated deoxyglucose uptake in 3T3-L1 fibroblasts about twofold, when the cells are assayed at confluence. During the course of differentiation, the stimulation increases to about 10- to 15-fold, due to a slight rise in the absolute rate of hormone-stimulatable uptake as well as a prominent decrease in basal flux (Resh, 1982; Garcia de Herreros and Birnbaum, 1989b; Tordjman et al., 1989; Weiland et al., 1990). The latter is a result of, at least in part, a redistribution of GLUT1 protein, which is expressed abundantly in both fibroblasts and adipocytes, from the plasma membrane to an intracellular compartment (Blok et al., 1988; Weiland et al., 1990). The onset of increased hormone responsiveness, as assayed with maximally effective concentrations of insulin, correlates perfectly with the appearance of GLUT4 mRNA and protein, both of which are undetectable in preadipocytes (Fig. 2) (Garcia de Herreros and Birnbaum, 1989b). Since insulin receptor mRNA and cell surface protein also increase with fatty conversion, it was important to determine the relative role of receptor number in the acquisition of marked hormone responsiveness. In fact, the increase in insulin-stimulatable hexose uptake precedes the increase in insulin receptors (Rubin et al., 1978; Resh, 1982; Garcia de Herreros and Birnbaum, 1989b). In addition, NIH-3T3 fibroblasts overexpressing human insulin receptors are capable of only a twofold increase in hexose uptake in response to insulin (Garcia de Herreros and Birnbaum, 1989b). Last, even though insulin-like growth factor I (IGF-I) receptors are present in sufficiently high numbers to produce biological responses in both fibroblasts and cultured adipocytes, only in the latter does IGF-I stimulate a 10-fold increase in deoxyglucose uptake (Weiland et al., 1990). Thus, it is reasonably clear that the increase in insulin receptors is not necessary for the maximal stimulation of hexose uptake, but allows the fat cell to respond to relatively low, physiological concentrations of the hormone. The striking correlation between the magnitude of insulin-stimulatable transport and the level of GLUT4 is consistent with but does not prove that the isoform is necessary for the full hormone response. Neither does it preclude the possibility that other adipose cell-specific factors in addition to GLUT4 are required for marked insulin-stimulated transport.

Thus, the picture has emerged of at least two facilitated hexose carriers, GLUT1 and GLUT4, expressed in insulin-responsive tissues, both of which are redistributed to the plasma membrane in response to insulin (Fig. 3). In the basal state, the ratio of intracellular to cell surface transporter is much larger for GLUT4 than GLUT1, and thus the former increases in the plasma membrane to a greater degree. In spite of the obvious appeal of such a model, a considerable effort over the past several years has been devoted to trying to ascertain whether it can quantitatively explain the complete activation of glucose transport. Obviously a technical prerequisite is the ability to accurately assess surface transporter number independent of cellular fractionation.

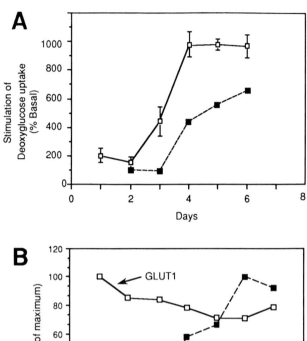

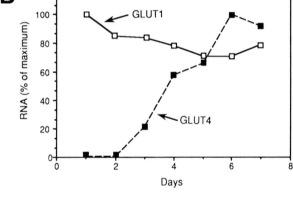

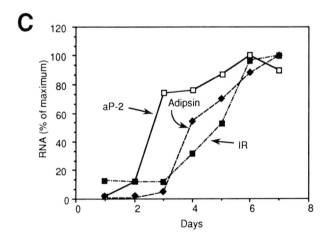

Extracellular   Extracellular

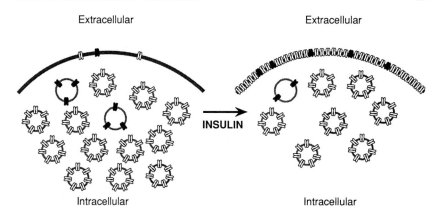

Intracellular   Intracellular

FIG. 3   The translocation of glucose transporters in rat adipocytes. GLUT1 is represented by closed symbols and GLUT4 by open symbols. The relative numbers of cell surface transporters for each of the two isoforms roughly correspond to that estimated by Holman *et al.* (1990).

D. Quantitative Considerations

Over the last 5 years, Holman had developed a series of reagents capable of photolabeling the exofacial surface of the facilitated hexose carriers (Holman *et al.*, 1986; Holman and Rees, 1987). The backbone of these compounds is two mannose residues linked at their C-4 units through a propyl bridge. This configuration reduces the affinity toward the internal site of the transporter while retaining binding to the exofacial site. In addition, the ligand is relatively membrane impermeable. The most recent version consists of a diazirine group linked to the middle carbon of the propyl moiety, resulting in a 2-*N*-4(1-azi-2,2,2-trifluoroethyl)benzoyl-1,3-

---

FIG. 2   Time course for the appearance of the highly insulin-stimulated deoxyglucose uptake in 3T3-L1 adipocytes. 3T3-L1 preadipocytes were grown to confluence, and induced to differentiate by treatment with dexamethasone, isobutyl methylxanthine, and 10% fetal bovine serum (day 0). (A) The stimulation of deoxyglucose uptake by 340 nM ($\square$) or 3.4 nM ($\blacksquare$) insulin is expressed as a percentage of nonstimulated uptake. Basal rates of deoxyglucose uptake on days 1 and 6 were 1.0 ± 0.13 and 0.1 ± 0.01 nmol/min/mg protein, respectively. (B and C) Expression of mRNA levels in differentiating 3T3-L1 adipocytes. Northern blots were quantitated by scanning densitometry, the data normalized to ribosomal RNA, and represented as a percentage of maximal mRNA concentration for each species of mRNA, as indicated. aP-2 Encodes the 13-kDa fatty acid binding protein of adipocytes, and IR is the insulin receptor. Note that adipocyte-specific proteins are induced asynchronously, but GLUT4 correlates well with insulin-stimulated transport. (From Garcia de Herreros and Birnbaum, 1989b.)

bis(D-mannos-4-yloxy)-2-propylamine, referred to as ATB-BMPA (Clark and Holman, 1990). Photolysis of this compound induces loss of N_2 and the generation of a reactive carbene. [^{3}H]ATB-BMPA has been used to examine the redistribution of transporters in freshly isolated rat adipocytes and in cultured murine 3T3-L1 adipocytes (Calderhead et al., 1990b; Holman et al., 1990). In order for values obtained in these studies to be reliable, the affinity and labeling efficiency of the compound must be similar for both transporter isoforms and must not be influenced by treatment of cells with hormone. In human erythrocytes as well as in basal and insulin-stimulated rat adipocytes, the K_i for the inhibition of hexose uptake by ATB-BMPA is about 300 μM, indicating that at least the first prerequisite is fulfilled. In addition, Holman et al. (1990) have established that photolabeling of GLUT1 and GLUT4 does not alter the ability of the transporters to be immunoprecipitated by antisera. Thus, it appears that the ATB-BMPA is an adequate photoreactive ligand for measuring transporters with an extracellularly disposed substrate binding site.

The level of GLUT1 protein in total membranes from rat adipocytes is about 5 to 10% of GLUT4 (Zorzano et al., 1989; Calderhead et al., 1990b). By labeling intact rat fat cells with [^{3}H]ATB-BMPA, the absolute ratio of GLUT4 to GLUT1 in plasma membranes has been determined to be greater than 2 (Holman et al., 1990). This is in contrast to a study of conventionally fractionated adipocytes, in which it was found that GLUT1 was the predominant cell surface transporter in nonhormonally stimulated cells (Zorzano et al., 1989). There are several possible reasons for this apparent discrepancy. First, experiments in intact cells and membrane fractions rely on photolabeling with ATB-BMPA or 3-iodo-4-azidophenethylamido-7-succinyldeacetyl forskolin (IAPS-forskolin), respectively, and there might be differences in labeling efficiency for the transporter isoforms. Second, the ATB-BMPA labeling was performed at 37°C, so any recycling of GLUT4 in the basal state would have the effect of artificially increasing the estimate of basal cell surface transporter. Last, the amount of radioactive ATB-BMPA incorporated into nonstimulated cells was quite low and thus subject to error.

Exposure of rat adipocytes to insulin resulted in an ~5-fold and 15- to 20-fold increase in the labeling of GLUT1 and GLUT4, respectively, by ATB-BMPA; in parallel, 3-O-MG uptake increased 20- to 30-fold (Holman et al., 1990). The ratio of plasma membrane GLUT4 to GLUT1 rose from 2 to 10. Thus, as noted in the previous section, the increase in hexose flux correlated much more closely with the change in GLUT4 than GLUT1, in spite of the presence of significant amounts of the latter isoform on the cell surface. This problem is more dramatic in 3T3-L1 adipocytes, in which there were about 3.5 to five times more total cellular GLUT1 than GLUT4 transporters (Calderhead et al., 1990b; Piper et al., 1991). Experiments in these cells analogous to those described above using ATB-BMPA as a probe for cell surface glucose transporters have yielded an estimate of the increase in plasma membrane GLUT1 and GLUT4 in response to insulin of 6- and 17-fold, respectively (Calderhead et al., 1990b). Again, the fold increase in GLUT4 much more closely correlated with the 21-fold augmentation of hexose uptake. However, the ratio of GLUT1 to GLUT4 in the plasma membrane changed from 4.6 to 1.6 from the basal to the hormone-stimulated state, again assuming equal efficiencies of photolabeling for the two isoforms. Thus, 3T3-L1 adipocytes present the paradox that the rate of hexose flux appears to correlate with the level of the less abundant transporter species. If the quantification of cell surface, catalytically active transporter using the Holman reagent is indeed reliable, then one of two explanations is likely to resolve this inconsistency: either GLUT4 is intrinsically more active in the transport of hexose than GLUT1, or there is a change in catalytic activity in one or both of the transporters accompanying insulin-induced translocation.

With the possible exception of the human erythrocyte glucose transporter, it has been impossible to unequivocally define kinetic parameters for each of the transporter isoforms in their native environments, due both to the small intracellular water space of a mammalian cell, and to the coexpression of multiple carrier species in the same cell. Overexpression by traditional gene transfer of cloned cDNAs into tissue culture cells is plagued by the same problems. As noted above, the $X.$ laevis oocyte expression system was used initially to prove that the GLUT4 cDNA encoded an authentic facilitated glucose carrier (Birnbaum, 1989). The oocyte system is almost ideal for the study of the function of transporters isolated from their native environments. In this technique, the cDNA of interest is cloned into a plasmid downstream of a RNA polymerase promoter, such as SP6 polymerase, which is then used to produce in vitro an mRNA transcript (Melton et al., 1984). Stage V oocytes, which are about 1 mm in diameter with an intracellular water space of about 0.5 μl, are harvested from gravid female $X.$ laevis and injected with 1 to 50 ng mRNA. Transport can be measured 2 to 5 days later. The amount of protein produced by such a system is at least 1000 times greater than a reticulocyte lysate in vitro translation product (Colman, 1984). Estimates of the quantity of GLUT1 transporter protein present in oocytes injected with in vitro synthesized mRNA have ranged from 0.2 to 7 ng/oocyte (Gould and Lienhard, 1989; Keller et al., 1989). A number of groups have utilized oocyte expression to study various aspects of glucose transport (Birnbaum, 1989; Gould and Lienhard, 1989; Janicot and Lane, 1989; Keller et al., 1989; Vera and Rosen, 1989, 1990; Gould et al., 1991; Janicot et al., 1991). In two independent studies, the apparent K_m for GLUT1 for influx of 3-O-MG under equilibrium exchange conditions was estimated to be

20 mM, and the turnover number about 0.5–2.2 × 10^3 sec (Gould and Lienhard, 1989; Keller *et al.*, 1989). It should be noted that these parameters were obtained at 20°C. In addition, the turnover number was calculated based on the assumption that all of the cellular GLUT1 was on the surface of the oocyte, which is almost certainly not the case. Keller *et al.*, (1989) also obtained a value of 1.8 mM for the apparent K_m for GLUT4 as measured under the same conditions.

One of the interesting issues in regard to *Xenopus* oocytes is their capacity to respond to insulin in terms of an increase in glucose uptake. In uninjected oocytes, insulin has been reported to produce diverse biological effects, including maturation and the stimulation of cyclic AMP phosphodiesterase and ribosomal protein S6 phosphorylation (Maller *et al.*, 1986; Korn *et al.*, 1987; Sadler and Maller, 1987). Several investigators have found that micromolar insulin increases deoxyglucose uptake about twofold in uninjected oocytes, most likely working through a *Xenopus* IGF-I receptor (Janicot and Lane, 1989; Vera and Rosen, 1989; Janicot *et al.*, 1991). The effects of insulin on oocytes expressing mammalian transporters are unclear. Vera and Rosen (1989, 1990) have reported that expression of GLUT1, GLUT2, and GLUT4 in oocytes leads to deoxyglucose uptake that can be stimulated about two- to threefold by insulin. When mRNA encoding the human insulin receptor was coinjected with transporter mRNA, the oocyte responded to nanomolar concentrations of insulin with an increase in uptake equivalent to that measured using the endogenous IGF-I receptor. Moreover, the change in uptake was attributed to a redistribution of hexose carriers, as determined by Western immunoblotting of a "plasma membrane" fraction. Several other groups have been unable to reproduce the effect of insulin on either 2-deoxyglucose or 3-*O*-MG uptake in oocytes injected with GLUT1 or GLUT4 mRNA (Gould and Lienhard, 1989; A. Garcia de Herreros, and M. J. Birnbaum, unpublished experiments). It is possible that these differences are a result of variability in the state of the oocytes used for these studies.

One can use the K_m values of 20 and 1.8 mM for GLUT1 and GLUT4, respectively, to arrive at a calculated increase in transport based on the values for surface transporter number obtained with ATB-BMPA (Holman *et al.*, 1990). At a 3-*O*-MG concentration of 50 μM, GLUT1 and GLUT4 would catalyze uptake at 0.25 and 2.7% of V_{max}, respectively. Thus, if the turnover numbers of the two transporters were equal, 96 and 99% of the uptake into adipocytes would be catalyzed by GLUT4 in basal and insulin-stimulated cells, respectively, based on the numbers of cell surface transporters cited above. In other words, the contribution of GLUT1 to the measured rate of hexose flux would be negligible, and uptake would be directly proportional to the number of GLUT4 transporters on the cell surface. If the measurement of surface GLUT4 were overestimated by 2-fold, then the increase in plasma membrane GLUT4 would account for a 40-fold increase in hexose uptake in rat adipocytes. If basal plasma membrane GLUT4 content were considerably lower than determined by the ATB-BMPA photolabel, then one would predict a decrease in apparent K_m concomitant with insulin treatment. As noted above, such a change has been reported by some investigators but remains controversial (Martz *et al.*, 1986; Toyoda *et al.*, 1987; Whitesell *et al.*, 1989).

The situation in 3T3-L1 adipocytes is more complex, due to the greater number of GLUT1 transporters (Calderhead *et al.*, 1990b). Again, assuming equal turnover numbers, 70% of basal uptake of 50 μM 3-O-MG would be catalyzed by GLUT4, and 86% of insulin-stimulated flux. This would result in a net 13-fold increase in transport. The reduction in apparent K_m for 3-O-MG in insulin-treated versus control murine adipocytes predicted by this model has been observed (Clancy *et al.*, 1991). Unlike the situation with rat adipocytes, if the turnover number for GLUT4 were greater than GLUT1, the calculated stimulation of transport would be even more than cited above. Thus, although a number of uncertainties remain, it is quite possible that virtually the entire increase in glucose transport in adipocytes in response to insulin may be explained by translocation of GLUT1 and GLUT4. Alternative mechanisms, which invoke the presence of another transporter or a change in the intrinsic activity of one the carriers accompanying insulin treatment, have not been ruled out. In fact, there is some experimental support for the latter possibility. Czech and collaborators have reported that treatment of 3T3-L1 adipocytes with inhibitors of protein synthesis slowly stimulates hexose uptake without a concomitant increase in total cellular or plasma membrane glucose transporter content (Clancy *et al.*, 1991). They interpret these data as indicating that a rapidly turning over protein binds to a glucose transporter in the plasma membrane, suppressing its activity. Moreover, they suggest that part of the stimulation of uptake induced by insulin might be the result of relief of such a chronic inhibition. It should be noted that other investigators have failed to observe similar effects of cycloheximide in isolated rat adipocytes (Jones and Cushman, 1989).

III. Intracellular Trafficking of the "Insulin-Sensitive" Glucose Transporter

The data reviewed in the previous section support a model in which, in the basal state, there is a significant intracellular pool of both GLUT1 and GLUT4 glucose transporter. Two important questions suggested by such a model concern (1) whether GLUT1 and GLUT4 reside in the same intracellular organelle, and (2) the nature of the compartment(s) in which the transporters exist in unstimulated insulin-responsive cells. Strategies that have been used for approaching the former question include the purification of transporter-containing vesicles by either biochemical or immunological methods, immunohistochemistry, and the expression of each of the transporters in similar cellular contexts by gene transfer of the respective cDNAs. The earliest attempts at purification of glucose transporter-containing vesicles were performed prior to the discovery of GLUT4 and therefore were concerned with enrichment of a GLUT1-containing fraction. Using either immunoadsorption to antibodies directed against GLUT1 or traditional biochemical techniques it was possible to partially purify a vesicle population enriched in GLUT1 from 3T3-L1 adipocytes or rat fat cells (Biber and Lienhard, 1986; James et al., 1987). Zorzano et al. (1989) have utilized 1F8, a monoclonal antibody directed against GLUT4, to immunopurify transporter-containing vesicles from the low-density microsome fraction of homogenates of isolated rat adipocytes. Vesicles adsorbed to 1F8 contained over 90% of GLUT4 but less than 5% of the cellular GLUT1. The IGF-II/mannose 6-phosphate receptor, which redistributes about twofold to the plasma membrane in response to insulin, remained predominantly in the nonadsorbed fraction (Oka and Czech, 1984; Wardzala et al., 1984). The idea that the IGF-II/mannose 6-phosphate receptor is located in a compartment separable from the bulk of intracellular glucose transporter is consistent with a study in which the recruitment of IGF-II/mannose 6-phosphate receptor and transporter were shown to differ kinetically (Appell et al., 1988). The immunopurified GLUT4-containing vesicles were composed of a collection of proteins distinct from total low-density microsomes, and GLUT4 represented at least 10% of the protein in the vesicle. The authors concluded the following: (1) GLUT1 and GLUT4 are located in distinct cellular compartments, and (2) GLUT4 probably resides in a distinct, differentiated organelle (Zorzano et al., 1989).

There have been two analogous studies in which GLUT4-containing vesicles were immunopurified from 3T3-L1 adipocytes. Adsorption by anti-peptide antisera directed against the carboxyl termini of either GLUT1 or GLUT4 removed almost all of the isoforms from the "low-speed" supernatant (Calderhead et al., 1990b). For example, adsorption of vesicles to fixed Staphylococcus aureus coated with α-GLUT4 precipitated essentially all the GLUT4 and about 85% of the GLUT1 from basal murine adipocytes. However, the starting material for the immunoadsorption was a 16,000 g supernatant containing 50% of both transporters, leaving open the possibility that a significant portion of the GLUT4-containing vesicles were excluded from the purification. The vesicles adsorbed were heterogeneous in size, averaging 75 nm in diameter. Fractionation of homogenized 3T3-L1 adipocytes by centrifugation through density sucrose gradients produced cosedimentation of GLUT1 and GLUT4, consistent with the idea that they reside in the same intracellular compartment.

An independent series of experiments performed by Piper *et al.* (1991) examined the immunoadsorption of glucose transporter vesicles from 3T3-L1 adipocytes to isoform-specific anti-peptide antisera linked to acrylamide beads. Adsorption of the GLUT4-containing vesicles from a partially purified preparation of low-density microsomes removed 90% of the GLUT4 but only 30% of the immunoreactive GLUT1 from solution. In addition, the distribution of vesicles containing each of the transporters was partially overlapping but distinct on equilibrium sucrose density centrifugation (Piper *et al.*, 1991). Thus, these data suggest that GLUT1 and GLUT4 are in different intracellular locations in murine adipocytes.

There are a number of possible explanations to resolve these seemingly contradictory results. One plausible idea is that there is partial overlap between the distribution of transporters, such that there exists a population of vesicles rich in GLUT1 but containing a few GLUT4 molecules, and vice versa. In that case, small differences in the efficiency of immunoadsorption might have minor effects on the precipitation of the isoform to which the antiserum is directed, but lead to profound reductions in adsorption of the other transporter species. In addition, the much higher ratio of GLUT1 to GLUT4 in 3T3-L1 adipocytes compared to rat fat cells might alter the apparent colocalization. Another possibility is that the degree of colocalization is dependent on the fragmentation during cell homogenization of a single organelle. A continuous endosomal reticulum with distinct, connected domains has been postulated; the degree of intermixing of proteins in such an organelle during cell disruption might be very dependent on the cell type (Hopkins *et al.*, 1990). Last, the differences might relate to the fact that 3T3-L1 adipocytes are an imperfect model for fat cells *in vivo*. This might lead to imprecise sorting, or the presence of a significant portion of the transporter in a biosynthetic compartment prior to the sorting step. Analogous inefficiencies in targeting of secretory products in cultured cell lines compared to authentic endocrine organs have been well documented (Burgess and Kelly, 1987).

An alternative strategy for assessing the colocalization of glucose transporter isoforms is morphological studies. At the light microscope level, immunofluorescent detection of GLUT4 in differentiated 3T3-L1 adipocytes revealed a perinuclear, punctate pattern, as well as some fine granules distributed throughout the cytoplasm (Garcia de Herreros and Birnbaum, 1989b). In the same cells, GLUT1 is located on the plasma membrane as well as in an intracellular, juxtanuclear distribution that at first glance resembles GLUT4 (Piper *et al.*, 1991). However, double-labeling experiments clearly revealed that GLUT1 and GLUT4 were localized in distinctly different compartments in the same region of the cell (Piper *et al.*, 1991). Immunoelectron microscopical studies of colocalization have been hampered by the lack of a suitable anti-GLUT1 antiserum; thus, there has not been a study evaluating the distribution of GLUT1 and GLUT4 in the same cryosections. The first ultrastructural study of glucose transporter was performed in 3T3-L1 cells prior to the cloning of GLUT4, therefore using an affinity-purified antiserum obtained from rabbits immunized with the human erythrocyte glucose transporter (Blok *et al.*, 1988). In retrospect, it is not clear whether this antiserum recognized exclusively GLUT1, or also had some reactivity toward GLUT4. Nonetheless, the intracellular distribution of immunoreactivity was predominantly in tubulovesicular structures, characterized as trans-Golgi network (TGN), as well as in the compartment of uncoupling of receptors and ligands (CURL). On exposure to insulin, the labeling of TGN/CURL decreased, while that in the plasma membrane increased about threefold (Blok *et al.*, 1988). In an important and elegant series of experiments, Slot *et al.* (1991) studied the distribution of transporter in brown adipose tissue by immunoelectron microscopy of ultrathin cryosections using an anti-peptide antiserum specific for GLUT4. Rats were fasted overnight and then placed at 4°C to deplete brown fat of triglyceride. In addition to the α-GLUT4 antibody, antisera against rat albumin and cathepsin D were used to identify early and late endosomes, respectively. In the basal state, GLUT4 was located exclusively intracellularly, predominantly in tubulovesicular structures both in the region of the plasma membrane and the trans-Golgi (Slot *et al.*, 1991). Exposure to insulin effected a major increase in cell surface immunolabeling for GLUT4, which was quantitated as being about 40-fold; this is consistent with the increase in hexose uptake in brown fat and the change in plasma membrane GLUT4 measured in white adipose tissue by photolabeling with ATB-BMPA. Concomitant with the increase in plasma membrane GLUT4, there was a decrease in the transporter in essentially all intracellular sites except early endosomes, in which GLUT4 increased significantly. The latter observation provides strong evidence that insulin effects the redistribution of GLUT4 by stimulating movement to the cell surface, as opposed to inhibiting internalization. Moreover, it suggests that GLUT4 continuously recycles between the plasma membrane and the cell interior in the sustained presence of insulin. Consistent with this idea, Slot *et al.* (1991) observed in brown adipose cells from insulin-treated animals early endosomes with tubular extensions, in which GLUT4 appeared to be enriched. Albumin, a marker for fluid phase endocytosis, was excluded from these tubular structures. This pattern of distribution is reminiscent of the recycling of plasma membrane receptors, which concentrate in tubular processes attached to early endosomes in the course of returning to the cell surface.

Another strategy to approach the problem of whether GLUT1 and GLUT4 are targeted to distinct cellular compartments is to express the two transporters in identical cellular contexts, and define their subcellular distributions. There have been a number of studies in which GLUT1 was overexpressed by DNA-mediated gene transfer in tissue culture cells (Asano et al., 1989; Gould et al., 1989; Harrison et al., 1990a,b). In these reports, transfection of the GLUT1 cDNA invariably resulted in an increase in basal 2-deoxyglucose uptake proportional to or, at high levels of expression, somewhat less than the increase in total GLUT1 protein. In 3T3-L1 adipocytes, overexpression of GLUT1 resulted in an increase in basal hexose uptake but did not alter the increment in flux in response to insulin (Harrison et al., 1990b). This is consistent with the model discussed above, in which GLUT4 is intrinsically more active at low substrate concentrations and represents the major species translocated in response to insulin. Expression of rat GLUT4 in fibroblastic tissue culture cells by conventional DNA or retrovirus-mediated gene transfer leads to results markedly different than GLUT1 (Hudson et al., 1991). Except at exceptionally high levels, expression of GLUT4 did not change basal or insulin-stimulated 2-deoxyglucose uptake. This is true even when GLUT4 is expressed in fibroblasts at a level 3 time greater than the endogenous GLUT1, or in cell lines also expressing greater than 10^6 insulin receptors. The obvious implication is that adipose cell-specific factors in addition to GLUT4 are required for the full stimulation in glucose transport in response to insulin. The lack of alteration in basal hexose flux in transfected cells was clarified by indirect immunofluorescence of infected cells. GLUT4 was found to have an intracellular, perinuclear punctate distribution, resembling that visualized in 3T3-L1 adipocytes (Garcia de Herreros and Birnbaum, 1989b; Hudson et al., 1991; Piper et al., 1991). In contrast, GLUT1 was present predominantly on the plasma membrane, and to a lesser extent in a juxtanuclear distribution. These data unequivocally demonstrate that the GLUT1 and GLUT4 proteins have within their primary sequences information that dictates targeting to distinct intracellular locales. A more detailed study of fibroblasts transfected with GLUT4 showed that the transporter was localized predominantly in early and late endosomes (Hudson et al., 1991). The latter observation suggests, but does not prove, that GLUT4 is inserted into the plasma membranes in fibroblasts, but is rapidly internalized.

How can one resolve these data regarding the intracellular distributions of the GLUT1 and GLUT4 glucose transporter isoforms? In general, the models fall within two classes (Fig. 4). In the first, GLUT4 resides in a mature storage organelle, whose presence is a characteristic feature of

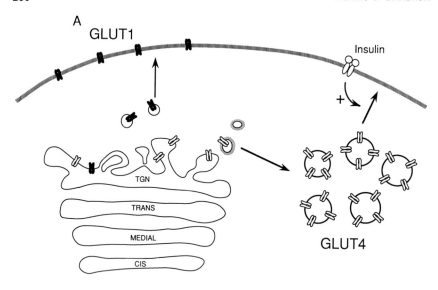

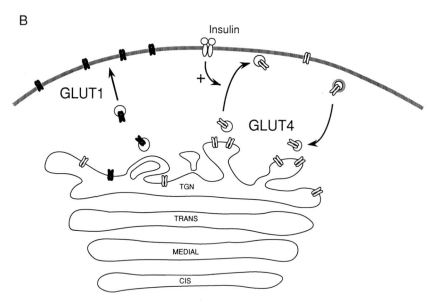

FIG. 4   Alternative models for the cellular trafficking of glucose transporters in rat adipocytes. GLUT1 is represented by closed symbols and GLUT4 by open symbols. (A) The "secretory" model, in which GLUT4 is sorted into a distinct storage organelle. (B) The "receptor" model, in which insulin regulates the rate of GLUT4 recycling though a ubiquitous organelle. See text for details.

insulin-responsive organs such as adipose tissue and muscle. By analogy to regulated secretion, this shall be referred to as the "secretory" model. Thus, one might imagine that GLUT4 is actively sorted at the trans-Golgi network into the equivalent of the regulated pathway, whereas GLUT1 follows a default route to the cell surface (Griffiths and Simons, 1986; Burgess and Kelly, 1987). This would account for the differences in relative distribution of the two isoforms between plasma membrane and intracellular compartments. The secretory model does not preclude a scenario in which synthesis of the insulin-responsive compartment proceeds directly from endocytosed proteins. As an example of such a process, the small synaptic vesicles (SSVs) of neurons are thought to be first synthesized by a regulated secretory pathway parallel to that of the large, dense core vesicles (LDCVs) (Cutler and Cramer, 1990; De Camilli and Jahn, 1990). Nonetheless, after depolarization-induced release of neurotransmitter, the integral membrane proteins of SSVs are sorted out of early endosomes and used to reconstruct synaptic vesicles, which exist as a mature organelle awaiting another exocytotic event. As is apparently the case with insulin-responsive vesicles, SSVs are capable of undergoing multiple rounds of exocytosis and resynthesis. In studies involving the immunadsorption of GLUT4 or its localization by immunoelectron microscopy, GLUT4 did not appear to reside in vesicles of uniform size resembling classic neuronal SSV (Calderhead et al., 1990b; De Camilli and Jahn, 1990; Slot et al., 1991). While this is evidence against a secretory model, it does not exclude one. De Camilli and co-workers have localized SSV markers in endocrine cells to tubulovesicular structures very much resembling the GLUT4-containing compartment of brown adipose tissue (Navone et al., 1986; Slot et al., 1991). Data suggest that this "neuronal-like" structure does indeed represent an authentic secretory compartment (Reetz et al., 1991). Thus, a precedent exists for a pleiomorphic tubulovesicular organelle that obeys the rules of regulated secretion.

Several additional recent experiments are consistent with a secretory model. Baldini et al. (1991) have studied the translocation of GLUT4 in α-toxin-permeabilized rat adipose cells. In such cells, nonhydrolyzable analogs of GTP mimicked insulin in inducing the translocation of GLUT4 from low-density microsomes to the plasma membrane. Small, Ras-related G proteins have been implicated a many vesicular fusion events (Balch, 1990; Hall, 1990). In particular, GTPγS stimulates secretion of regulatory pathway products in neuroendocrine cells (Knight and Baker, 1985; Barrowman et al., 1986; Bittner et al., 1986; Cockcroft et al., 1987; Vallar et al., 1987). Nonetheless, there are numerous sites in the insulin signal transduction pathway at which guanine nucleotides might influence translocation, and thus the experiments of Baldini et al. do not prove the existence of a small G protein associated with the GLUT4-containing vesicle. For example, nonhydrolyzable GTP analogs stimulate the production of diacylglycerol, which can activate glucose transport via protein kinase C (Farese *et al.*, 1985; Strålfors, 1988; Bader *et al.*, 1989) (see Section VI,A below), and GTPγS inhibits the fusion of endosomes, which might increase the abundance of GLUT4 in the plasma membrane by slowing its internalization (Chavrier *et al.*, 1990; Gorvel *et al.*, 1991). Nonetheless, the simplest and most likely explanation is that a small G protein is associated with the insulin-responsive compartment and is required for fusion to the plasma membrane. If true, this would provide a potential site of regulation that could be modulated by phosphorylation or dephosphorylation. Phosphorylation of the endosomal small G protein Rab4 by $p34^{cdc2}$ has been postulated to cause the inhibition of vesicular fusion during mitosis (Bailly *et al.*, 1991). Neuronal cells possess a unique small G protein, Rab3, which is localized on the cytoplasmic surface of synaptic vesicles, and dissociates during exocytosis (Darchen *et al.*, 1990; Fischer von Mollard *et al.*, 1990, 1991). It will be interesting to determine whether an analogous protein is expressed in insulin-responsive tissues.

The major alternative model of intracellular trafficking of GLUT4 is based on the recycling of plasma membrane receptors, and will therefore be referred to as the "receptor" model (Fig. 4). According to this hypothesis, GLUT4 constitutively recycles through the endocytotic pathway, in the basal state residing predominantly in an intracellular location. The important feature is that at all times GLUT4 is located in nonspecialized cellular compartments, i.e., either the plasma membrane or components of the endosomal system. Insulin would alter the distribution by increasing the rate constant determining extraction of GLUT4 from the endosome to the cell surface. This hypothesis engenders specific predictions about the mechanism of insulin-stimulated transport. For example, according to the receptor model, insulin must initiate a chain of events that leads to an alteration in GLUT4 protein, either by covalent modification or allosteric interaction with another protein or regulatory molecule, instructing it to exit the endosomes and move to the plasma membrane. The secretory model requires no such alteration in GLUT4; once the transporter has been correctly sorted to its resident vesicle, insulin has only to effect the fusion of that organelle with the cell surface. Another prediction different for each of the models relates to the nature of the adipose cell factors lacking in fibroblasts and required for translocation. In the secretory model, the most likely deficiency in nondifferentiated cells is the information to construct an insulin-responsive organelle; thus, it should be possible to demonstrate differences between the location of GLUT4 in transfected fibroblasts and adipocytes. According to the receptor model, transporter might well reside in precisely the same compartment in adipocytes and fibroblasts, and the latter lack an essential component of the signal transduction pathway.

IV. Glucose Transport in Muscle

Thus far, the discussion has primarily concerned the uptake of glucose in adipocytes. The preponderance of biochemical data concerning the mechanism by which insulin activates hexose transport has been obtained through the study of fat cells. This is a result of the existence of two exquisitely hormone-sensitive isolated cell systems, rat adipocytes and 3T3-L1 cells (Rodbell, 1964; Green and Kehinde, 1973). Nonetheless, numerous studies have established that muscle is responsible for the majority of glucose disposal *in vivo,* particularly in the adsorptive state. Daniel *et al.* (1975) found that within 1 min following injection of a 4-mmol/kg load of glucose, 25% of the sugar had entered muscle cells. More recently, investigators have made more sophisticated efforts at quantitating glucose disposal. For example, DeFronzo *et al.* (1981) have utilized the euglycemic, hyperinsulinemic clamp technique combined with calorimetry to assess the contributions of storage and oxidation to intravenously administered glucose. At a serum insulin level of 100 μU/ml, about 85% of infused glucose was taken up by skeletal muscle, and most was converted to glycogen. Even at elevated insulin levels, glucose transport is the rate-limiting step for hexose metabolism in muscle, except possibly during exceptionally severe hyperglycemia (Goodman *et al.,* 1983; Yki-Järvinen *et al.,* 1987; Katz *et al.,* 1988). An informative way to approach *in vivo* glucose uptake is to separate it into noninsulin-mediated and insulin-mediated components; these can be defined functionally as glucose disposal during somatostatin-induced insulinopenia and hyperinsulinemia, respectively (Baron *et al.,* 1988). In normal hyperinsulinemic humans, 75% of glucose uptake was into muscle during euglycemia and 95% during hyperglycemia. Under basal or insulinopenic conditions, the central nervous system consumed most of the glucose. Surprisingly, even in the absence of significant circulating insulin, virtually all of the increase in glucose disposal associated with hyperglycemia was accounted for by muscle. In other words, the central nervous and other constitutive users of glucose are saturated at basal conditions, and muscle is responsible for any increment in glucose usage (Baron *et al.,* 1988). Thus, although glucose flux in adipocytes is more approachable experimentally, any comprehensive model of physiologically relevant hormone-regulated glucose transport must address muscle.

R. Levine was among the first to suggest that the major action of insulin was to increase the permeability of the muslce plasma membrane to glucose (Levine et al., 1949). Levine suggested that "insulin acts upon the cell membranes of certain tissues (skeletal muscle, etc.) in such a way that the transfer of hexoses . . . from the extracellular fluid into the cell is facilitated"(Levine et al., 1949; Park et al., 1955; Park and Johnson, 1955; Gey, 1956). That the determinant of permeability in skeletal muscle was a "mobile carrier" was established by experiments demonstrating saturability, sensitivity to inhibitors (e.g., phloretin), stereospecificity, and countertransport, i.e., the shifting outward of nonmetabolizable sugar in the presence of increasing extracellular levels of glucose (Morgan et al., 1964; Park et al., 1968). As described above for adipocytes, most, but not all, investigators have found that the predominant effect of insulin on the kinetics of transport is to increase the V_{max} with little or no change in the K_m (Table I) (Narahara and Özand, 1963; Chaudry and Gould, 1969; Gottesman et al., 1982; Shanahan, 1984; Young et al., 1986; Ploug et al., 1987; Sternlicht et al., 1988). The study of insulin-regulated glucose uptake in muscle is complicated by heterogeneity among muscle fiber types. In general, the magnitude of the maximal response to insulin is greatest in slow-twitch oxidative muscle (e.g., rat soleus), less so in fast-twitch oxidative–glycolytic (rat red gastrocnemius), and least in fast-twitch glycolytic (rat white gastrocnemius) (Ariano et al., 1973; James et al., 1985a; Ploug et al., 1987). As expected, muscles in the rat that are composed of mixed fiber type, such as extensor digitorum longus, are of intermediate responsiveness to insulin (Ariano et al., 1973; James et al., 1985a; Shoji, 1988). In addition, the responsiveness of most muscles groups appears to decrease with advancing age in the rat (Goodman et al., 1983).

In skeletal muscle, glucose transport is also increased by electrically stimulated contractions in vitro or exercise in vivo (Park et al., 1968). This was initially demonstrated in the frog sartorius muscle by Holloszy and Narahara (1965), who reported that the increase was mediated exclusively by an increase in the V_{max} with no change in K_m for 3-O-MG. The stimulation of transport by exercise varies among different muscle groups, but roughly parallels that due to insulin (James et al., 1985b). Stimulation of transport by work in muscle does not require the continued presence of insulin (Ploug et al., 1984; Wallberg-Henriksson and Holloszy, 1984; Nesher et al., 1985). In fact, most investigators have found that the effects on skeletal muscle of brief exposure to a maximal concentration of insulin and exercise are additive (Nesher et al., 1985; Idström et al., 1986; Ploug et al., 1987; Constable et al., 1988). Nonetheless, there is abundant evidence in favor of an interaction between the effects of insulin and exercise. For example, exercise increases the sensitivity of skeletal muscle to stimulation by insulin, even when the muscle is excised and incubated in vitro for several hours before exposure to the hormone (Young *et al.*, 1987; Cartee *et al.*, 1989; Gulve *et al.*, 1990).

The discovery of redistribution of glucose transporters in isolated rat adipocytes was inevitably followed by attempts to demonstrate an analogous phenomenon in muscle. However, subcellular fractionation of skeletal muscle is fraught with much greater difficulties than fat cells and quantitation suffers from the same lack of an independent marker for the intracellular compartment containing glucose transporters. Cytochalasin B photolabels a protein of apparent M_r ~45,000 in rat skeletal muscle (Klip *et al.*, 1983). Wardzala and Jeanrenaud (1981, 1983) studied fractionated insulin-treated rat diaphragm and found a twofold increase in glucose-inhibitable cytochalasin B binding in plasma membrane and a concomitant decrease in transporters in low-density microsomes. Klip and colleagues have devoted a significant effort toward developing a subcellular fractionation protocol effective for skeletal muscle (Burdett *et al.*, 1987; Klip *et al.*, 1987). They were able to achieve a moderate enrichment but poor recovery of plasma membrane markers. Under conditions in which insulin stimulated glucose uptake in perfused hindquarter fivefold, transporter increased in the plasma membrane fraction about twofold (Klip *et al.*, 198 /). Again, this was accompanied by a decrease in transporter sedimenting at a higher density thought to include low-density microsomes. Grimditch *et al.* (1985), using an alternative purification scheme, reported that insulin increased the V_{max} for glucose uptake in sarcolemmal vesicles from gastrocnemius-plantaris and quadriceps muscle 3.2-fold, but the transporter number as determined by cytochalasin B binding rose less than twofold. This was interpreted as indicating a combination of recruitment and activation of individual plasma membrane carriers, but could also be explained by contamination of the basal plasma membrane fraction with leaky low-density microsomes vesicles (Sternlicht *et al.*, 1988).

Exercise-induced glucose uptake has been associated with a twofold increase in cytochalasin B binding in a plasma membrane fraction, or a 3.2-fold greater V_{max} for glucose transport in sarcolemmal vesicles (Hirshman *et al.*, 1988; Sternlicht *et al.*, 1989). Douen *et al.* (1989), using Klip's fractionation method, found that exercise or insulin stimulated glucose uptake into perfused hindlimb two- and fivefold, respectively, and increased cytochalasin B binding sites in the plasma membrane fraction twofold, although only insulin produced a concomitant decrease in intracellular transporter. Fushiki *et al.* (1989), using both Sternlicht's and Klip's protocol for preparations of subcellular fractions, found that exercise led to a measurable increase in plasma membrane and a reciprocal decrease in intracellular cytochalasin B binding in membranes isolated by either procedure. Thus, the weight of evidence favors a model in which recruitment of glucose transporters to the plasma membrane is important to the activation of transport in skeletal muscle by contraction or insulin. The cloning of cDNAs encoding the GLUT4 glucose transporter led to the observation that this isoform was expressed at high levels in cardiac and skeletal muscle as well as in white and brown adipose tissue (Birnbaum, 1989; Charron *et al.*, 1989; Fukumoto *et al.*, 1989; James *et al.* 1989b; Kaestner *et al.*, 1989). Moreover, James *et al.* (1989b) reported greater levels of GLUT4 protein in membranes from red (oxidative) than white (glycolytic) skeletal muscle. In a later study, GLUT4 protein content was quantitated in several rat muscles of varying fiber type, and this compared to the rates of 2-deoxyglucose uptake in the basal state and following exposure to insulin and/or exercise (Henriksen *et al.*, 1990). The muscles studied were epitrochlearis (15% type I, slow-twitch oxidative; 20% type IIa, fast-twitch oxidative–glycolytic; 65% type IIb, fast-twitch glycolytic), soleus (84–16–0%), extensor digitorum longus (EDL, 3–57–40%), and flexor digitorum brevis (FDB, 7–92–1%) (Ariano *et al.*, 1973; Henriksen *et al.*, 1990). Insulin-stimulated hexose uptake was greatest in soleus (about sevenfold) followed by, in order, FDB, EDL, and epitrochlearis. Contraction was most potent as a stimulus for hexose uptake in FDB, where it stimulated the highest rate of uptake, followed by EDL, soleus, and epitrochlearis. Contractile activity and insulin treatment were additive on deoxyglucose uptake, the order of maximal transport rates being FDB > soleus > EDL > epitrochlearis. Most striking in these experiments was the remarkable correlation between total cellular GLUT4 protein and the capacity of the muscle for maximal transport ($r = 0.992$). Kern *et al.* (1990) also reported a correlation between insulin responsiveness and GLUT4 levels, with maximal hexose transport, and GLUT4 mRNA and protein, all two- to fivefold greater in red than white muscle. This suggests that GLUT4 catalyzes both insulin- and exercise-stimulated hexose uptake and is consistent with the idea that GLUT4, as opposed to some auxiliary protein(s) involved in translocation, is a primary determinant of the capacity for transport. Further evidence for this is derived from experiments correlating transport activity after exercise training with the levels of GLUT4. Ten weeks of endurance swimming training in rats increased the effects of contraction in red gastrocnemius muscle and insulin in soleus on 3-O-MG transport by about one-third (Ploug *et al.*, 1990). Exercise training also increased the level of both GLUT1 and GLUT4 mRNA twofold and the encoded proteins about 30% in red gastrocnemius. In an independent series of experiments, rats exposed to volitional exercise on a wheel cage for 6 weeks had a higher concentration of GLUT4 protein in plantaris muscle, and in soleus an increase in GLUT4 proportional to total mass of the muscle (Rodnick *et al.*, 1990). Training resulted in higher levels of GLUT4 mRNA in white quadriceps, plantaris, and red quadriceps (Wake *et al.*, 1991). Again, these data are consistent with the stimulatory effects of training on glucose flux into skeletal muscle (James *et al.*, 1984, 1985c; Kraegen *et al.*, 1989). Nonetheless, it remains unproven whether GLUT4 plays a primary role in determining maximal rates of glucose transport. It is also possible, for example, that training coordinately regulates a set of muscle genes, including GLUT4, and other proteins which may influence the activation of hexose flux. However, the observation that exercise training also augments insulin-stimulated glucose transport in adipocytes, and accomplishes this via an expansion of the intracellular pool of carriers (almost certainly representing GLUT4), favors a pivotal role for GLUT4 in establishing maximal transport rates (Vinten and Galbo, 1984; Vinten *et al.*, 1985; Hirshman *et al.*, 1989).

Over the last several years, the translocation of glucose transporters in muscle has been studied using isoform-specific antisera. Hirshman *et al.* (1990) have reported that, under conditions in which insulin stimulated glucose uptake in perfused hindlimb greater than threefold, the number of GLUT4 transporters in the plasma membrane fraction increased 3.3-fold, accompanied by a concomitant decrease in GLUT4 in the intracellular pool. Interestingly, the ratio of basal intracellular to plasma membrane GLUT4 was 11, whereas the ratio of glucose-inhibitable cytochalasin B binding sites was 1.6 (Hirshman *et al.*, 1990). This is reminiscent of the disparity between relative transporter number in subcellular fractions of adipocytes as measued by reactivity with α-GLUT1 versus cytochalasin B binding (see above) and suggests that a significant portion of the basal cell surface transporters in skeletal muscle is not GLUT4. The data of Douen *et al.* (1990a,b) are consistent with GLUT1 playing a role as the "basal" glucose transporter. They found that both insulin injection *in vivo* and treadmill exercise increased the GLUT4 in the plasma membrane fraction; only exposure to insulin decreased the GLUT4 in the intracellular fraction, consistent with their previous studies measuring total transporter by cytochalasin B binding (Douen *et al.*, 1989, 1990b). GLUT1 was detected virtually exclusively in the plasma membrane fraction and did not change on injection of insulin (Douen *et al.*, 1990b). When exercise was followed by insulin, glucose uptake was additive, but the abundance of GLUT4 in the plasma membrane was not greater than that due to insulin or exercise alone, about 2.5 fold (Douen *et al.*, 1990b). Moreover, intracellular transporters did not decrease any more than that with insulin alone.

GLUT4 has been localized in human skeletal muscle by immunoelectron microscopy of ultrathin cryosections (Friedman *et al.*, 1991). The transporter was not visualized on the plasma membrane, but was on transverse tubules and terminal cisternae of the triad, as well as other ill defined intracellular sites. The presence of GLUT4 on transverse tubules and not on the sarcolemma was also demonstrated by fractionation of rat muscle membranes, confirming a previous report of cytochalasin B binding

(Burdett *et al.*, 1987). There was a modest increase in transverse tubule labeling in one subject after insulin injection; however, the ultrastructure in this study was not clear enough to allow a distinction between transverse tubule and intracellular components of the triad (Friedman *et al.*, 1991). Slot *et al.* (1990) have studied the ultrastructure of rat muscle and the distribution of GLUT4. In nonstimulated cardiac and skeletal muscle, GLUT4 was localized predominantly to tubular and vesicular structures in the region of the Golgi and scattered throughout the cytoplasm. In heart, there was also GLUT4 at the Z-line level, often near the transverse tubule.

Thus, the weight of evidence favors the idea that both exercise and insulin increase glucose transport in skeletal muscle by enhancing the redistribution of GLUT4 to the plasma membrane. It is likely that many of the contradictory results and internally inconsistent data are a function of the difficulty inherent in the purification of subcellular fractions from muscle, in which the recovery of marker enzymes is typically less than 10% (Hirshman *et al.*, 1990). The question of whether exercise and insulin recruit different intracellular pools has yet to be resolved (Douen *et al.*, 1990a,b). Alternatively, insulin and contraction may regulate distinct steps, such as exocytosis and internalization, controllng the steady state distribution of the same pool of hexose carriers. Of note, hypoxia also stimulates redistribution of glucose transporters in skeletal muscle, and is additive to the action of insulin but not contractile activity (Cartee *et al.*, 1991).

Like skeletal muscle, cardiac muscle displays increased glucose transport in response to insulin (Morgan *et al.*, 1961). This has been reported to be exclusively an effect on V_{max} with little change in apparent K_m although, as for adipose tissue, this finding is controversial (Cheung *et al.*, 1978; Zaninetti *et al.*, 1988). Activation of transport is associated with a redistribution of transporter from a subcellular fraction thought to represent internal membranes to a fraction enriched in plasma membrane markers. This has been demonstrated in heart using reconstitution, cytochalasin B binding, and immunoreactivity with an antiserum specific for GLUT1 (Watanabe *et al.*, 1984; Wheeler, 1988; Zaninetti *et al.*, 1988). There is, to this author's knowledge, no equivalent study using α-GLUT4 antisera, although cytochalasin B detects a larger pool of basal intracellular transporters than can be accounted for by GLUT1 (Wheeler, 1988). It has been estimated that cardiac muscle contains about fourfold more total GLUT4 protein than GLUT1 (Calderhead *et al.*, 1990b).

Continuous cell lines derived from muscle or tumors have been advocated as model systems in which to study insulin-stimulated glucose transport (Klip *et al.*, 1984). The BC3H-1, C2, and C8 cells do not contain detectable GLUT4 and, in BC$_3$H-1 cells, most of the 2.5-fold stimulation in transport can be accounted for by the recruitment of GLUT1 protein

(Calderhead *et al.*, 1990a). This has led to the questioning of these cell lines as useful, physiologically relevant models (Calderhead *et al.*, 1990a). L_6 myoblasts express low levels of GLUT4, which increase during differentiation; the ratio of GLUT4 to GLUT1 protein has not been quantitated (Mitsumoto *et al.*, 1991). Translocation of glucose transporter, as assayed by cytochalasin B binding to subcellular fractions, correlates well with the 1.5- to 2-fold stimulation of transport by insulin in differentiated L_6 cells (Ramlal *et al.*, 1988). It has been reported that chronic, 24-hr treatment of L_6 muscle cells with insulin increases the fraction of GLUT4 in the plasma membrane (Koivisto *et al.*, 1991).

V. Long-Term Regulation of Insulin Responsiveness

All forms of diabetes mellitus involve insulin resistance as a part of the disease. In type II, or non-insulin-dependent diabetes mellitus (NIDDM), it is a prominent and possibly causal component. Over the last 15 years, there have been numerous studies evaluating the role of glucose transporters in the abnormal action of insulin. The data pertaining to animal models of diabetes and other altered metabolic states is considered briefly, followed by a discussion of the role of glucose transport proteins in human diabetes.

A. Animal Models

One of the oldest and most frequently used models for diabetes mellitus is animals tested with streptozotocin, a relatively specific poison for β cells of the pancreas (Junod *et al.*, 1967). Injection of the drug produces a syndrome resembling type I, insulin-dependent diabetes mellitus in that it is an insulin-deficient state. On the other hand, animals treated with streptozotocin do not develop ketoacidosis, and, if they survive the period of hypoglycemia 8–10 hr after the drug is administered, can live for substantial periods of time (Junod *et al.*, 1967). It has been long recognized that rats made diabetic with streptozotocin display decreased responsiveness to the action of insulin (Kasuga *et al.*, 1978; Kobayashi and Olefsky, 1978, 1979; Wieringa, and Krans, 1978). Two parameters that have been frequently measured are the dose–response curve for insulin and the maximal rate of hormone-regulated transport. This article conforms to the convention of referring to the changes in the half-maximal value for insulin action as variations in insulin ''sensitivity,'' and alterations in the stimulation of glucose transport by maximum concentrations of hormone as insulin ''responsiveness.'' Thus, streptozotocin diabetes in rats does not change or increases slightly the insulin sensitivity of fat cells, while substantially reducing the responsiveness (Kasuga et al., 1978; Kobayashi and Olefsky, 1978, 1979). The earliest studies found that diabetes reduced the V_{max} for 2-deoxyglucose uptake in both basal and acutely insulin-treated adipocytes, with no alteration in apparent K_m (Kobayashi and Olefsky, 1978). More recent studies have shown a greater stimulation by acute insulin and little or no effect of streptozotocin treatment on basal rates of transport (Karnieli et al., 1981a; Garvey et al., 1989).

Karnieli et al. (1981a) were the first to propose a mechanism for the aberrant insulin responsiveness in adipocytes from streptozotocin-treated rats: a selective depletion of intracellular glucose transporters. Basal transporter number as assayed by cytochalasin B binding was unchanged in plasma membranes but decreased 45% in basal low-density microsomes. This resulted in a greatly diminished number of transporters available to move to the plasma membrane fraction after exposure of the cells to insulin. Interestingly, when the diabetic rats were treated with insulin, the rate of hexose transport in response to a maximal concentration of insulin transiently surpassed that in untreated adipocytes, reaching about threefold normal levels at 7–8 days after the initiation of insulin therapy (Kahn and Cushman, 1987). The augmented responsiveness could not be accounted for by the repletion of intracellular hexose carriers, leading the authors to suggest that insulin deficiency followed by replacement leads to a change in the intrinsic activity of one of the fat cell transporters.

Investigators have experienced somewhat more variable results when studying hexose transport in muscle from streptozotocin diabetic animals, although most data would favor an impairment of both basal and insulin-stimulated flux. A number of early communications reported little or no decrement in maximum insulin-stimulated hexose transport in diabetic muscle (Le Marchand-Brustel and Freychet, 1979; Chiasson et al., 1984; Dall'Aglio et al., 1985). Subsequently, Maegawa et al. (1986) showed that muscle required a longer period of diabetes than adipose tissue to develop insulin resistance, so the interval between injection of streptozotocin and sacrifice of the animal is critical. In rats treated with 50 mg/kg streptozotocin, insulin-stimulatable hexose transport was unimpaired at 1 week, or decreased 36 and 60% at 2 and 4 weeks after injection, respectively. Thus, studies in which streptozotocin was injected less than 1 week prior to measurement of muscle glucose uptake are difficult to interpret in the context of chronic insulin deficiency (Wallberg-Henriksson et al., 1987). Barnard et al. (1990) prepared ''sarcolemmal'' vesicles from hindlimb muscles 10 weeks after streptozotocin treatment of rats, and found significant decreases in basal and insulin-stimulated transporter, as assayed either by hexose flux or cytochalasin B binding. Intracellular transporters are also depleted from diabetic skeletal muscle (Bourey *et al.*, 1990). It is likely that streptozotocin-treated animals develop an impairment in basal and insulin-stimulated transport in cardiac muscle as well (Almira *et al.*, 1986; Eckel *et al.*, 1990).

Shortly after the cloning of the GLUT4 cDNA, a number of groups reported experiments investigating the levels of glucose transporter isoforms and their mRNAs during streptozotocin-induced diabetes. All workers agreed that the insulin-deficient state produced a marked suppression of GLUT4 mRNA in adipocytes, with little or no change in GLUT1 (Garvey *et al.*, 1989; Kahn *et al.*, 1989; Sivitz *et al.*, 1989). GLUT4 mRNA was also decreased in cardiac and skeletal muscle, although to a less dramatic degree (Garvey *et al.*, 1989; Bourey *et al.*, 1990; Eckel *et al.*, 1990). There was a depletion of GLUT4 protein equal to or greater than the change in mRNA that was apparent in both the basal low-density microsomes and the plasma membranes from acutely insulin-treated cells (Berger *et al.*, 1989; Garvey *et al.*, 1989; Kahn *et al.*, 1989). In muscle, however, the plasma membrane fraction appeared to be responsible for most of the loss in total cellular GLUT4 transporters (Klip *et al.*, 1990). These data led to the obvious conclusion that the insulin resistance of streptozotocin-induced insulin deficiency in rodents, and possibly some diabetes in humans, was caused by a selective loss of intracellular GLUT4 protein, itself a result of abnormal expression of the *GLUT4* gene. Inconsistent with this simple view, however, are the experiments of Kahn *et al.* (1991b), in which after 1 week of diabetes, there was no change in either GLUT1 or GLUT4 protein in spite of a 50% impairment in *in vivo* insulin-stimulated glucose uptake. If the immunodetectable transporter in these experiments does indeed reflect the total cellular content in muscle, these data could be interpreted as indicating a defect either in the catalytic activity or the efficiency of translocation of the GLUT4 transporter in muscle.

Several interesting points are relevant to this issue. Garvey *et al.* (1989) gave daily injections of insulin for the second week of 2 weeks of streptozotocin-induced diabetes in rats, and found commensurate increases in adipocyte GLUT4 mRNA, GLUT4 protein located in basal low-density microsomes and maximal transport rates to about two-fold that measured in fat cells from normal, untreated animals. Several other studies are qualitatively in agreement with this observation (Berger *et al.*, 1989; Sivitz *et al.*, 1989; Bourey *et al.*, 1990). Furthermore, Sivitz *et al.* (1990) has reported that the insulin-mimetic agent vanadate corrects the insulin resistance of experimental diabetes by restoring GLUT4 mRNA levels. On the other hand, Kahn and Flier (1990) have argued that the repletion of GLUT4 protein is insufficient to account for the augmented response of transport to insulin exposure *in vitro*, and invoke a change in intrinsic activity of the transporter. A similar conclusion derived from experiments aimed at resolving whether the hypoinsulinemia or the hyperglycemia of experimental diabetes was the signal for abnormal GLUT4 gene expression. Sivitz *et al.* (1990) found that correction of the elevated serum glucose level by poisoning renal tubular reabsorption of glucose with phlorizin did not restore GLUT4 levels in fat cells unless insulin was also injected. Kahn and co-workers (1991a) obtained similar results after inducing diabetes by partial pancreatectomy, but also measured parameters of insulin responsiveness in the phlorizin-treated animals. They found that correction of hyperglycemia restored insulin-stimulated 3-*O*-MG transport in adipocytes and glucose disposal *in vivo*, but GLUT4 protein in basal low-density microsomes remained low. Again, they suggest that glucose levels are modulating the catalytic activity of the transporter independent of total carrier number.

Another approach to understanding the effects of insulin on glucose transport *in vivo* has been to study animals during fasting, which like diabetes is characterized by hypoinsulinemia, but differs in that blood glucose levels are low normal. It has been long recognized that starvation leads to *in vivo* insulin resistance, i.e., a decreased rate of glucose disposal at an equivalent concentration of circulating insulin (Olefsky, 1976; Ruderman *et al.*, 1977; Newman and Brodows, 1983; Pénicaud *et al.*, 1985). In adipocytes from fasted rats, there was a blunted hexose transport response to insulin accompanied by a decrease in basal intracellular transporters, as measured by cytochalasin B binding (Kahn *et al.*, 1988). This is accounted for by a depletion of GLUT4 mRNA and protein in adipose tissue (Berger *et al.*, 1989; Sivitz *et al.*, 1989; Charron and Kahn, 1990). On the other hand, adaptation to starvation in muscle is a complicated process. During a fast, insulin was more efficacious in stimulating hexose transport into both perfused hindquarter and muscle explanted from animals that showed significant insulin resistance *in vivo* (Ruderman *et al.*, 1977; Le Marchand-Brustel and Freychet, 1979; Brady *et al.*, 1981; Stirewalt *et al.*, 1985; Issad *et al.*, 1987b). Some of the former effect is caused by an increase in the number of insulin receptors, but there is also a more efficient activation of transport by comparable levels of insulin receptor occupancy. Consistent with the *in vitro* data but contrary to what is measured *in vivo*, GLUT4 protein is increased in muscle, and GLUT4 mRNA is either increased or unchanged, depending on the study (Charron *et al.*, 1989; Bourey *et al.*, 1990). Two conclusions can be drawn from these experiments: (1) under conditions of starvation, GLUT4 is regulated in fat cells in a manner opposite to that in skeletal muscle, and (2) the *in vivo* response of muscle to insulin is contrary to what is found in *in vitro* systems and what would be predicted on the basis of GLUT4 levels. The simplest interpretation is that during fasting there exists a circulating factor whose effect as an antagonist of insulin-stimulated glucose transport is dominant *in vivo*. Treatment of normal rats with daily injections of insulin for 1 week has been reported to increase GLUT4 mRNA and protein in soleus, whereas in a different study continuous infusion of insulin for 4 days decreased GLUT4 mRNA and protein in tibialis muscle and GLUT4 mRNA in diaphragm (Bourey *et al.*, 1990; Cusin *et al.*, 1990). In the latter series of experiments, depletion of GLUT4 was associated with a decrease in basal and insulin-stimulated glucose utilization (Cusin *et al.*, 1990). Under the same conditions, white adipose tissue had a higher level of GLUT4 and a heightened responsiveness to acute challenge with insulin. Thus, at least one study found that pharmacologically elevated insulin with or without hypoglycemia initiated changes in GLUT4 opposite to that seen during fasting (Cusin *et al.*, 1990).

Another example of reciprocal changes in insulin-responsiveness in muscle and adipose tissue is the Zucker "fatty" rat. This animal model of hereditary obesity results from an autosomal recessive mutation leading to excessive adipose deposition even if hyperphagia is eliminated by pair-feeding with lean controls (Bray, 1977). Rats with the *fa/fa* genotype are visibly bigger than their heterozygous lean littermates by 4 weeks, although their fat pads are larger earlier. The Zucker rat meets the classical criterion for insulin resistance: normal or near normal glycemia coincident with significant hyperinsulinemia (Zucker and Antoniades, 1972). There are numerous studies demonstrating abnormal insulin-stimulated glucose transport and disposal in muscles of all fiber types in fatty rats, and several reports of normal GLUT4 mRNA and protein (Kemmer *et al.*, 1979; Crettaz *et al.*, 1980; Sherman *et al.*, 1988; Ivy *et al.*, 1989; Friedman *et al.*, 1990; Hainault *et al.*, 1991; Yamamoto *et al.*, 1991). Nonetheless, Zucker rats are capable of increasing their muscle GLUT4 and insulin responsiveness after exercise training (Friedman *et al.*, 1990). The augmentation of hexose uptake in heart in response to either insulin or increased pressure load is blunted in the Zucker rat; this is due to a loss of cytochalasin B binding sites in both plasma membranes and intracellular membranes, although GLUT4 has not been measured directly (Zaninetti *et al.*, 1983, 1989; Eckel *et al.*, 1985). Interestingly, however, there was an increased responsiveness of glucose transport to insulin in adipose tissue from young Zucker rats (Czech *et al.*, 1978; Pénicaud *et al.*, 1987; Krief *et al.*, 1988). This was associated with an increase in GLUT4 protein, predominantly in the low-density microsome fraction, accompanied by an elevation in the level of GLUT4 mRNA (Guerre-Millo *et al.*, 1985; Ezaki, 1989; Hainault *et al.*, 1991). Thus, the overall constellation of insulin responsiveness is opposite to that in fasting, and somewhat similar to hyperinsulinemia

(Cusin *et al.*, 1990). Recently, however, it was shown that the changes in GLUT4 in adipocytes of Zucker rats precede the elevation in serum insulin, making the latter an unlikely candidate for the primary causative event (Hainault *et al.*, 1991).

There are several other animal models of insulin resistance for which the role of alteration in glucose transport has been evaluated. *ob* and *db* are autosomal recessive mutations at different loci in mice that produce essentially the same phenotype. In the C57BL/6J inbred strain, both mutants lead to marked obesity, hyperphagia, transient hyperglycemia, and chronically elevated serum insulin (Coleman, 1978). On the C57BL/KsJ background, the disease is much more severe, leading ultimately to a fall in insulin levels with attendant marked diabetes mellitus. Skeletal muscle from animals homozygous for either mutation display reduced glucose transport activity in the presence of a maximal concentration of insulin (Cuendet *et al.*, 1976; Chan and Dehaye, 1981; Chan and Tatoyan, 1984). However, GLUT4 protein is not reduced in cardiac or skeletal muscle, at least in *db/db* mice (Koranyi *et al.*, 1990). Large adipocytes from aged, obese rats have increased basal 3-*O*-MG but a severely blunted maximal response to insulin associated with a decrease in adipocyte GLUT4 protein (Hissin *et al.*, 1982a; Ezaki *et al.*, 1990). Ablation of the ventromedial hypothalamus, either surgically or by parenteral administration of gold thioglucose, results in hyperphagia and insulin resistance (Debons *et al.*, 1977; Morrison, 1977). Like the Zucker rat, this is characterized by a transient increase in responsiveness in adipose tissue, but insulin resistance in muscle (Pénicaud *et al.*, 1989). Four to 6 months after treatment of mice with gold thioglucose, there was a decrease of insulin-stimulatable glucose transport into muscle, accompanied by a depletion of GLUT4 protein in brown and white fat, but no change in muscle (Tanti *et al.*, 1989; Le Marchand-Brustel *et al.*, 1990). Interestingly, the thermogenic agent BRL 26830A, which is an adrenergic agonist specific for the β_3 subtype of receptor, corrects GLUT4 levels in both types of adipocyte.

It has been long recognized that a diet high in fat and low in carbohydrate causes insulin resistance and a depletion of basal low-density microsome glucose transporter in rat adipocytes, as measured by cytochalasin B binding (Hissin *et al.*, 1982b). A special time of physiological high-fat diet is during the suckling period. At the suckling–weaning transition in rats, there was a marked increase in insulin responsiveness, in terms of the inhibition of hepatic glucose output as well as the stimulation of peripheral glucose utilization (Issad *et al.*, 1987a, 1988). The response of glucose transport to insulin was blunted in the suckling rat, both *in vivo* and *in vitro* in isolated adipocytes (Issad *et al.*, 1989). When rats were weaned to a high carbohydrate diet, there was a significant increase in the content of GLUT4 mRNA and protein in both skeletal muscle and white adipose tissue (Leturque *et al.*, 1991). Insulin-stimulated translocation of GLUT4 proteins occurred in adipocytes from both suckling and weaned rodents, but in the former a smaller pool of basal intracellular transporters contributed to the insulin response. Weaning on a high-fat diet prevented the increase in insulin responsiveness and GLUT4 in both muscle and fat. The induction of GLUT4 mRNA in response to carbohydrate was remarkably rapid, with a threefold increase 12 hr after feeding. It is not known whether insulin plays a role in this regulation, although the level of circulating hormone clearly rose on weaning to a high carbohydrate diet (Leturque *et al.*, 1991).

B. Diabetes Mellitus

All forms of diabetes mellitus in humans are associated with some degree of peripheral insulin resistance. In type I diabetes, it is unclear whether the insensitivity to the actions of insulin on glucose transport is a direct result of low circulating insulin or a response to hyperglycemia (Yki-Järvinen *et al.*, 1990). As a consequence of numerous studies implicating glucose utilization as a primary causal event in type II, or non-insulin-dependent diabetes (NIDDM), the role of glucose transport in this form of the disease has received considerable attention (Hansen and Bodkin, 1986; Lillioja *et al.*, 1987b; Garvey, 1988). In patients with type II diabetes, as well as in many obese subjects, there is a decrease in the maximal stimulation of *in vivo* glucose disposal, as measured by the euglycemic glucose-clamp technique (Kolterman *et al.*, 1980, 1981). The decrease correlates well with a defect in the ability of all concentrations of insulin to augment 3-*O*-MG uptake in adipocytes prepared from obese, insulin-resistant subjects or patients with type II diabetes (Ciaraldi *et al.*, 1981, 1982; Kashiwagi *et al.*, 1983). Similarly, in an *in vitro* preparation from rectus abdominus muscle, the normal twofold increase in 3-*O*-MG transport was blunted or absent in muscle strips from obese subjects or patients with NIDDM, respectively (Dohm *et al.*, 1988). Garvey *et al.* (1988) found that the basal low-density microsome fraction of adipocytes from insulin-resistant humans was depleted of transporter, as assayed by the glucose-inhibitable binding of cytochalasin B. Thus, as in experimental streptozotocin-induced diabetes in rodents, a brief exposure of isolated fat cells to insulin resulted in a smaller increment in plasma membrane transporter number than in fat cells from normal subjects. In a more recent study, Garvey and collaborators (1991) reported a loss of GLUT4 protein primarily from intracellular fractions of subcutaneous adipocytes from patients with obesity or NIDDM; in type II diabetes, the depletion of GLUT4 was greater and appeared to be from the plasma membranes as well. In all insulin-resistant populations, the depletion of GLUT4 protein was correlated with a similarly loss of GLUT4 mRNA, which was reduced in diabetic patients to 14% of that measured in lean controls (Garvey *et al.*, 1991). Similar results have been reported by Sinha *et al.* (1991). The situation in muscle, which, as noted above, is responsible for most of the insulin-stimulated glucose disposal *in vivo*, is much less clear. There have been two reports of a lack of alteration in GLUT4 mRNA and protein in biopsies from vastus lateralis skeletal muscle from obese subjects or patients with NIDDM (Handberg *et al.*, 1990; Pedersen *et al.*, 1990). This is consistent with the apparent absence of linkage of the GLUT4 gene to type II diabetes mellitus in blacks (Matsutani *et al.*, 1990). On the other hand, Dohm *et al.* (1991) have reported modest decreases in GLUT4 protein in biopsies from rectus abdominus and vastus lateralis from patients with morbid obesity and NIDDM. Thus, the current evidence suggests that obesity-related insulin resistance in human muscle is accompanied by no depletion of GLUT4 or a loss that appears small in relation to the decrement in maximal glucose transport. This is reminiscent of several conditions in rats described above, in which adipose cell and muscle GLUT4 are regulated independently and divergently. Nonetheless, care should be taken in interpreting the early results of muscle biopsies. Lillioja *et al.*, (1987a) have shown that fiber type and capillary density are highly correlated with insulin resistance in humans. Thus, sampling error could obscure significant changes and there is the theoretical possibility that the most resistant muscles are inaccessible to biopsy. Nonetheless, it seems most likely that the etiology of insulin-resistant glucose transport in NIDDM is unrelated to changes in total GLUT4 mRNA or protein.

VI. Regulators of Insulin-Responsive Glucose Transport

A. Protein Kinase C

A topic of enormous controversy has been the role of protein kinase C in the actions of insulin in general and on glucose transport in particular. Within the context of this article, the data are briefly summarized without attempting either to be comprehensive or to arrive at resolution of seemingly disparate results. Moreover, the discussion is limited to experiments performed in classical insulin target tissues or reasonable *in vitro* models thereof. One fact on which virtually all investigators can agree is that activators of protein kinase C, such as phorbol 12-myristate 13-acetate (PMA), are capable of increasing glucose transport. PMA stimulated glucose transport in rat adipocytes, 3T3-L1 adipocytes, and rat skeletal muscle to approximately 30, 10, and 10–50%, respectively, of that produced by a maximal concentration of insulin (Kirsch *et al.*, 1985; Sowell *et al.*, 1988; Henriksen *et al.*, 1989; Tanti *et al.*, 1989; Ishizuka *et al.*, 1990; Gibbs *et al.*, 1991). Typically, the activation by PMA was slower than that of insulin (Kirsch *et al.*, 1985; Saltis *et al.*, 1991). In adipocytes, PMA has been reported to either lower or leave unchanged the apparent K_m for hexose transport, although in both studies the predominant effect was to increase the V_{max} (Martz *et al.*, 1986; Suzuki, 1988). The addition of PMA to maximal concentrations of insulin is either nonadditive or frankly inhibitory (Kirsch *et al.*, 1985; Henriksen *et al.*, 1989; Tanti *et al.*, 1989). This experiment is complicated by a direct negative effect of protein kinase C on the insulin receptor (Jacobs *et al.*, 1983; Bollag *et al.*, 1986). The issue of whether insulin physiologically activates protein kinase C is more confusing. Farese and colleagues (1985) have been the primary, although certainly not exclusive, proponents of the idea that insulin significantly activates protein kinase C. They studied the incorporation of [^{3}H]glycerol into lipids of rat adipocytes, soleus muscle, and diaphragm, and found an increased accumulation of [^{3}H]diacylglycerol, the physiological activator of protein kinase C, after treatment with insulin (Ishizuka *et al.*, 1990; Hoffman *et al.*, 1991). Egan *et al.* (1990) reported an increase in protein kinase C activity and translocation to the plasma membrane, an event usually associated with activation of the enzyme, after exposure of rat adipocytes to insulin. Insulin also stimulated redistribution of protein kinase C from the soluble to the particulate fraction of soleus muscle (Ishizuka *et al.*, 1990). In contrast, Henriksen *et al.*, (1989) reported that insulin was without effect on the subcellular distribution of protein kinase C in epitrochlearis muscle under conditions in which PMA and phospholipase C clearly induced translocation of the kinase. Interestingly, these authors suggested that protein kinase C may be a mediator of the augmented glucose uptake in response to contractile activity in skeletal muscle. This is supported by independent experiments in which repetitive, electrically induced contractions stimulated the translocation of protein kinase C in skeletal muscle (Cleland *et al.*, 1989). Nonetheless, even in these experiments the activation of the kinase alone appeared insufficient to increase transport.

Blackshear and collaborators have devoted significant energy to debunking protein kinase C as a mediator of insulin action. Their strategies have been twofold: showing the persistence of the effects of insulin after "downregulation" of protein kinase C by chronic exposure of cells to PMA, and demonstrating the lack of insulin-stimulatable phosphorylation of the myristoylated, alanine-rich C kinase substrate (MARCKS) protein (Blackshear *et al.*, 1985, 1991; Stumpo and Blackshear, 1986). To confuse matters more, Strålfors (1988) has suggested that diacylglycerol can increase glucose transport in rat adipocytes in the absence of changes in the activity of protein kinase C. This has been challenged based on the observation of redistribution of immunoreactive C kinase to the particulate fraction after treatment of cells with diolein (Ishizuka *et al.*, 1990).

Perhaps the clearest studies have been those aimed at determining the mechanism of PMA-stimulated glucose transport in terms of the redistribution of hexose carriers. Initially, Mühlbacher *et al.* (1988) reported that PMA was equivalent to insulin in stimulating translocation of transporter, but less efficacious as an agonist of glucose flux. They suggested a two-step model in which newly arrived plasma membrane transporters required activation. The assay for glucose transporter in these experiments was cytochalasin B binding, so the model required modification after several groups used isoform-specific antisera and subcellular fractionation of rat adipocytes to address the problem. Saltis *et al.* (1991) found that both insulin and PMA increased GLUT1 in the plasma membranes about 1.5-fold, but the latter effected only ~35% of the six fold stimulation by insulin of GLUT4 translocation. Vogt et al. (1990, 1991) reported no effect of PMA on GLUT1 but a 2.5-fold stimulation of the redistribution of GLUT4. In both studies, the PMA-dependent change in transport roughly correlated with the translocation of GLUT4 to the cell surface. The quantitative aspects of this issue have been addressed with much more clarity by use of the impermeant glucose transporter photolabel, ATB-BMPA. Holman *et al.* (1990) found that in rat adipocytes, PMA stimulated approximately four- and five fold increases in cell surface GLUT4 and GLUT1, respectively, concomitant with an approximate threefold increase in 3-*O*-MG uptake. Thus if, as discussed above, GLUT4 catalyzes most of the 3-*O*-MG uptake in this assay, there is no need to invoke anything other than translocation to explain the augmentation in hexose flux in response to activation of protein kinase C. A similar mechanism appears operative in 3T3-L1 adipocytes: PMA stimulates a 1.7- to 2.5-fold increase in hexose transport concomitant with 2.5- and 1.7-fold increases in cell surface GLUT1 and GLUT4, respectively (Gibbs *et al.*, 1991). In any case, it is difficult to imagine how protein kinase C could be the sole mediator of the actions of insulin, given the disparity in translocation of glucose transporters induced by PMA and insulin.

B. Cyclic AMP and Phosphorylation

A topic that has received some attention due to its obvious physiological relevance is the regulation of insulin-stimulated glucose transport by catecholamines. Several years ago, it was reported that isoproterenol, in the presence of adenosine deaminase, inhibited the stimulation of glucose transport by insulin; moreover, the inhibition was preserved in isolated plasma membranes, in which there was a ~50% decrease in V_{max} for hexose transport and no alteration in the number of total transporters (Joost et al., 1986). This was interpreted as indicating a decrease in the intrinsic catalytic activity of the plasma membrane transporters (Joost et al., 1987). There is some evidence that the inhibitory effect of β-adrenergic agents is not mediated by increases in cyclic AMP, but is related to direct interactions of G proteins with the transporter or an associated regulatory protein (Kuroda et al., 1987). Schürmann et al. (1989) have reported that GTPγS inhibits hexose uptake into liposomes reconstituted with glucose transporter isolated from either plasma membranes or low-density microsomes prepared from insulin-treated but not basal adipocytes. In contrast to these observations, James et al. (1989a) have argued that the inhibitory effects of β-adrenergic agents on glucose transport in rat adipocytes are a result of a cyclic AMP-dependent phosphorylation of GLUT4. In metabolically labeled rat adipocytes, isoproterenol stimulated the incorporation of ^{32}P into GLUT4 about twofold in both plasma membranes and low-density microsomes. This was mimicked by analogs of cyclic AMP, and occurred in broken cells in the presence of the catalytic subunit of cyclic AMP-dependent protein kinase and $[\gamma^{32}P]ATP$ (James et al., 1989a). Catecholamines did not alter the subcellular distribution of GLUT4. The amino acid phosphorylated in response to an elevation in cyclic AMP has been mapped to Ser^{488}, located at the carboxyl terminus of the protein (Lawrence et al., 1990a). Clancy and Czech (1990) have reported a stimulation of glucose transport in 3T3-L1 adipocytes exposed to cholera toxin or dibutyryl cyclic AMP for 4 to 24 hr; this was accompanied by an increase in total cellular GLUT1 and plasma membrane GLUT4. However, the increase in cell surface transporter did not appear sufficient to explain the higher rate of 2-deoxyglucose uptake.

A role for phosphorylation in regulating glucose transport activity in insulin-responsive cells has been suggested by the stimulation of 2-deoxyglucose uptake in rat adipocytes by okadaic acid, an inhibitor of protein phosphatase-1 and -2A, (Haystead et al., 1989). In a careful study, Lawrence et al. (1990b) showed that okadaic acid was about 50% as potent as insulin in stimulating hexose uptake in adipocytes, and inhibited slightly the stimulation by a maximal concentration of insulin when the two agents were present concurrently. Moreover, both the stimulatory and inhibitory effects of the phosphatase inhibitor correlated reasonably well with the amount of GLUT4 protein in the plasma membrane as determined by subcellular fractionation of adipocytes. Okadaic acid increased the phosphorylation of GLUT4 in the low-density microsome fraction by about 60%, whereas the $^{32}P_i$ incorporated into plasma membrane transporter was increased three- to fourfold. The overall stoichiometry of phosphorylation changed from ~0.2 mol phosphate/mol GLUT4 to 0.6–0.8 mol/mol, all at the carboxyl terminus of the protein. Both phosphatase 1 and 2A removed phosphate from GLUT4 *in vitro* (Lawrence *et al.*, 1990b). Corvera *et al.* (1991) obtained qualitatively similar results concerning the stimulation of hexose transport in rat adipocytes by okadaic acid. However, they found a poor correlation between the inhibition of insulin-stimulated transport and plasma membrane transporters, leading to the suggestion that okadaic acid exerts its primary effect on a step other than translocation. Okadaic acid is also insulin mimetic in soleus muscle (Tanti *et al.*, 1991). Interestingly, both basal and okadaic acid-stimulated hexose uptake are inhibited proportionally in muscle from animals made obese by gold thioglucose injection.

Thus, there is abundant evidence that the GLUT4 is a phosphoprotein, although there is little direct information on the effects of phosphorylation on catalytic activity or intracellular trafficking. Investigators have failed to detect a significant insulin-stimulated change in phosphorylation of total cellular GLUT4 protein in either rat adipocytes or 3T3-L1 adipocytes (James *et al.*, 1989a; A. Garcia de Hererros and M. J. Birnbaum, unpublished observations). Lawrence *et al.* (1990b) have argued for a lower stoichiometry of phosphorylation of GLUT4 protein in basal low-density microsomes than in plasma membrane, although this is a difficult measurement due of the scarcity of cell surface transporters in nonstimulated cells.

C. Other Regulators

Given the likely importance of GLUT4 in determining peripheral insulin responsiveness, the long-term regulation of GLUT4 expression has been studied in some detail. In addition to its modulation by the alterations in metabolic state discussed above, hyperthyroidism has been reported to increase basal deoxyglucose uptake and GLUT4 protein levels in skeletal muscle about twofold (Casla *et al.*, 1990). In cultured murine and freshly isolated rat adipocytes, surprisingly few agents have been identified that alter the expression of GLUT4. In particular, insulin does not appear to change total GLUT4 mRNA or protein in 3T3-L1 adipocytes under conditions in which GLUT1 gene transcription is clearly activated (Garcia de Herreros and Birnbaum, 1989a; Tordjman *et al.*, 1989). This observation suggests that either the absence of insulin is not the signal acting directly on adipocytes in streptozotocin diabetic rodents, or 3T3-L1 cells are not an adequate model for the study of *in vivo* GLUT4 gene expression. Similarly, addition of the sulfonylurea tolbutamide or growth hormone to cultured murine adipocytes increased or decreased, respectively, hexose uptake with congruent changes in the levels of GLUT1 mRNA and protein but with no alteration in GLUT4 (Tordjman *et al.*, 1989; Tai *et al.*, 1990).

The stimulation of 2-deoxyglucose uptake during long-term glucose star-vation of 3T3-L1 adipocytes was associated with increases in GLUT1 mRNA and protein and GLUT4 mRNA, but no change in total GLUT4 protein (Tordjman *et al.*, 1990). Perhaps the only stimulus convincingly demonstrated to alter GLUT4 gene expression *in vitro* is an elevation in intracellular cyclic AMP (Kaestner *et al.*, 1991). Kaestner *et al.* have reported that a 16-hr treatment of 3T3-L1 adipocytes with either forskolin or 8-bromo-cyclic AMP resulted in a 70% decrease in GLUT4 mRNA and protein. As reported previously for other cell types, cyclic AMP increases the abundance of GLUT1 mRNA and protein (Hiraki *et al.*, 1989; Ka-estner *et al.*, 1991). These changes correlated well with an augmentation of basal hexose transport and an inhibition of insulin-stimulated flux (Ka-estner *et al.*, 1991). The depletion of mRNA in response to cyclic nucleo-tides was entirely explained by inhibition of gene transcription. The sug-gestion by the authors that increases in intracellular cyclic AMP might be responsible for the loss of GLUT4 in adipose tissue during diabetes and starvation is supported by a study in which the β-adrenergic blocking agent propranolol partially prevented insulin resistance in skeletal muscle from streptozotocin diabetic rats (Bostrom *et al.*, 1989). However, there are numerous other explanation for the effects of β-receptor blockade, includ-ing a reduction in the level of circulating free fatty acids, so that a role for cyclic AMP in the insulin resistance of diabetes remains to be proven directly.

References

Almira, E. C., Garcia, A. R., and Boshell, B. R. (1986). *Am. J. Physiol.* **250,** E402–E406.
Andersson, L., and Lundahl, P. (1988). *J. Biol. Chem.* **263,** 11414–11420.
Appell, K. C., Simpson, I. A., and Cushman, S. W. (1988). *J. Biol. Chem.* **263,** 10824–10829.
Ariano, M. A., Armstrong, R. B., and Edgerton, V. R. (1973). *J. Histochem. Cytochem.* **21,** 51–55.
Asano, T., Shibasaki, Y., Ohno, S., Taira, H., Lin, J. L., Kasuga, M., Kanazawa, Y., Akanuma, Y., Takaku, F., and Oka, Y. (1989). *J. Biol. Chem.* **264,** 3416–3420.
Bader, M. F., Sontag, J. M., Thierse, D., and Aunis, D. (1989). *J. Biol. Chem.* **264,** 16426–16434.
Bailly, E., McCaffrey, M., Touchot, N., Zahraoui, A., Goud, B., and Bornens, M. (1991). *Nature (London)* **350,** 715–718.
Balch, W. E. (1990). *Trends Biochem. Sci.* **15,** 473–477.
Baldini, G., Hohman, R., Charron, M. J., and Lodish, H. F. (1991). *J. Biol. Chem.* **266,** 4037–4040.
Baldwin, S. A., and Henderson, P. J. F. (1989). *Annu. Rev. Physiol.* **51,** 459–471.
Barnard, R. J., Youngren, J. F., Kartel, D. S., and Martin, D. A. (1990). *Endocrinology (Baltimore)* **126,** 1921–1926.
Baron, A. D., Brechtel, G., Wallace, P., and Edelman, S. V. (1988). *Am. J. Physiol.* **255,** E769–E774.

288   MORRIS J. BIRNBAUM

Barrowman, M. M., Cockcroft, S., and Gomperts, B. D. (1986). *Nature (London)* **319**, 504–507.

Bell, G. I., Murray, J. C., Nakamura, Y., Kayano, T., Eddy, R. L., Fan, Y. S., Byers, M. G., and Shows, T. B. (1989). *Diabetes* **38**, 1072–1075.

Bell, G. I., Kayano, T., Buse, J. B., Burant, C. F., Takeda, J., Lin, D., Fukumoto, H., and Seino, S. (1990). *Diabetes Care* **13**, 198–208.

Berger, J., Biswas, C., Vicario, P. P., Strout, H. V., Saperstein, R., and Pilch, P. F. (1989). *Nature (London)* **340**, 70–72.

Biber, J. W., and Lienhard, G. E. (1986). *J. Biol. Chem.* **261**, 16180–16184.

Birnbaum, M. J. (1989). *Cell (Cambridge, Mass.)* **57**, 305–315.

Birnbaum, M. J., Haspel, H. C., and Rosen, O. M. (1986). *Proc. Natl. Acad. Sci. U.S.A.* **83**, 5784–5788.

Bittner, M. A., Holz, R. W., and Neubig, R. R. (1986). *J. Biol. Chem.* **261**, 10182–10188.

Blackshear, P. J., Witters, L. A., Girard, P. R., Kuo, J. F., and Quamo, S. N. (1985). *J. Biol. Chem.* **260**, 13304–13315.

Blackshear, P. J., Haupt, D. M., and Stumpo, D. J. (1991). *J. Biol. Chem.* **266**, 10946–10952.

Blok, J., Gibbs, E. M., Lienhard, G. E., Slot, J. W., and Geuze, H. J. (1988). *J. Cell Biol.* **106**, 69–76.

Bollag, G. E., Roth, R. A., Beaudoin, J., Mochly-Rosen, D., and Koshland, D. E. (1986). *Proc. Natl. Acad. Sci. U.S.A.* **83**, 5822–5824.

Bostrom, M., Nie, Z., Goertz, G., Henriksson, J., and Wallberg, H. H. (1989). *Diabetes* **38**, 906–910.

Bourey, R. E., Koranyi, L., James, D. E., Mueckler, M., and Permutt, M. A. (1990). *J. Clin. Invest.* **86**, 542–547.

Brady, L. J., Goodman, M. N., Kalish, F. N., and Ruderman, N. B. (1981). *Am. J. Physiol.* **240**, E184–E190.

Bray, G. A. (1977). *Fed. Proc., Fed. Am. Soc. Exp. Biol.* **36**, 148–153.

Burdett, E., Beeler, T., and Klip, A. (1987). *Arch. Biochem. Biophys.* **253**, 279–286.

Burgess, T. L., and Kelly, R. B. (1987). *Annu. Rev. Cell Biol.* **3**, 243–293.

Calderhead, D. M., and Lienhard, G. E. (1988). *J. Biol. Chem.* **263**, 12171–12174.

Calderhead, D. M., Kitagawa, K., Lienhard, G. E., and Gould, G. W. (1990a). *Biochem. J.* **269**, 597–601.

Calderhead, D. M., Kitagawa, K., Tanner, L. I., Holman, G. D., and Lienhard, G. E. (1990b). *J. Biol. Chem.* **265**, 13801–13808.

Carruthers, A. (1990). *Physiol. Rev.* **70**, 1135–1176.

Cartee, G. D., Young, D. A., Sleeper, M. D., Zierath, J., Wallberg, H. H., and Holloszy, J. O. (1989). *Am. J. Physiol.* **256**, E494–E499.

Cartee, G. D., Douen, A. G., Ramlal, T., Klip, A., and Holloszy, J. O. (1991). *J. Appl. Physiol.* **70**, 1593–1600.

Carter, S. C., Pessin, J. E., Mora, R., Gitomer, W., and Czech, M. P. (1982). *J. Biol. Chem.* **257**, 5419–5425.

Casla, A., Rovira, A., Wells, J. A., and Dohm, G. L. (1990). *Biochem. Biophys. Res. Commun.* **171**, 182–188.

Chan, T. M., and Dehaye, J. P. (1981). *Diabetes* **30**, 211–218.

Chan, T. M., and Tatoyan, A. (1984). *Biochim. Biophys. Acta* **798**, 325–332.

Charron, M. J., and Kahn, B. B. (1990). *J. Biol. Chem.* **265**, 7994–8000.

Charron, M. J., Brosius, F. C., Alper, S. L., and Lodish, H. F. (1989). *Proc. Natl. Acad. Sci. U.S.A.* **86**, 2535–2539.

Chaudry, I. H., and Gould, M. K. (1969). *Biochim. Biophys. Acta* **177**, 527–536.

Chavrier, P., Parton, R. G., Hauri, H. P., Simons, K., and Zerial, M. (1990). *Cell (Cambridge, Mass.)* **62**, 317–329.

Cheung, J. Y., Conover, C., Regen, D. M., Whitfield, C. F., and Morgan, H. E. (1978). *Am. J. Physiol.* **234**, E70–E78.

Chiasson, J. L., Germain, L., Srivastava, A. K., and Dupuis, P. (1984). *Metab., Clin. Exp.* **33**, 617–621.

Ciaraldi, T. P., Kolterman, O. G., and Olefsky, J. M. (1981). *J. Clin. Invest.* **68**, 875–880.

Ciaraldi, T. P., Kolterman, O. G., Scarlett, J. A., Kao, M., and Olefsky, J. M. (1982). *Diabetes* **31**, 1016–1022.

Clancy, B. M., and Czech, M. P. (1990). *J. Biol. Chem.* **265**, 12434–12443.

Clancy, B. M., Harrison, S. A., Buxton, J. M., and Czech, M. P. (1991). *J. Biol. Chem.* **266**, 10122–10130.

Clark, A. E., and Holman, G. D. (1990). *Biochem. J.* **269**, 615–622.

Cleland, P. J., Appleby, G. J., Rattigan, S., and Clark, M. G. (1989). *J. Biol. Chem.* **264**, 17704–17711.

Cockcroft, S., Howell, T. W., and Gomperts, B. D. (1987). *J. Cell Biol.* **105**, 2745–2750.

Coleman, D. L. (1978). *Diabetologia* **14**, 141–148.

Colman, A. (1984). *In* "Transcription and Translation: A Practical Approach" (B. D. Hames and S. J. Higgins, eds.), pp. 49–70. IRL Press, Oxford.

Constable, S. H., Favier, R. J., Cartee, G. D., Young, D. A., and Holloszy, J. O. (1988). *J. Appl. Physiol.* **64**, 2329–2332.

Corvera, S., Jaspers, S., and Pasceri, M. (1991). *J. Biol. Chem.* **266**, 9271–9275.

Crettaz, M., Prentki, M., Zaninetti, D., and Jeanrenaud, B. (1980). *Biochem. J.* **186**, 525–534.

Crofford, O. B., and Renold, A. E. (1965a). *J. Biol. Chem.* **240**, 14–21.

Crofford, O. B., and Renold, A. E. (1965b). *J. Biol. Chem.* **240**, 3237–3244.

Csary, T. Z., and Wilson, J. E. (1956). *Biochim. Biophys. Acta* **22**, 185–186.

Cuendet, G. S., Loten, E. G., Jeanrenaud, B., and Renold, A. E. (1976). *J. Clin. Invest.* **58**, 1078–1088.

Cushman, S. W., and Wardzala, L. J. (1980). *J. Biol. Chem.* **255**, 4758–4762.

Cusin, I., Terrettaz, J., Rohner, J. F., Zarjevski, N., Assimacopoulos, J. F., and Jeanrenaud, B. (1990). *Endocrinology* (*Baltimore*) **127**, 3246–3248.

Cutler, D. F., and Cramer, L. P. (1990). *J. Cell Biol.* **110**, 721–730.

Czech, M. P. (1976). *J. Biol. Chem.* **251**, 2905–2910.

Czech, M. P., Lynn, D. G., and Lynn, W. S. (1973). *J. Biol. Chem.* **248**, 3636–3641.

Czech, M. P., Lawrence, J. C., and Lynn, W. S. (1974). *J. Biol. Chem.* **249**, 5421–5427.

Czech, M. P., Richardson, D. K., Becker, S. G., Walters, C. G., Gitomer, W., and Heinrich, J. (1978). *Metab., Clin. Exp.* **12**, 1967–1981.

Dall'Aglio, E., Chang, H., Hollenbeck, C. B., Mondon, C. E., Sims, C., and Reaven, G. M. (1985). *Am. J. Physiol.* **249**, E312–E316.

Daniel, P. M., Love, E. R., and Pratt, O. E. (1975). *J. Physiol.* (*London*) **247**, 273–288.

Darchen, F., Zahraoui, A., Hammel, F., Monteils, M. P., Tavitian, A., and Scherman, D. (1990). *Proc. Natl. Acad. Sci. U.S.A.* **87**, 5692–5696.

Debons, A. F., Krimsky, I., Maayan, M. L., Fani, K., and Jimenez, F. A. (1977). *Fed. Proc., Fed. Am. Soc. Exp. Biol.* **36**, 143–147.

De Camilli, P., and Jahn, R. (1990). *Annu. Rev. Physiol.* **52**, 625–645.

De Fronzo, R. A., Jacot, E., Jequier, E., Maeder, E., Wahren, J., and Felber, J. P. (1981). *Diabetes* **30**, 1000–1007.

Denton, R. M., Yorke, R. E., and Randle, P. J. (1966). *Biochem. J.* **100**, 407–419.

Dohm, G. L., Tapscott, E. B., Pories, W. J., Dabbs, D. J., Flickinger, E. G., Meelheim, D., Fushiki, T., Atkinson, S. M., Elton, C. W., and Caro, J. F. (1988). *J. Clin. Invest.* **82**, 486–494.

Dohm, G. L., Elton, C. W., Friedman, J. E., Pilch, P. F., Pories, W. J., Atkinson, S. M. J., and Caro, J. F. (1991). *Am. J. Physiol.* **260**, E459–E463.

Douen, A. G., Ramlal, T., Klip, A., Young, D. A., Cartee, G. D., and Holloszy, J. O. (1989). *Endocrinology (Baltimore)* **124**, 449–454.

Douen, A. G., Ramlal, T., Rastogi, S., Bilan, P. J.. Cartee, G. D., Vranic, M., Holloszy, J. O., and Klip, A. (1990a). *J. Biol. Chem.* **265**, 13427–13430.

Douen, A. G., Ramlal, T., Cartee, G. D., and Klip, A. (1990b). *FEBS Lett.* **261**, 256–260.

Douen, A. G., Burdett, E., Ramlal, T., Rastogi, S., Vranić, M., and Klip, A. (1991). *Endocrinology (Baltimore)* **128**, 611–616.

Eckel, J., Wirdeier, A., Herberg, L., and Reinauer, H. (1985). *Endocrinology (Baltimore)* **116**, 1529–1534.

Eckel, J., Gerlach, E. E., and Reinauer, H. (1990). *Biochem. J.* **272**, 691–696.

Egan, J. J., Saltis, J., Wek, S. A., Simpson, I. A., and Londos, C. (1990). *Proc. Natl. Acad. Sci. U.S.A.* **87**, 1052–1056.

Ezaki, O. (1989). *Diabetologia* **32**, 290–294.

Ezaki, O., and Kono, T. (1982). *J. Biol. Chem.* **257**, 14306–14310.

Ezaki, O., Kasuga, M., Akanuma, Y., Kuniaki, T., Hirano, H., Fujita-Yamaguchi, Y., and Kasahara, M. (1986). *J. Biol. Chem.* **261**, 3295–3305.

Ezaki, O., Fukuda, N., and Itakura, H. (1990). *Diabetes* **39**, 1543–1549.

Farese, R. V., Standaert, M. L., Barnes, D. E., Davis, J. S., and Pollet, R. J. (1985). *Endocrinology (Baltimore)* **116**, 2650–2655.

Fischer von Mollard, G., Mignery, G. A., Baumert, M., Perin, M. S., Hanson, T. J., Burger, P. M., Jahn, R., and Sudhof, T. C. (1990). *Proc. Natl. Acad. Sci. U.S.A.* **87**, 1988–1992.

Fischer von Mollard, G., Sudhof, T. C., and Jahn, R. (1991). *Nature (London)* **349**, 79–81.

Flier, J. S., Mueckler, M., MacCall, A. L., and Lodish, H. F. (1987). *J. Clin. Invest.* **79**, 657–661.

Friedman, J. E., Sherman, W. M., Reed, M. J., Elton, C. W., and Dohm, G. L. (1990). *FEBS Lett.* **268**, 13–16.

Friedman, J. E., Dudek, R. W., Whitehead, D. S., Downes, D. L., Frisell, W. R., Caro, J. F., and Dohm, G. L. (1991). *Diabetes* **40**, 150–154.

Fukumoto, H., Seino, S., Imura, H., Seino, Y., and Bell, G. I. (1988). *Diabetes* **37**, 657–661.

Fukumoto, H., Kayano, T., Buse, J. B., Edwards, Y., Pilch, P. F., Bell, G. I., and Seino, S. (1989). *J. Biol. Chem.* **264**, 7776–7779.

Fushiki, T., Wells, J. A., Tapscott, E. B., and Dohm, G. L. (1989). *Am. J. Physiol.* **256**, E580–E587.

Garcia de Herreros, A., and Birnbaum, M. J. (1989a). *J. Biol. Chem.* **264**, 9885–9890.

Garcia de Herreros, A., and Birnbaum, M. J. (1989b). *J. Biol. Chem.* **264**, 19994–19999.

Garvey, W. T. (1988). *In* "Pathogenesis of Non-insulin Dependent Diabetes Mellitus" (V. Grill and S. Efendic, eds.), pp. 171–200. Raven Press, New York.

Garvey, W. T., Huecksteadt, T. P., Matthaei, S., and Olefsky, J. M. (1988). *J. Clin. Invest.* **81**, 1528–1536.

Garvey, W. T., Huecksteadt, T. P., and Birnbaum, M. J. (1989). *Science* **245**, 60–63.

Garvey, W. T., Maianu, L., Huecksteadt, T. P., Birnbaum, M. J., Molina, J. M., and Ciaraldi, T. P. (1991). *J. Clin. Invest.* **87**, 1072–1081.

Gey, K. F. (1956). *Biochem. J.* **64**, 145–150.

Gibbs, E. M., Lienhard, G. E., and Gould, G. W. (1988). *Biochemistry* **27**, 6681–6685.

Gibbs, E. M., Calderhead, D. M., Holman, G. D., and Gould, G. W. (1991). *Biochem. J.* **275**, 145–150.

Goodman, M. N., Dluz, S. M., McElaney, M. A., Belur, E., and Ruderman, N. B. (1983). *Am. J. Physiol.* **244**, E93–E100.

Gorvel, J. P., Chavrier, P., Zerial, M., and Gruenberg, J. (1991). *Cell (Cambridge, Mass.)* **64**, 915–925.

Gottesman, I., Mandarino, L., Verdonk, C., Rizza, R., and Gerich, J. (1982). *J. Clin. Invest.* **70,** 1310–1314.

Gould, G. W., and Bell, G. I. (1990). *Trends Biochem. Sci.* **15,** 18–23.

Gould, G. W., and Lienhard, G. E. (1989). *Biochemistry* **28,** 9447–9452.

Gould, G. W., Derechin, V., James, D. E., Tordjman, K., Ahern, S., Gibbs, E. M., Lienhard, G. E., and Mueckler, M. (1989). *J. Biol. Chem.* **264,** 2180–2184.

Gould, G. W., Thomas, H. M., Jess, T. J., and Bell, G. I. (1991). *Biochemistry* **30,** 5139–5145.

Green, H., and Kehinde, O. (1973). *Cell (Cambridge, Mass.)* **1,** 113–116.

Griffiths, G., and Simons, K. (1986). *Science* **234,** 438–443.

Grimditch, G. K., Barnhard, R. J., Kaplan, S. A., and Sternlicht, E. (1985). *Am. J. Physiol.* **249,** E398–E408.

Guerre-Millo, M., Lavau, M., Horne, J. S., and Wardzala, L. J. (1985). *J. Biol. Chem.* **260,** 2197–2201.

Gulve, E. A., Cartee, G. D., Zierath, J. R., Corpus, V. M., and Holloszy, J. O. (1990). *Am. J. Physiol.* **259,** E685–E691.

Hainault, I., Guerre, M. M., Guichard, C., and Lavau, M. (1991). *J. Clin. Invest.* **87,** 1127–1131.

Hall, A. (1990). *Science* **249,** 635–640.

Handberg, A., Vaag, A., Damsbo, P., H., B.-N., and Vinten, J. (1990). *Diabetologia* **33,** 625–627.

Hansen, B. C., and Bodkin, N. C. (1986). *Diabetologia* **29,** 713–719.

Harrison, S. A., Buxton, J. M., Helgerson, A. L., MacDonald, R. G., Chlapowski, F. J., Carruthers, A., and Czech, M. P. (1990a). *J. Biol. Chem.* **265,** 5793–5801.

Harrison, S. A., Buxton, J. M., Clancy, B. M., and Czech, M. P. (1990b). *J. Biol. Chem.* **265,** 20106–20116.

Haspel, H. C., Rosenfeld, M. G., and Rosen, O. M. (1988). *J. Biol. Chem.* **263,** 398–403.

Haystead, T. A., Sim, A. T., Carling, D., Honnor, R. C., Tsukitani, Y., Cohen, P., and Hardie, D. G. (1989). *Nature (London)* **337,** 78–81.

Henriksen, E. J., Rodnick, K. J., and Holloszy, J. O. (1989). *J. Biol. Chem.* **264,** 21536–21543.

Henriksen, E. J., Bourey, R. E., Rodnick, K. J., Koranyi, L., Permutt, M. A., and Holloszy, J. O. (1990). *Am. J. Physiol.* **259,** E593–E598.

Hiraki, Y., McMorrow, I. M., and Birnbaum, M. J. (1989). *Mol. Endocrinol.* **3,** 1470–1476.

Hirshman, M. F., Wallberg-Henriksson, H., Wardzala, L. J., Horton, E. D., and Horton, E. S. (1988). *FEBS Lett.* **238,** 235–239.

Hirshman, M. F., Wardzala, L. J., Goodyear, L. J., Fuller, S. P., Horton, E. D., and Horton, E. S. (1989). *Am. J. Physiol.* **257,** E520–E530.

Hirshman, M. F., Goodyear, L. J., Wardzala, L. J., Horton, E. D., and Horton, E. S. (1990). *J. Biol. Chem.* **265,** 987–991.

Hissin, P. J., Foley, J. E., Wardzala, L. J., Karnielli, E., Simpson, I. A., Salans, L. B., and Cushman, S. W. (1982a). *J. Clin. Invest.* **70,** 780–790.

Hissin, P. J., Karnieli, E., Simpson, I. A., Salans, L. B., and Cushman, S. W. (1982b). *Diabetes* **31,** 589–592.

Hoffman, J. M., Ishizuka, T., and Farese, R. V. (1991). *Endocrinology (Baltimore)* **128,** 2937–2948.

Holloszy, J. O., and Narahara, H. T. (1965). *J. Biol. Chem.* **240,** 3493–3500.

Holman, G. D., and Rees, W. D. (1987). *Biochim. Biophys. Acta* **897,** 395–405.

Holman, G. D., Parkar, B. A., and Midgley, P. J. (1986). *Biochim. Biophys. Acta* **855,** 115–126.

Holman, G. D., Kozka, I. J., Clark, A. E., Flower, C. J., Saltis, J., Habberfield, A. D., Simpson, I. A., and Cushman, S. W. (1990). *J. Biol. Chem.* **265,** 18172–18179.

Hopkins, C. R., Gibson, A., Shipman, M., and Miller, K. (1990). *Nature (London)* **346,** 335–339.

Hudson, A. W., Ruiz, M. L., and Birnbaum, M. J. (1992). *J. Cell Biol.* **116,** 785–797.

Idström, J. P., Rennie, M. J., Scherstén, T., and Bylund-Fellenius, A. C. (1986). *Biochem. J.* **233,** 131–137.

Ishizuka, T., Cooper, D. R., Hernandez, H., Buckley, D., Standaert, M., and Farese, R. V. (1990). *Diabetes* **39,** 181–190.

Issad, T., Penicaud, L., Ferré, P., Kande, J., Baudon, M. A., and Girard, J. (1987a). *Biochem. J.* **246,** 241–244.

Issad, T., Coupé, C., Ferré, P., and Girard, J. (1987b). *Am. J. Physiol.* **253,** E142–E148.

Issad, T., Coupé, C., Pastor-Anglada, M., Ferré, P., and Girard, J. (1988). *Biochem. J.* **251,** 685–690.

Issad, T., Ferre, P., Pastor, A. M., Baudon, M. A., and Girard, J. (1989). *Biochem. J.* **264,** 217–222.

Ivy, J. L., Brozinick, J. T. J., Torgan, C. E., and Kastello, G. M. (1989). *J. Appl. Physiol.* **66,** 2635–2641.

Jacobs, S., Sahyoun, N. E., Saltiel, A. R., and Cuatrecasas, P. (1983). *Proc. Natl. Acad. Sci. U.S.A.* **80,** 6211–6213.

James, D. E., Kraegen, E. W., and Chisholm, D. J. (1984). *J. Appl. Physiol.* **56,** 1217–1222.

James, D. E., Jenkins, A. B., and Kraegen, E. W. (1985a). *Am. J. Physiol.* **248,** E567–E574.

James, D. E., Kraegen, E. W., and Chisholm, D. J. (1985b). *Am. J. Physiol.* **248,** E575–E580.

James, D. E., Kraegen, E. W., and Chisholm, D. J. (1985c). *J. Clin. Invest.* **76,** 657–666.

James, D. E., Lederman, L., and Pilch, P. F. (1987). *J. Biol. Chem.* **262,** 11817–11824.

James, D. E., Brown, R., Navarro, J., and Pilch, P. F. (1988). *Nature (London),* **333,** 183–185.

James, D. E., Hiken, J., and Lawrence, J. C. J. (1989a). *Proc. Natl. Acad. Sci. U.S.A.* **86,** 8368–8372.

James, D. E., Strube, M., and Mueckler, M. (1989b). *Nature (London)* **338,** 83–87.

Janicot, M., and Lane, M. D. (1989). *Proc. Natl. Acad. Sci. U.S.A.* **86,** 2642–2646.

Janicot, M., Flores, R. J. R., and Lane, M. D. (1991). *J. Biol. Chem.* **266,** 9382–9391.

Jones, M. N., and Nickson, J. K. (1981). *Biochim. Biophys. Acta* **650,** 1–20.

Jones, T. L., and Cushman, S. W. (1989). *J. Biol. Chem.* **264,** 7874–7877.

Joost, H. G., Weber, T. M., Cushman, S. W., and Simpson, I. A. (1986). *J. Biol. Chem.* **261,** 10033–10036.

Joost, H. G., Weber, T. M., Cushman, S. W., and Simpson, I. A. (1987). *J. Biol. Chem.* **262,** 11261–11267.

Joost, H. G., Weber, T. M., and Cushman, S. W. (1988). *Biochem. J.* **249,** 155–161.

Junod, A., Lambert, A. E., Orci, L., Pictet, R., Gonet, A. E., and Renold, A. E. (1967). *Proc. Soc. Exp. Biol. Med.* **126,** 201–205.

Juranka, P. F., Zastawny, R. L., and Ling, V. L. (1989). *FASEB J.* **3,** 2583–2592.

Kaestner, K. H., Christy, R. J., McLenithan, J. C., Braiterman, L. T., Cornelius, P., Pekala, P. H., and Lane, M. D. (1989). *Proc. Natl. Acad. Sci. U.S.A.* **86,** 3150–3154.

Kaestner, K. H., Christy, R. J., and Lane, M. D. (1990). *Proc. Natl. Acad. Sci. U.S.A.* **87,** 251–255.

Kaestner, K. H., Flores, R. J. R., McLenithan, J. C., Janicot, M., and Lane, M. D. (1991). *Proc. Natl. Acad. Sci. U.S.A.* **88,** 1933–1937.

Kahn, B. B., and Cushman, S. W. (1987). *J. Biol. Chem.* **262,** 5118–5124.

Kahn, B. B., and Flier, J. S. (1990). *Diabetes Care* **13,** 548–564.

Kahn, B. B., Simpson, I. A., and Cushman, S. W. (1988). *J. Clin. Invest.* **82,** 691–699.

Kahn, B. B., Charron, M. J., Lodish, H. F., Cushman, S. W., and Flier, J. S. (1989). *J. Clin. Invest.* **84,** 404–411.

Kahn, B. B., Shulman, G. I., DeFronzo, R. A., Cushman, S. W., and Rossetti, L. (1991a). *J. Clin. Invest.* **87**, 561–570.

Kahn, B. B., Rossetti, L., Lodish, H. F., and Charron, M. J. (1991b). *J. Clin. Invest.* **87**, 2197–2206.

Karnieli, E., Hissin, P. J., Simpson, I. A., Salans, L. B, and Cushman, S. W. (1981a). *J. Clin. Invest.* **68**, 811–814.

Karnieli, E., Zarnowski, M. J., Hissin, P. J., Simpson, I. A., Salans, L. B., and Cushman, S. W. (1981b). *J. Biol. Chem.* **256**, 4772–4777.

Kasanicki, M. A., and Pilch, P. F. (1990). *Diabetes Care* **13**, 219–227.

Kashiwagi, A., Verso, M. A., Andrews, J., Vasquez, B., Reaven, G., and Foley, J. E. (1983). *J. Clin. Invest.* **72**, 1246–1254.

Kasuga, M., Akanuma, Y., Iwamoto, Y., and Kosaka, K. (1978). *Am. J. Physiol.* **235**, E175–E182.

Katz, A., Nyomba, B. L., and Bogardus, C. (1988). *Am. J. Physiol.* **255**, E942–E945.

Keller, K., Strube, M., and Mueckler, M. (1989). *J. Biol. Chem.* **264**, 18884–18889.

Kemmer, F. W., Berger, M., Herberg, L., Gries, F. A., Wirdeier, A., and Becker, K. (1979). *Biochem. J.* **178**, 733–741.

Kern, M., Wells, J. A., Stephens, J. M., Elton, C. W., Friedman, J. E., Tapscott, E. B., Pekala, P. H., and Dohm, G. L. (1990). *Biochem. J.* **270**, 397–400.

Kirsch, D., Obermaier, B., and Haring, H. U. (1985). *Biochem. Biophys. Res. Commun.* **128**, 824–832.

Kletzien, R. F., and Perdue, J. F. (1973). *J. Biol. Chem.* **248**, 711–719.

Klip, A., and Paquet, M. R. (1990). *Diabetes Care* **13**, 228–243.

Klip, A., Walker, D., Ransome, K. J., Schroer, D. W., and Lienhard, G. E. (1983). *Arch. Biochem. Biophys.* **226**, 198–205.

Klip, A., Li, G., and Logan, W. J. (1984). *Am. J. Physiol.* **247**, E291–E296.

Klip, A., Ramlal, T., Young, D. A., and Holloszy, J. O. (1987). *FEBS Lett.* **224**, 224–230.

Klip, A., Ramlal, T., Bilan, P. J., Cartee, G. D., Gulve, E. A., and Holloszy, J. O. (1990). *Biochem. Biophys. Res. Commun.* **172**, 728–736.

Knight, D. E., and Baker, P. F. (1985). *FEBS Lett.* **189**, 345–349.

Kobayashi, M., and Olefsky, J. M. (1978). *J. Clin. Invest.* **62**, 73–81.

Kobayashi, M., and Olefsky, J. M. (1979). *Diabetes* **28**, 87–95.

Koivisto, U. M., Martinez, V. H., Bilan, P. J., Burdett, E., Ramlal, T., and Klip, A. (1991). *J. Biol. Chem.* **266**, 2615–2621.

Kolterman, O. G., Insel, J., Saekow, M., and Olefsky, J. M. (1980). *J. Clin. Invest.* **65**, 1272–1284.

Kolterman, O. G., Gray, R. S., Griffin, J., Burstein, P., Insel, J., Scarlett, J. A., and Olefsky, J. M. (1981). *J. Clin. Invest.* **68**, 957–969.

Kono, T., Robinson, F. W., Blevins, T. L., and Ezaki, O. (1982). *J. Biol. Chem.* **257**, 10942–10947.

Koranyi, L., James, D., Mueckler, M., and Permutt, M. A. (1990). *J. Clin. Invest.* **85**, 962–967.

Korn, L. J., Siebel, C. W., McKormick, F., and Rothe, R. A. (1987). *Science* **236**, 840–843.

Kraegen, E. W., Storlien, L. H., Jenkins, A. B., and James, D. E. (1989). *Am. J. Physiol.* **256**, E242–E249.

Krief, S., Bazin, R., Dupuy, F., and Lavau, M. (1988). *Am. J. Physiol.* **254**, E342–E348.

Kuri-Harcuch, W., and Green, H. (1977). *J. Biol. Chem.* **252**, 2158–2160.

Kuroda, M., Honnor, R. C., Cushman, S. W., Londos, C., and Simpson, I. A. (1987). *J. Biol. Chem.* **262**, 245–253.

Landau, B. R., Ship, A. G., and Levine, H. J. (1958). *Am. J. Physiol.* **193**, 461–465.

Lawrence, J. C. J., Hiken, J. F., and James, D. E. (1990a). *J. Biol. Chem.* **265**, 2324–2332.

Lawrence, J. C. J., Hiken, J. F., and James, D. E. (1990b). *J. Biol. Chem.* **265**, 19768–19776.

Le Marchand-Brustel, Y., and Freychet, P. (1979). *J. Clin. Invest.* **64**, 1505–1515.

Le Marchand-Brustel, Y., Olichon, B. C., Gremeaux, T., Tanti, J. F., Rochet, N., and Van, O. E. (1990). *Endocrinology (Baltimore)* **127**, 2687–2695.

Leturque, A., Postic, C., Ferré, P., and Girard, J. (1991). *Am. J. Physiol.* **260**, E588–E593.

Levine, R., and Goldstein, M. S. (1955). *Recent Prog. Horm. Res.* **11**, 343–380.

Levine, R., Goldstein, M., Klein, S., and Huddlestun, B. (1949). *J. Biol. Chem.* **179**, 985–986.

Lienhard, G. E., Kim, H. H., Ransome, K. J., and Gorga, J. C. (1982). *Biochem. Biophys. Res. Commun.* **105**, 1150–1156.

Lillioja, S., Mott, D. M., Zawadzki, J. K., Young, A. A., Abbott, W. G. H., Knowler, W. C., H., B. P., Moll, P., and Bogardus, C. (1987a). *Diabetes* **36**, 1329–1335.

Lillioja, S., Young, A. A., Culter, C. L., Ivy, J. L., Abbott, W. G. H., Zawadzki, J. K., Yki-Järvinen, H., Christian, L., Secomb, T. W., and Bogardus, C. (1987b). *J. Clin. Invest.* **80**, 415–424.

Loten, E. G., and Jeanrenaud, J. (1974). *Biochem. J.* **140**, 185–192.

Ludvigsen, C., and Jarett, L. (1980). *Diabetes* **29**, 373–378.

Mackall, J. C., Student, A. K., Polakis, S. E., and Lane, M. D. (1976). *J. Biol. Chem.* **251**, 6462–6464.

Maegawa, H., Kobayashi, M., Watanabe, N., Ishibashi, O., Takata, Y., Kitamura, E., and Shigeta, Y. (1986). *Metab., Clin. Exp.* **35**, 499–504.

Maller, J. L., Pike, L. J., Freidenberg, G. R., Cordera, R., Stith, B. J., Olefsky, J. M., and Krebs, E. G. (1986). *Nature (London)* **230**, 459–461.

Martz, A., Mookerjee, B. K., and Jung, C. Y. (1986). *J. Biol. Chem.* **261**, 13606–13609.

Matsutani, A., Koranyi, L., Cox, N., and Permutt, M. A. (1990). *Diabetes* **39**, 1534–1542.

May, J. M., and Mikulecky, D. C. (1982). *J. Biol. Chem.* **257**, 11601–11608.

McKeel, D. W., and Jarett, L. (1970). *J. Cell Biol.* **44**, 417–432.

Melton, D. A., Krieg, P. A., Rebagliati, M. R., Maniatis, T., Zinn, K., and Green, M. R. (1984). *Nucleic Acids Res.* **12**, 7035–7056.

Mitsumoto, Y., Burdett, E., Grant, A., and Klip, A. (1991). *Biochem. Biophys. Res. Commun.* **175**, 652–659.

Mizel, S. B., and Wilson, L. (1972). *J. Biol. Chem.* **247**, 4102–4105.

Morgan, H. E., Cadenus, E., Regan, D. M., and Park, C. R. (1961). *J. Biol. Chem.* **236**, 253–261.

Morgan, H. E., Regen, D. M., and Park, C. R. (1964). *J. Biol. Chem.* **239**, 369–374.

Morrison, S. D. (1977). *Fed. Proc., Fed. Am. Soc. Exp. Biol.* **36**, 139–142.

Mueckler, M. (1990). *Diabetes* **39**, 6–11.

Mueckler, M., Caruso, C., Baldwin, S. A., Panico, M., Blench, I., Morris, H. R., Allard, W. J., Lienhard, G. E., and Lodish, H. F. (1985). *Science* **229**, 941–945.

Mühlbacher, C., Karnielli, E., Schaff, P., Obermaier, B., Mushack, J., Rattenhuber, E., and Häring, H. U. (1988). *Biochem. J.* **249**, 865–870.

Narahara, H. T., and Özand, P. (1963). *J. Biol. Chem.* **238**, 40–49.

Narahara, H. T., Özand, P., and Cori, C. F. (1960). *J. Biol. Chem.* **235**, 3370–3378.

Navone, F., Jahn, R., Di, G. G., Stukenbrok, H., Greengard, P., and De, C. P. (1986). *J. Cell Biol.* **103**, 2511–2527.

Nesher, R., Karl, I. E., and Kipnis, D. M. (1985). *Am. J. Physiol.* **249**, C226–C232.

Newman, W. P., and Brodows, R. G. (1983). *Metab., Clin. Exp.* **32**, 590–596.

Oka, Y., and Czech, M. P. (1984). *J. Biol. Chem.* **259**, 8125–8133.

Oka, Y., Asano, T., Shibasaki, Y., Kasuga, M., Kanazawa, Y., and Takaku, F. (1988). *J. Biol. Chem.* **263**, 13432–13439.

Olefsky, J. M. (1975). *J. Clin. Invest.* **56**, 1499–1508.

Olefsky, J. M. (1976). *J. Clin. Invest.* **58**, 1450–1460.
Olefsky, J. M. (1978). *Biochem. J.* **172**, 137–145.
Park, C. R., and Johnson, L. H. (1955). *Am. J. Physiol.* **182**, 17–23.
Park, C. R., Bornstein, J., and Post, R. L. (1955). *Am. J. Physiol.* **182**, 12–16.
Park, C. R., Crofford, O. B., and Kono, T. (1968). *J. Gen. Physiol.* **52**, 296S–318S.
Pedersen, O., Bak, J. F., Andersen, P. H., Lund, S., Moller, D. E., Flier, J. S., and Kahn, B. B. (1990). *Diabetes* **39**, 865–870.
Pénicaud, L., Kandé, J., Le Magnen, J., and Girard, J. R. (1985). *Am. J. Physiol.* **249**, E514–E518.
Pénicaud, L., Ferre, P., Terretaz, J., Kinebanyan, M. F., Leturque, A., Dore, E., Girard, J., Jeanrenaud, B., and Picon, L. (1987). *Diabetes* **36**, 626–631.
Pénicaud, L., Kinebanyan, M. F., Ferré, P., Morin, J., J., K., Smadja, C., Marfaing-Jallat, P., and Picon, L. (1989). *Am. J. Physiol.* **257**, E255–E260.
Pessin, J. E., Tillotson, L. G., Yamada, K., Gitomer, W., Carter, S. C., Mora, R., Isselbacher, K. J., and Czech, M. P. (1982). *Proc. Natl. Acad. Sci. U.S.A.* **79**, 2286–2290.
Piper, R. C., Hess, L. J., and James, D. E. (1991). *Am. J. Physiol.* **260**, C570–C580.
Ploug, T., Galbo, H., and Richter, E. A. (1984). *Am. J. Physiol.* **247**, E726–E731.
Ploug, T., Galbo, H., Vinten, J., Jorgensen, M., and Richter, E. A. (1987). *Am. J. Physiol.* **253**, E12–E20.
Ploug, T., Stallknecht, B. M., Pedersen, O., Kahn, B. B., Ohkuwa, T., Vinten, J., and Galbo, H. (1990). *Am. J. Physiol.* **259**, E778–E786.
Ramlal, T., Sarabia, V., Bilan, P. J., and Klip, A. (1988). *Biochem. Biophys. Res. Commun.* **157**, 1329–1335.
Reed, B. C., Kaufmann, S. H., Mackall, J. C., Student, A. K., and Lane, M. D. (1977). *Proc. Natl. Acad. Sci. U.S.A.* **74**, 4876–4880.
Reetz, A., Solimena, M., Matteoli, M., Folli, F., Takei, K., and De Camilli, P. (1991). *EMBO J.* **10**, 1275–1284.
Resh, M. D. (1982). *J. Biol. Chem.* **27**, 6978–6986.
Rodbell, M. (1964). *J. Biol. Chem.* **239**, 375–380.
Rodnick, K. J., Holloszy, J. O., Mondon, C. E., and James, D. E. (1990). *Diabetes* **39**, 1425–1429.
Rubin, C. S., Hirsch, A., Fung, C., and Rosen, O. M. (1978). *J. Biol. Chem.* **253**, 7570–7578.
Ruderman, N. B., Goodman, M. N., Berger, M., and Hagg, S. (1977). *Fed. Proc., Fed. Am. Soc. Exp. Biol.* **36**, 171–176.
Sadler, S. E., and Maller, J. L. (1987). *J. Biol. Chem.* **262**, 10644–10650.
Saltis, J., Habberfield, A. D., Egan, J. J., Londos, C., Simpson, I. A., and Cushman, S. W. (1991). *J. Biol. Chem.* **266**, 261–267.
Schroer, D. W., Frost, S. C., Kohanski, R. A., Lane, M. D., and Lienhard, G. E. (1986). *Biochim. Biophys. Acta* **885**, 317–326.
Schürmann, A., Rosenthal, W., Hinsch, K. D., and Joost, H. G. (1989). *FEBS Lett.* **255**, 259–264.
Shanahan, M. F. (1984). *Mol. Cell. Endocrinol.* **38**, 171–178.
Shanahan, M. F., Olson, S. A., Weber, M. J., Lienhard, G. E., and Gorga, J. C. (1982). *Biochem. Biophys. Res. Commun.* **107**, 38–43.
Sherman, W. M., Katz, A. L., Cutler, C. L., Withers, R. T., and Ivy, J. L. (1988). *Am. J. Physiol.* **255**, E374–E382.
Shibata, Y., Flanagan, J. E., Smith, M. M., Robinson, F. W., and Kono, T. (1987). *Biochim. Biophys. Acta.* **902**, 154–158.
Shoji, S. (1988). *Comp. Biochem. Physiol. A* **91**, 363–365.
Simpson, I. A., Yver, D. R., Hissin, P. J., Wardzala, L. J., Karnieli, E., Salans, L. B., and Cushman, S. W. (1983). *Biochim. Biophys. Acta* **763**, 393–407.

Sinha, M. K., Raineri, M. C., Buchanan, C., Pories, W. J., Carter-Su, C, Pilch, P. F., and Caro, J. F. (1991). *Diabetes* **40**, 472–477.

Sivitz, W. I., DeSautel, S. L., Kayano, T., Bell, G. I., and Pessin, J. E. (1989). *Nature (London)* **340**, 72–74.

Sivitz, W. I., DeSautel, S. L., Kayano, T., Bell, G. I., and Pessin, J. E. (1990). *Mol. Endocrinol.* **4**, 583–588.

Slot, J. W., Moxley, R., Geuze, H. J., and James, D. E. (1990). *Nature (London)* **346**, 369–371.

Slot, J. W., Geuze, H. J., Gigengack, S., Lienhard, G. E., and James, D. E. (1991). *J. Cell Biol.* **113**, 123–135.

Smith, M. M., Robinson, F. W., Watanabe, T., and Kono, T. (19840. *Biochim. Biophys. Acta* **775**, 121–128.

Sowell, M. O., Treutelaar, M. K., Burant, C. F., and Buse, M. G. (1988). *Diabetes* **37**, 499–506.

Sternlicht, E., Barnard, R. J., and Grimditch, G. K. (1988). *Am. J. Physiol.* **254**, E633–E638.

Sternlicht, E., Barnard, R. J., and Grimditch, G. K. (1989). *Am. J. Physiol.* **256**, E227–E230.

Stirewalt, W. S., Low, R. B., and Slaiby, J. M. (1985). *Biochem. J.* **227**, 355–362.

Strålfors, P. (1988). *Nature (London)* **335**, 554–556.

Stumpo, D. J., and Blackshear, P. J. (1986). *Proc. Natl. Acad. Sci. U.S.A.* **83**, 9453–9457.

Suzuki, K. (1988). *J. Biol. Chem.* **263**, 12247–12252.

Suzuki, K., and Kono, T. (1980). *Proc. Natl. Acad. Sci. U.S.A.* **77**, 2542–2545.

Tai, P. K., Liao, J. F., Chen, E. H., Dietz, J., Schwartz, J., and Carter, S. C. (1990). *J. Biol. Chem.* **265**, 21828–21834.

Tanti, J. F., Rochet, N., Grémeaux, T., Van, O. E., and LeMarchand-Brustel, Y. (1989). *Biochem. J.* **258**, 141–146.

Tanti, J.-F., Grémeaux, T., Van, O. E., and Le Marchand-Brustel, Y. (1991). *J. Biol. Chem.* **266**, 2099–2103.

Taylor, L. P., and Holman, G. D. (1981). *Biochim. Biophys. Acta* **642**, 325–335.

Tordjman, K. M., Leingang, K. A., James, D. E., and Mueckler, M. M. (1989). *Proc. Natl. Acad. Sci. U.S.A.* **86**, 7761–7765.

Tordjman, K. M., Leingang, K. A., and Mueckler, M. (1990). *Biochem. J.* **271**, 201–207.

Toyoda, N., Robinson, F. W., Smith, M. M., Flanagan, J. E., and Kono, T. (1986). *J. Biol. Chem.* **261**, 2117–2122.

Toyoda, N., Flanagan, J. E., and Kono, T. (1987). *J. Biol. Chem.* **262**, 2737–2745.

Vallar, L., Biden, T. J., and Wollheim, C. B. (1987). *J. Biol. Chem.* **262**, 5049–5056.

Vera, J. C., and Rosen, O. M. (1989). *Mol. Cell. Biol.* **9**, 4187–4195.

Vera, J. C., and Rosen, O. M. (1990). *Mol. Cell. Biol.* **10**, 743–751.

Vilaro, S., Palacin, M., Pilch, P. F., Testar, X., and Zorzano, A. (1989). *Nature (London)* **342**, 798–800.

Vinten, J. (1978). *Biochim. Biophys. Acta* **511**, 259–273.

Vinten, J., and Galbo, H. (1984). *Am. J. Physiol.* **244**, E129–E134.

Vinten, J., Gliemann, J., and Østerling, K. (1976). *J. Biol. Chem.* **251**, 794–800.

Vinten, J., Petersen, N., Sonne, B., and Galbo, H. (1985). *Biochim. Biophys. Acta* **841**, 223–227.

Vogt, B., Mushack, J., Seffer, E., and Haring, H. U. (1990). *Biochem. Biophys. Res. Commun.* **168**, 1089–1094.

Vogt, B., Mushack, J., Seffer, E., and Haring, H. U. (1991). *Biochem. J.* **275**, 597–600.

Wake, S. A., Sowden, J. A., Storlien, L. H., James, D. E., Clark, P. W., Shine, J., Chisholm, D. J., and Kraegen, E. W. (1991). *Diabetes* **40**, 275–279.

Wallberg-Henriksson, H., and Holloszy, J. O. (1984). *J. Appl. Physiol.* **57**, 1045–1049.

Wallberg-Henriksson, H., Zetan, N., and Henriksson, J. (1987). *J. Biol. Chem.* **262**, 7665–7671.

Wang, C. (1985). *Proc. Natl. Acad. Sci. U.S.A.* **82**, 3621–3625.

Wang, C. (1987). *J. Biol. Chem.* **262**, 15689–15695.

Wardzala, L. J., and Jeanrenaud, B. (1981). *J. Biol. Chem.* **256**, 7090–7093.

Wardzala, L. J., and Jeanrenaud, B. (1983). *Biochim. Biophys. Acta* **730**, 49–56.

Wardzala, L. J., Cushman, S. W., and Salans, L. B. (1978). *J. Biol. Chem.* **253**, 8002–8005.

Wardzala, L. J., Simpson, I. A., Rechler, M. M., and Cushman, S. W. (1984). *J. Biol. Chem.* **259**, 8378–8383.

Watanabe, T., Smith, M. M., Robinson, F. W., and Kono, T. (1984). *J. Biol. Chem.* **259**, 13117–13122.

Weiland, M., Schurmann, A., Schmidt, W. E., and Joost, H. G. (1990). *Biochem. J.* **270**, 331–336.

Wheeler, T. J. (1988). *J. Biol. Chem.* **263**, 19447–19454.

Wheeler, T. J., Simpson, I. A., Sogin, D. C., Hinkle, P. C., and Cushman, S. W. (1982). *Biochem. Biophys. Res. Commun.* **105**, 89–95.

Whitesell, R. R., and Abumrad, N. A. (1985). *J. Biol. Chem.* **260**, 2894–2899.

Whitesell, R. R., and Abumrad, N. A. (1986). *J. Biol. Chem.* **261**, 15090–15096.

Whitesell, R. R., and Gliemann, J. (1979). *J. Biol. Chem.* **254**, 5276–5283.

Whitesell, R. R., Regen, D. M., and Abumrad, N. A. (1989). *Biochemistry* **28**, 6937–6943.

Wieringa, T., and Krans, H. M. (1978). *Biochim. Biophys. Acta* **538**, 563–570.

Williams, S. A., and Birnbaum, M. J. (1988). *J. Biol. Chem.* **263**, 19513–19518.

Yamamoto, T., Fukumoto, H., Koh, G., Yano, H., Yasuda, K., Masuda, K., Ikeda, H., Imura, H., and Seino, Y. (1991). *Biochem. Biophys. Res. Commun.* **175**, 995–1002.

Yki-Järvinen, H., Young, A. A., Lamkin, C., and Foley, J. E. (1987). *J. Clin. Invest.* **79**, 1713–1719.

Yki-Järvinen, H., Sahlin, K., Ren, J. M., and Koivisto, V. A. (1990). *Diabetes* **39**, 157–167.

Young, D. A., Wallberg, H. H., Sleeper, M. D., and Holloszy, J. O. (1987). *Am. J. Physiol.* **253**, E331–E335.

Young, D. G., Uhl, J. J., Cartee, G. D., and Holloszy, J. O. (1986). *J. Biol. Chem.* **261**, 16049–16053.

Zaninetti, D., Crettaz, M., and Jeanrenaud, B. (1983). *Diabetologia* **25**, 525–529.

Zaninetti, D., Greco-Perotto, R., Assimacopoulos-Jeannet, F., and Jeanrenaud, B. (1988). *Biochem. J.* **250**, 277–283.

Zaninetti, D., Greco, P. R., Assimacopoulos, J. F., and Jeanrenaud, B. (1989). *Diabetologia* **32**, 56–60.

Zorzano, A, Wilkinson, W., Kotliar, N., Thoidis, G., Wadzinkski, B. E., Ruoho, A. E., and Pilch, P. F. (1989). *J. Biol. Chem.* **264**, 12358–12363.

Zucker, L. M., and Antoniades, H. N. (1972). *Endocrinology (Baltimore)* **90**, 1320–1330.

Molecular Genetics of Yeast Ion Transport

Richard F. Gaber

Department of Biochemistry, Molecular Biology and Cell Biology
Northwestern University
Evanston, Illinois 60208

I. Introduction

This article is presented as a forum from which to call for a closer relationship between what have in the past been considered separate disciplines: molecular genetics on one side of the aisle and physiology/biochemistry on the other. It has been over a decade since Borst-Pauwels (1981) published a comprehensive review on the energetics and kinetics of ion transport in yeast. In that time dramatic advances in ion transport physiology have been made not so much with microbial systems as with the cells and tissues of multicellular eukaryotes. The reasons for this go beyond the simple explanation that there are far fewer investigators working in the microbial ion transport fields. Several factors have vexed physiological studies of ion transport in yeast in particular. First, unlike the abundance of pharmacological agents that are available to block transport activities in higher organisms, specific inhibitors of yeast ion transport are few. Second, until recently, the smaller cell size and presence of the cell wall precluded electrophysiological approaches in yeast. Finally, we have only recently uncovered what, to some, is a disturbing fact of fungal life: extensive functional redundancy. Not only do yeast cells harbor multiple genes encoding the transporters for H^+, K^+, and Ca^{2+}, but the transporters of one ion can, under some circumstances, transport others well enough to greatly complicate physiological analyses.

In the last 5 years, however, advances in two areas have regenerated excitement and expectation in the field of yeast ion transport. First, the barriers to electrophysiological analysis have been successfully breached. Collaborative experiments from the Kung and Culbertson laboratories (Gustin *et al.*, 1986, 1988) have shown that the powerful patch-clamping technique can be used successfully in yeast. Other groups are now joining in the use of this approach to further our understanding of fungal ion channel activities (Ramirez et al., 1989; Bertl and Slayman, 1990). Second, molecular and genetic approaches have led to the identification and manipulation of the genes encoding yeast ion transporters. Geneticists have successfully identified many of the principal ion transporters in this tiny eukaryote and have generated novel cell types in which one or more of these transporters have been functionally altered. The contributions of these advances make up the primary focus of this article. Obviously, molecular genetics alone cannot be the vehicle to a full understanding of ion transport. Nor can purely physiological or biochemical studies. In what is hoped to be a harbinger to things to come, this article includes some of the recent (but few) physiological studies that have exploited the powerful new tools generated by yeast geneticists: precisely simplified cells in which specific transport components have been altered or even abolished.

II. Plasma Membrane Ion Transport

A. H$^+$ Transport

The H$^+$-ATPase is the principal electrogenic ion pump in the plasma membranes of plants and fungi. Generating both an inward-facing proton gradient and an internal-negative electrical potential, the plasma membrane ATPase can produce a 10,000-fold difference between the concentration of intracellular and extracellular protons. The electrochemical energy of this gradient drives the import of nutrients and other ions (Foury and Goffeau, 1975; Boutry et al., 1977; Foury et al., 1977; Serrano, 1984; Cid et al., 1987; Ulaszewski et al., 1987b). In addition to providing the energy for nutrient uptake, the proton pump also controls the intracellular pH (Serrano, 1984; Eraso et al., 1987; Ulaszewski et al., 1987b). Evidence suggests that this function may be intimately involved in progression of the cell cycle (Cid et al., 1987; McCusker et al., 1987). The H$^+$-ATPase is a very abundant protein; it constitutes as much as 20–40% of the total plasma membrane protein and accounts for more than 90% of the total plasma membrane ATPase activity (Serrano, 1984). The structure–function relationships that allow ion pumps to transfer the chemical energy of ATP into the translocation of an ion across the plasma membrane have been the focus of much attention. Our understanding of the structure and function of the H$^+$-ATPase has been greatly enhanced through the molecular and genetic analyses described below.

Since the field of proton transport via the mitochondrial H$^+$-ATPase constitutes a large field that is not yeast specific or even fungal specific, it has not been included in this article.

1. Genetic Identification of *PMA1*

PMA1, the gene that encodes the H^+-ATPase, was first described as a locus that gave rise to mutations conferring multiple drug resistance (Ulaszewski *et al.*, 1983, see Table I). Mutations conferring *in vitro* resistance to vanadate (Ulaszewski *et al.*, 1987b), and *in vivo* resistance to Dio9 (Roon *et al.*, 1978; Ulaszewski *et al.*, 1983), the amino glycoside hygromycin B (McCusker *et al.*, 1987), cesium (M. Gustin, D. Conklin, and M. Culbertson, unpublished results), and polymyxin B (G. Boguslawski, unpublished results), have been shown to map at *PMA1*. The molecular bases of some of these drug resistance phenotypes are discussed below.

2. Cloning of *PMA1*

Although the existence of the fungal H^+ pump was proposed over 40 years ago (Conway and Downey, 1951), much of what we know about the structure of the plasma membrane H^+-ATPase has been determined from the molecular genetic analysis of *PMA1* genes. *PMA1* from *Saccharomyces cerevisiae* was cloned by "reverse genetics" (Serrano *et al.*, 1986) and by the isolation of a DNA fragment that complemented the tightly linked *leu1* mutation (Ulaszewski *et al.*, 1987b). Plasmids carrying the *PMA1* gene from *Schizosaccharomyces pombe* were subsequently identified by hybridization to the *S. cerevisiae PMA1* gene (Ghislain *et al.*, 1987). Molecular analysis inferred that the *PMA1* genes from *S. cerevisiae* and *S. pombe* encode 100-kDa proteins in agreement with the size determined from biochemical experiments (Serrano, 1984; Ulaszewski *et al.*, 1987a). Underscoring its central role in the transport biology of yeast, disruption of *PMA1* results in cell death (Serrano *et al.*, 1986).

3. Structure–Function Model of the H^+-ATPase

Through the electrogenic pumping of protons the plasma membrane ATPase plays both regulatory and energetic roles in yeast cell biology. The H^+-ATPase is a member of the E1–E2, E–P, or P-type class of ion pumps because it alternates between E1 and E2 conformations and undergoes autophosphorylation during ion transport. Prior to the switch from E1 to E2, the enzyme is believed to bind a proton and an ATP molecule. By an unknown mechanism, chemical energy in the bound ATP molecule is

TABLE I

PMA1 Mutations

Allele	Mutation	Effect	Ref.
S.c.[a] pma1-155	A→G at −39	60% of normal PMA1 abundance; strongly inhibited by vanadate[b]	Perlin et al. (1989)
S.c. Δpma1-241	Δ(Met-1–Ala-27)	No discernible effect	Portillo and Serrano (1989)
S.c. Δpma1-242	Δ(Met-1–Asp-61)	Lethal	Portillo and Serrano (1989)
S.c. Δpma1-243	Δ(Met-1–Pro-84)	Lethal	Portillo and Seranno (1989)
S.c. pma1-203	Asp-91→Tyr	Temperature sensitivity for growth	Cid et al. (1987)
	Glu-92→Lys		
S.c. pma1-210	Glu-129→Gln	Normal activity; normal inhibition by DCCD	Portillo and Serrano (1988)
S.c. pma1-211	Glu-129→Leu	Normal activity; normal inhibition by DCCD	Portillo and Serrano (1988)
S.c. pma1-230	Asp-143→Asn	No discernible effect	Portillo and Serrano (1989)
S.c. pma1-114	Gly-158→Asp	ATP hydrolysis and H⁺ transport may be uncoupled	Perlin et al. (1989)
S.c. pma1-212	Asp-200→Asn	Lethal; 20% of wild-type specific activity	Portillo and Serrano (1988)
S.c. pma1-202	Thr-212→Ile	Temperature sensitivity for growth	Cid et al. (1987)
S.c. pma1-231	Asp-226→Asn	Partially phosphorylation defective; wild-type levels of ATPase activity but 65% H⁺ transport; ATPase/H⁺ transport uncoupled	Portillo and Serrano (1989)
S.c. pma1-232	Ser-228→Ala	No discernible effect	Portillo and Serrano (1989)
S.c. pma1-233	Thr-231→Gly	45% of wild-type ATPase activity; dephosphorylation inhibited; completely resistant to vanadate	Portillo and Serrano (1989)
S.c. pma1-213	Glu-233→Gln	Lethal; 50% of wild-type specific activity; accumulates higher levels of phosphorylated intermediate: dephosphorylation activity greatly diminished; ATPase/H⁺ transport partially uncoupled	Portillo and Serrano (1988)

S.c. *pmal-234*	Ser-234→Ala	290% of wild-type ATPase activity; slightly decreased rate of proton transport; growth rate 31% of wild type; decreased intracellular ATP levels; ATPase/H^+ transport uncoupled	Portillo and Serrano (1989)
S.p.[a] *pmal-2*	Lys-250→Thr	Reduced stimulation by phosphate; increased stimulation by DMSO; vanadate resistance; prevents activation by glucose	M. Ghislain and A. Goffeau (personal communication), Goffeau and de Meis (1990)
S.c. *pmal-201*	Gly-254→Ser	Temperature sensitivity for growth	Cid *et al.*, (1987)
S.p. *pmal-1*	Gly-268→Asp	Increased stimulation by phosphate; decreased inhibition by TFP; vanadate insensitive; glucose constitutive K_m phenotype	Ulaszewski *et al.* (1986, 1987), Ghislain *et al.* (1987), Goffeau and de Meis (1990)
S.c. *pmal-214*	Arg-271→Thr	No discernible effect	Portillo and Serrano (1988)
S.p. *pmal*	Gly-268→Asp	Vanadate insensitivity	Ulaszewski *et al.* (1987a,b)
S.c. *pmal-101*	unknown	Hygromycin resistant; confers conditional cell cycle phenotype	Perlin *et al.* (1989)
S.c. *pmal-215*	Pro-335→Ala	Lethal; two to three times less enzyme in membrane	Portillo and Serrano (1988)
S.c. *pmal-105*	Ser-368→Phe	Shows sharp decline in K_m and V_{max} below pH 6.5; vanadate insensitive	Perlin *et al.* (1989)
S.c. *pmal-141*	unknown		
S.c. *pmal-235*	Cys-376→Leu	65% of wild-type ATPase activity; partially phosphorylation defective; partially resistant to vanadate	Portillo and Serrano (1989 No. 56)
S.c. *pmal-216*	Asp-378→Asn	Lethal or dominant negative (see text)	Portillo and Serrano (1988), J. Haber (personal communication)
S.c. *pmal-217*	Asp-378→Glu	Probably lethal (see text)	Portillo and Serrano (1988), J. Haber (personal communication)

(continued)

303

TABLE I
(Continued)

Allele	Mutation	Effect	Ref.
S.c. *pmal-218*	Asp-378→Thr	Probably lethal (see text)	Portillo and Serrano (1988). J. Haber (personal communication)
S.c. *pmal-236*	Lys-379→Gln	70% of wild-type ATPase activity; hypersensitive to vanadate	Portillo and Serrano (1989)
S.c. *pmal-219*	Lys-474→Gln	Lethal; 10% of wild-type specific activity	Portillo and Serrano (1988)
S.c. *pmal-237*	Lys-474→Arg	30% of wild-type ATPase activity; partially phosphorylation defective; FITC resistant (does not bind)	Portillo and Serrano (1989)
S.c. *pmal-238*	Lys-474→His	10% of wild-type ATPase activity; no phosphorylation detected	Portillo and Serrano (1989)
S.c. *pmal-220*	Asp-534→Asn	(Dominant?) lethal; five times less enzyme in membrane; nucleotide specificity decreased	Portillo and Serrano (1988)
S.c. *pmal-200*	Ala-547→Val	Temperature sensitivity for growth; confers glucose-independent shift in kinetic properties[b]	Cid and Serrano (1988)
S.c. *pmal-221*	Asp-560→Asn	Lethal; 25% of wild-type specific activity; nucleotide specificity decreased	Portillo and Serrano (1988)
S.c. *pmal-1*	Ala-608→Thr	Dio9 resistant; decreased ATPase activity; increased K_m; vanadate resistant	Ulaszewski et al. (1983, 1987a), Van Dyck et al. (1990)
S.c. *pmal-239*	Asp-634→Asn	15% of wild-type ATPase activity; dephosphorylation inhibited: slower turnover demonstrated; partially resistant to vanadate	Portillo and Serrano (1989)
S.c. *pmal-222*	Asp-638→Asn	Lethal; 15% of wild-type specific activity; nucleotide specificity decreased	Portillo and Serrano (1988)
S.c. *pmal-147*	Pro-640→Leu	Vanadate insensitive	McCusker et al. (1987)
S.c. *pmal*	Glu-703→Gln	No discernible effect	Portillo et al. (1989)
S.c. *pmal-240*	Asp-730→Asn	Lethal; dephosphorylation inhibited	Portillo and Serrano (1989)
S.c. *pmal-223*	Asn-848→Asp	No discernible effect	Portillo and Serrano (1988)

S.c. species[a]	Amino acid changes	Effect	Reference
S.c. *pma1*	Glu-803→Gln	No discernible effect	Portillo et al. (1989)
S.c. Δ*pma1-249*	Ala-813→Ser Gly-814→Ser Pro-815→Ala Phe-816→stop (deletes 103 amino acids)	Lethal	Portillo et a. (1989)
S.c. Δ*pma1-248*	Asp-833→Ala Ile-834→Ser Ile-835→Arg Ala-836→Ala Thr-837→stop (deletes 81 amino acids)	Lethal	Portillo et al. (1989)
S.c. Δ*pma1-247*	Glu-847→stop (deletes 71 amino acids)	Lethal	Portillo et al. (1989)
S.c. *pma1*	Asn-848→Asp	No discernible effect	Serrano and Portillo (1990)
S.c. Δ*pma1-246*	Tyr-873→stop (deletes 46 amino acids)	No discernible effect	Portillo et al. (1989)
S.c. Δ*pma1-245*	Glu-901→Ala Asp-902→Ser Phe-903→Arg Met-904→Ala Ala-905→stop (deletes 18 amino acids)	Normal amount of enzyme in membrane; confers glucose-independent shift in kinetic properties[b]	Portillo et al. (1989)
S.c. Δ*pma1-244*	Gln-908→stop (deletes 11 amino acids)	Normal amount of enzyme in membrane; confers glucose-independent shift in kinetic properties[c]	*Portillo et al.* (1989)

[a] S.c., *Saccharomyces cerevisiae*; S.p. *Schizosaccharomyces pombe*.

[b] Vanadate sensitivity or resistance refers to the effect of vanadate on enzymatic activity measured *in vitro*.

[c] Glucose fermentation confers a 4-fold decrease in K_m for ATP, a 3-fold increase in V_{max}, a 20-fold increase in vanadate sensitivity, and a shift in pH optimum to a more alkaline value.

transformed into mechanical energy for the translocation of the bound proton across the membrane.

Members of the H^+-ATPase family found in plants, fungi, and parasites share about 30% amino acid identity. Many of the basic structure–function relationships determined for the H^+-ATPases are also conserved among the Na^+/K^+-, K^+/H^+-, and Ca^{2+}-ATPases found in animal cells. Not only do all members of these P-type ATPases exhibit the E1–E2 conformational changes and form a phosphorylated intermediate during ion transport, but they also contain multiple membrane-spanning domains (from 8 to 10) and share significant amino acid identities with each other (Goffeau *et al.*, 1989; Nakamoto *et al.*, 1989; Serrano, 1989; Goffeau and Green, 1990).

The first structural model of the proton pump, based only on its inferred amino acid sequence and similarities to the ion pumps from other organisms (Serrano *et al.*, 1986), described 4 principal functional domains of the yeast H^+-ATPases: (1) 10 putative membrane-spanning domains and the regions immediately adjacent to them (some of which may constitute the proton translocation domain), (2) the phosphorylation domain containing the completely conserved aspartate residue shown to be phosphorylated in other ion pumps (Walderhaug *et al.*, 1985), (3) the ATP binding domain, identified by its sequence similarity to the region of the Na^+/K^+-ATPase that is labeled by fluorescein 5'-isothiocyanate (Ohta *et al.*, 1986), and (4) a region defined as the energy-transducing domain consisting of the only conserved transmembrane domain among all of the P-type ATPases and the cytoplasmic region adjacent to it. Contributions by a number of laboratories have allowed a more detailed model of the proton pump to be presented (Fig. 1).

4. *PMA1* Mutational Analyses

Although the overall structures of P-type ATPases from different organisms have been evolutionarily conserved, few of the pharmacological agents that can be used to inhibit these enzymes in animals are effective against the plant and fungal proton pumps. The genetic analysis available in yeast has, however, more than compensated for the lack of pharmacological probes. Genetic analysis of the yeast proton pump has been facilitated in two ways. First, mutations in *PMA1* have been isolated by direct selection, usually for resistance to drugs, including hygromycin B and Dio9 (referenced above). Second, diverse phenotypes, both *in vivo* and *in vitro*, resulting from site-directed mutagenesis of *PMA1*, have permitted the assignment of particular functions to specific structural domains (Table I). Two findings should serve as a general cautionary warning, however, when interpreting the outcome of experiments based on molecular manipulations of gene structure. First, dominant negative mutations, although

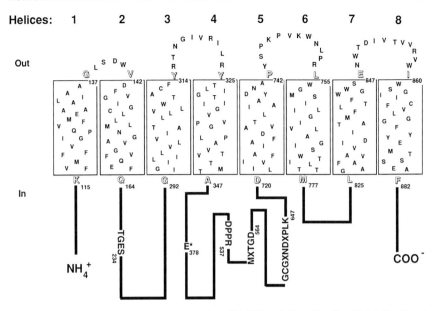

FIG. 1   Membrane-spanning model of the plasma membrane H^+-ATPase. Amino acid residues likely to lie adjacent to the membrane are shown in outlined font with their positions subscripted. Those cytoplasmic domains that are highly conserved between P-type ATPases and are mentioned in the text are shown in larger font with their positions subscripted. E^*_{378} is the site of autophosphorylation during E_1 to E_2 transition. Adapted from figures provided by D. Perlin and R. Serrano.

expected to be quite rare, can be generated *in vitro* but overlooked *in vivo* due to repair of the mutation by gene conversion (J. Haber, personal communication; see Section II.A.4.c). Second, the ability to conditionally express mutant versions of genes is a powerful technique that allows one to examine the effect of lethal and sublethal mutations. However, sometimes the growth conditions in which the cells must be placed to repress conditionally a particular gene construct can lead to unexpected and perhaps global effects on gene expression. Nevertheless, because of the advantages inherent in both approaches, *in vitro* and *in vivo* mutagenesis strategies have proved fruitful and have allowed a more sophisticated understanding of the H^+-ATPase structure and function.

a. In Vivo Mutant Isolation   Not surprisingly, mutations that alter the activity of what has been called the "master enzyme" of the cell (Serrano, 1989) cause a variety of *in vivo* phenotypes. These phenotypes include cell death (Serrano *et al.*, 1986), heat sensitivity (Cid *et al.*, 1987; Cid and Serrano, 1988) and cold sensitivity (McCusker *et al.*, 1987) for growth, slow growth (Serrano *et al.*, 1986; Ulaszewski *et al.*, 1986, 1987a), hypersensitivity to weak acids and to ammonium ion (McCusker *et al.*, 1987), resistance to hygromycin (McCusker *et al.*, 1987) and to Dio9 (Ulaszewski *et al.*, 1983, 1986, 1987a), and alteration of cell cycle (McCusker *et al.*, 1987) and growth (Portillo and Serrano, 1989) regulation.

Apparently any mutation that decreases the activity of the proton pump can, if severe enough, result in hypersensitivity to weak acids, resistance to hygromycin and Dio9, and perhaps sensitivity to NH_4^+. These phenotypes are likely to result from a decreased membrane potential and decreased ability to extrude protons from the cytoplasm (see this Section). McCusker *et al.* (1987) showed that most of their low pH-sensitive, hygromycin-resistant mutants had reduced levels of ATPase activity (as low as 25% of wild-type levels). They postulated that hypersensitivity to low pH was due to intracellular acidification in the presence of a net increased H^+ influx and that resistance to hygromycin resulted from decreased uptake of the drug due to a decreased membrane potential. Subsequent experiments (Perlin *et al.*, 1988) confirmed that the membrane potential in these mutants is reduced. Vallejo and Serrano (1989) also showed that mutations conferring decreased expression levels of *PMA1* lead to hygromycin resistance by decreasing the ability of the cell to take up the drug. Molecular analysis of two *pma1* mutations revealed that hygromycin resistance does not require structural alteration of the H^+-ATPase but merely reduced activity of the enzyme. *pma1-155* contains an alteration in the 5′-untranslated region reducing the expression level of the gene to 61% of wild-type protein levels (Perlin *et al.*, 1989). Similarly, a deletion mutation removing an upstream activation sequence (UAS) required for full *PMA1* expression confers hygromycin resistance (Capieaux *et al.*, 1989).

Some hygromycin-resistant *pma1* mutants are hypersensitive to NH_4^+ (McCusker *et al.*, 1987). Although the basis of this inhibition remains unknown, it is not due to an inability of these mutants to utilize extracellular NH_4^+. NH_4^+ hypersensitivity of some *pma1* mutants may be a reflection of the normal ion traffic across the plasma membrane. The rate of proton pumping is significantly increased in the presence of extracellular K^+ due to the ability of K^+ to act as a counterion (Conway and Brady, 1950). The activity of the proton pump, negatively regulated by the membrane potential, is increased on the temporary depolarization that results from the influx of K^+ (Serrano, 1983, 1984). The hypersensitivity of hygromycin-resistant *pma1* mutants to NH_4^+ could reflect the combined effects of a decreased membrane potential and competitive inhibition of K^+ uptake by NH_4^+ (Borst-Pauwels, 1981). Since K^+ is an essential physiological ion, the combination of these two effects may starve the cell for potassium when its extracellular concentration is sufficiently low.

Consistent with this interpretation, McCusker *et al.* (1987) reported that increased concentrations of extracellular K^+ suppress the NH_4^+ hypersensitivity of these *pma1* mutants.

Hygromycin resistance has also been shown to result from mutations that affect several of the distinct activities carried out by the plasma membrane ATPase during proton pumping. The details of some of these mutations are discussed below.

Genetic experiments involving hygromycin resistance have led to speculation that the H^+-ATPase functions as a multimer. McCusker *et al.* (1987) found that some recessive *pma1* mutants complemented each other for the hygromycin-resistance phenotype. However, alternative conclusions can be drawn from intragenic complementation tests. For example, in a heterozygote in which one of the mutant H^+-ATPases is defective for dephosphorylation and the other is defective for some other aspect of pump function, the former might be dephosphorylated by the latter. If so, H^+ pumping itself may actually be accomplished by monomers.

As mentioned above, direct selection for *pma1* mutations was also achieved by selecting for resistance to Dio9 (Roon *et al.*, 1978; Ulaszewski *et al.*, 1983, 1986, 1987a). DNA sequence analysis of the Dio9-resistant alleles *pma1-1* (Gly-268→Asp) and *pma1-2* (Lys-250→Thr), combined with the observation that these mutations confer decreased sensitivity *in vitro* to vanadate and modified stimulation by phosphate anion (Ghislain *et al.*, 1987; Goffeau and de Meis, 1990), suggests that these residues make up part of the P_i-binding domain of the enzyme.

b. Targeting Structural Domains of the Plasma Membrane ATPase by Mutations Constructed in Vitro Because it is an essential gene, direct replacement of *PMA1* in haploid cells with *in vitro*-generated mutant alleles does not yield viable cells if the mutation confers a severe defect in function. To circumvent this difficulty, three independent approaches were developed to investigate the effects of *pma1* mutations generated *in vitro*. Serrano and colleagues (Cid *et al.*, 1987) developed a system in which expression of the wild-type *PMA1* gene was placed under the control of a conditional promoter *(GAL1)*. Growing the cells on glucose represses the *GAL1::PMA1* construct, forcing the cell to rely solely on the expression of mutant *pma1* alleles carried on a plasmid. Although residual wild-type ATPase activity present in cells and cell extracts following the shift to the glucose medium obscured the activity of some of their weakest mutants, this system allowed the analysis of mutations generated by site-directed mutatgenesis and by hydroxylamine mutagenesis (including the first temperature-sensitive allele). As alluded to above, growth of yeast cells on galactose as the sole carbon source can complicate the interpretation of the activity levels of some of the *pma1* mutants that were analyzed with this system, since growth on galactose results in lower activity levels of the wild-type H⁺-ATPase (A. Goffeau, personal communication). Nakamoto *et al.* (1991) modified this general approach by including a *sec6* mutation that blocks secretion at a step prior to vesicle–membrane fusion (Novick *et al.*, 1981). The expression of plasmid-borne mutant *pma1* alleles (from a heat-shock promoter) results in the accumulation of secretory vesicles containing only the mutant H⁺-ATPase. This method has the potential to yield tightly sealed vesicles containing only the mutant ATPase. Gene replacement methodologies have been exploited to generate diploid cells that are heterozygous for specific *pma1* mutations made *in vitro* (S. Harris and J. Haber, personal communication). Sporulation of this diploid results in four ascospores (two containing only the mutant copy of the H⁺-ATPase gene) that can be analyzed directly. The results of these studies are discussed below and a list of mutants is contained in Table I.

5. *In Vitro* Analysis of Mutant Proton Pumps

In vitro analysis of the H⁺-ATPase purified from *pma1* mutants has identified distinct defects in enzyme function. These include reduced (McCusker *et al.*, 1987; Portillo and Serrano, 1988, 1989) or increased (Portillo and Serrano, 1989) specific activity, defects in phosphorylation of the intermediate (Portillo and Serrano, 1989), decreased dephosphorylation activity (Portillo and Serrano, 1988, 1989), defective H⁺ translocation (Portillo and Serrano, 1989), vanadate resistance (Ghislain *et al.*, 1987; Ulaszewski *et al.*, 1987a,b; Portillo and Serrano, 1988, 1989; Perlin *et al.*, 1989; Goffeau *et al.*, 1990), vanadate hypersensitivity (Portillo and Serrano, 1989), glucose-independent changes in kinetic parameters of the enzyme (Cid and Serrano, 1988; Portillo *et al.*, 1989; Goffeau *et al.*, 1990), increased and decreased inhibition by phosphate anion and vanadate (Goffeau and de Meis, 1990), decreased specificity in nucleotide binding activity (Portillo and Serrano, 1988), uncoupling of ATPase activity from H⁺ translocation (Portillo and Serrano, 1988; Perlin *et al.*, 1989; Portillo *et al.*, 1989), and decreased abundance of the enzyme in the membrane (Portillo and Serrano, 1988; Perlin *et al.*, 1989; Portillo *et al.*, 1989; see Table I).

6. Correlating Changes in H⁺-ATPase Structure with Function

The combination of *in vivo* and *in vitro* analysis of many *pma1* mutations has led to a fairly sophisticated model of H⁺-ATPase structure and function. Studies of mutations that were found to alter clearly definable functions of the H⁺-ATPase are discussed below.

a. Transmembrane Domains   The transmembrane domains of the H^+-ATPase serve two primary functions. First, they anchor the transporters to the plasma membrane and, second, they form the polar pathway that protons take in their journey across the membrane. Several of the more highly conserved amino acids located within the putative transmembrane domains of the pump have been mutationally altered to assess their role(s) in proton transport.

Pro-335 is completely conserved among all of the P-type ATPases and resides in the middle of the transmembrane segment that is itself the most highly conserved among these ion pumps (Nakamoto *et al.*, 1989). This segment has been postulated to consist of part of the energy transduction pathway (Shull *et al.*, 1985). It has been proposed (Brandl *et al.*, 1986) that cis–trans isomerization of intramembrane prolines may be involved in important conformational changes during ion transport. Portillo and Serrano (1988) mutationally altered Pro-335 and found that, although a Pro-335→Ala mutation was lethal due to insufficient levels of enzyme in the membrane, it had only moderate effects on specific activity of the residual enzyme. In contrast, mutation of the homologous mutation in the Ca^{2+}-ATPase of the sarcoplasmic reticulum was found to modify the affinity of the enzyme for Ca^{2+} (Vilsen *et al.*, 1989).

Since the plasma membrane ATPase of yeast is a cation-transporting pump, acidic residues within transmembrane domains could be involved in direct interactions with the permeant protons. Asp-143 and Asp-730 (Fig. 1) are the only acidic residues that are conserved in the transmembrane domains of the fungal P-type ATPases. Although an Asp-143→Asn mutation had no discernable effect on enzyme function, an Asp-730→Asn mutation was lethal (Portillo and Serrano, 1989).

b. Dephosphorylation/H^+-Transduction Domain   The fourth transmembrane segment and the region containing the conserved motif TGES (amino acids 231–234) are considered to be part of the proton transduction domain since mutations in these regions can uncouple ATP hydrolysis from H^+ transport. For example, a Ser-234→Ala mutation exhibited levels of ATPase activity significantly above wild-type levels, yet exhibited a reduced rate of proton transport that resulted in a slow growth phenotype (Portillo and Serrano, 1989). Similarly, the Pro-335→Ala mutation mentioned above decreased H^+ transport more than it did ATP hydrolysis (Portillo and Serrano, 1988). The Glu-129 site in *Neurospora* was shown by Sussman *et al.* (1987) to be the residue modified by dicyclohexylcarbodiimide (DCCD), which prevents H^+ translocation. However, Portillo and Serrano (1988) showed that in *S. cerevisiae* Glu-129→Gln and Glu-129→Leu mutants exhibited normal ATPase activity and remained sensitive to DCCD, indicating that Glu-129 may not be critical for transduction in the yeast enzyme.

The ability of the P-type ATPases to recognize the covalently bound phosphate and catalyze its release from the enzyme is crucial for completion of the E1 to E2 conformational cycle during ion translocation. On phosphate release, the phosphate-recognition domain is empty, a condition that leads to vanadate sensitivity, one of the strongest *in vitro* characteristics of these enzymes. It is thought that sensitivity to vanadate originates from the structural similarity of vanadate to phosphate, a similarity that results in the binding of vanadate to the phosphate binding site with such high affinity that further dephosphorylation activity of the enzyme is inhibited (Cantley *et al.*, 1978; Pick, 1982; Huang and Askari, 1984). In the presence of inhibitory concentrations of vanadate an accumulation of the phosphorylated intermediate occurs.

Mutations in the dephosphorylation domain can have different effects on enzyme activity. Vanadate sensitivity can be decreased by mutants that have decreased affinity for the inhibitor yet remain partially dephosphorylation competent. Included among these mutations are Thr-231→Gly, Cys-376→Leu, and Asp-634→Asn. In contrast, the Lys-379→Gln mutation actually increases vanadate sensitivity (Portillo and Serrano, 1988, 1989), probably by reducing the affinity of the enzyme for phosphate more than its affinity for vanadate. Other mutations, such as Thr-231→Gly, Glu-233→Gln, and Cys-376→Leu (Portillo and Serrano, 1989) inhibit the dephosphorylation activity of the enzyme to the point where the steady state level of the phosphorylated intermediate increases relative to wild type. All of these mutations lie within the highly conserved TGES motif. The region encompassing Asp-634 and Asp-730 may be part of the dephosphorylation domain since a mutation to Asn at either position inhibits dephosphorylation of the enzyme (Portillo and Serrano, 1989). Asp-226 is also likely to make up part of the dephosphorylation domain since an Asp-226→Asn mutation confers a faster rate of dephosphorylation than that observed for the wild-type enzyme (Portillo and Serrano, 1989).

Goffeau and de Meis (1990) have proposed that the region from Lys-250 and Gly-268 in the *S. pombe* enzyme is part of the low-affinity binding site for the phosphate anion HPO_4^{2-}. Their suggestion is based on the observation that mutations making this region more hydrophobic (Lys-250→Thr) result in decreased ATPase stimulation by phosphate while mutations that increase the hydrophilicity of this region (Gly-268→Asp) confer increased phosphate stimulation.

The conformational change in the H^+-ATPase that accompanies phosphate binding can be detected experimentally by measuring the ability of phosphate to protect the enzyme against trypsin inactivation. Perlin *et al.* (1989) assayed mutations that altered phosphate binding ability and found that the vanadate-resistant mutants Ser-368→Phe and Pro-640→Leu were not as well protected by phosphate as were wild-type and mutant vanadate-sensitive enzymes. Based on these results, they suggested that Ser-368 and Pro-640 are part of the phosphate binding domain. The Ser-368 site has also been mutated to Ala, Thr, Cys, Val, and Leu. *In vitro* analysis of these mutant enzymes indicates that vanadate resistance directly corresponds to the relative size of the amino acid at this position (S. Harris and J. Haber, personal communication). In addition, mutations at the completely conserved Glu-367 can suppress an adjacent Ser-368→Phe mutation, evidently by reducing the steric mass in this region (S. Harris and J. Haber, personal communication).

c. ATP Binding and Kinase Domain

Like all P-type ATPases, the proton pump undergoes autophosphorylation at the completely conserved Asp-378 residue. The results of initial experiments involving the mutagenesis of this site (Portillo and Serrano, 1989) suggested that substitution of Asp-378 with Glu or Thr abolished phosphorylation of the proton pump but had little effect on the ability of the mutant enzyme to hydrolyze ATP (Serrano and Portillo, 1990). Serrano has indicated that the Asp-378→Asn mutation (Portillo and Serrano, 1989) had reverted to wild type and that experiments performed in his laboratory show that this mutation confers an inactive enzyme (R. Serrano, personal communication). In contrast, Haber's group claims that the Asp-378→Asn mutation actually functions as a dominant negative allele: heterozygotes containing both the mutant and wild-type allele exhibit an H^+-ATPase-defective phenotype (J. Haber, personal communication). To what degree these contradictory claims might reflect strain background differences, including differences in *PMA2* function (see Section II.A.12), will have to await further analysis. Haber has indicated that mutations such as Asp-378→Asn can revert at high frequency, due, most likely, to gene conversion with the resident *PMA1* wild-type allele (J. Haber, personal communication). More moderate effects are seen with mutations near the Asp-378 site. For example, the Cys-376→Leu mutant is only partially phosphorylation defective (Portillo and Serrano, 1989).

Other phosphorylation-defective *pma1* mutants are defective in ATP binding, ATP hydrolysis, or transfer of the high-energy phosphate bond to Asp-378. Among mutations that define the ATP binding domain are those found in the highly conserved DPPR (Asp-534→Asn), MXTGD (Asp-560→Asn), and GDGXNDXPXLK (Asp-638→Asn) domains (Table I).

These mutations diminish the level of ATPase activity and confer a significant decrease in the specificity of nucleotide binding (Portillo and Serrano, 1988).

One of the advantages of a site-directed approach in analyzing the functional role of specific residues was underscored by the analysis of a critical region of the ATP binding domain containing Lys-474. Lys-474 was first shown to be essential for function by replacing it with Gln (*pma1-219,* (Portillo and Serrano, 1988). The mutant enzyme exhibited less than 10% of the wild-type specific activity. Subsequently, more conservative changes were made (Portillo and Serrano, 1989). The Lys-474→Arg mutation retains 30% of wild-type specific activity and the mutant enzyme fails to bind fluorescein isothiocyanate (FITC). Since ATP protects against both FITC inhibition and FITC binding, Lys-474 likely constitutes part of the ATP binding site. Although the Lys-474→Arg mutation reduces ATP hydrolysis, it does so without altering nucleotide binding specificity, suggesting that this region is also important in the postbinding phosphorylation reaction. Asp-378, Lys-474, Asp-534, Asp-560, and Asp-638 are fully conserved in the P-type ATPase family. Consistent with their central roles in ATP binding, mutations at these sites lead to a decrease in nucleotide binding specificity (Table I).

Biochemical and genetic analyses have occasionally combined to yield exquisite detail regarding structure and function. A striking example is the analysis of Asp-634 within the ATP binding domain of the plasma membrane ATPase. Chemical labeling of Asp-634 with an ATP analog (Ovchinnikov *et al.,* 1987) revealed that this region is likely to be in close proximity to the γ-phosphate group of ATP. Retention of specificity of nucleotide binding of an Asp-634→Asn mutant (Portillo and Serrano, 1989) indicated that Asp-634 is not involved in binding to the adenine moiety. The Asp-634→Asn mutation conferred a low level of vanadate resistance and was defective in turnover of the phosphorylated intermediate, suggesting that this residue may also interact with the phosphorylated intermediate as part of the dephosphorylation domain. Other mutations, including Ala-574→Val (Portillo and Serrano, 1989) and Ala-608→Thr (Van Dijck, 1990, No. 206), alter the K_m for ATP hydrolysis, suggesting that these residues are also important in ATP binding.

d. Glucose Regulation Domain   When yeast cells are grown in glucose prior to purification of the H^+-ATPase the K_m for ATP decreases 4-fold, the V_{max} increases 3-fold, vanadate sensitivity increases 20-fold, and the pH optimum shifts to a slightly more alkaline value compared to wild type (Serrano, 1983). Deletion of the last 11 carboxy-terminal amino acids confers a shift in the enzymatic activity that mimicks this glucose effect (Portillo *et al.*, 1989). Thus, the carboxy terminus of the proton pump contains a negative regulatory domain, the function of which is alleviated by glucose metabolism.

e. Other Regions   The principal function of the amino terminus of the proton pump remains poorly defined. Portillo and Serrano (1989) showed that although deletion of the first 26 amino acids had no discernible effect, deletion up to Asp-61 confers lethality. Some amino acids have been identified as structurally important although their functional role remains unknown. For example, Asp-91→Tyr and Glu-92→Lys mutations confer heat sensitivity to the enzyme (Cid *et al.*, 1987) and the Pro-335→Ala and Asp-534→Asn mutations reduce the amount of enzyme in the membrane but have little effect on its specific activity. Presumably the latter mutations cause structural alterations that prevent proper localization.

Mutations at Asp-200 and Glu-233, the only conserved acidic residues among all of the P-type ATPases, demonstrated that both residues are required for normal function of the proton pump. Portillo and Serrano (1988) found that a mutation of Glu-233 to Gln decreased the specific activity of the enzyme to 50% of wild type. The Asp-200→Asn mutation was lethal, decreasing the specific activity of the mutant ATPase to less than 20% of wild type. Indeed, all H^+-ATPase mutations identified thus far that yield less than 20% of wild-type activity are lethal. Although McCusker *et al.* (1987) were able to isolate hygromycin-resistant mutants precisely because of decreased enzyme activity, all of the mutants obtained exhibited greater than 20% wild-type activity (Perlin *et al.*, 1988).

7. Reversion Analysis

The ability to select for revertants of a mutation can often yield important information regarding how a particular protein functions and/or the identity of other proteins with which it interacts. In the case of mutations at *PMA1*, negative phenotypes such as sensitivity to low pH offer the ability to select directly for revertants. In experiments designed to isolate low pH-resistant revertants from low pH-sensitive *pma1* cells, Haber's group obtained an unusually high frequency of true revertants that arose due to interchromosomal gene conversion between the mutant *pma1* gene and the wild-type *PMA2* gene (J. Haber, personal communication). Goffeau and colleagues have shown that the *PMA1* and *PMA2* genes are 90% homologous (Schlesser *et al.*, 1988). Because of the extensive homology between these genes, the frequency at which gene convertants are obtained is approximately two orders of magnitude higher than the frequency at which intragenic revertants are obtained (S. Harris and J. Haber, personal communication).

a. Intragenic Revertants   Some intragenic revertants of *pma1* mutations have been isolated. For example, 13 independent intragenic revertants of *pma1-105* (Ser-368→Phe) have been isolated, half of which map to putative transmembrane regions (S. Harris and J. Haber, personal communication). The original Ser-368→Phe mutation results in aberrant regulation of a putative K^+ channel function (Ramirez *et al.*, 1989). Second-site suppressor mutations that partially restore normal function could decrease the increased K^+ influx conferred by the original mutation. Thus, the suppressor mutations seem to alter the structure and function of the pathway that ions take to get into or out of the cell via PMA1.

b. Extragenic Revertants   Thus far, extragenic suppressors of *pma1* mutations have not been isolated. However, with the discovery of *PMA2* and its subsequent deletion (see Section II,A,12), strains better engineered to yield such suppressors can now be constructed. Extragenic mutations that affect plasma membrane ATPase activity have, however, been identified. Starting with hygromycin-resistant *pma1* mutants and selecting for increased resistance to the drug, Haber's group has isolated mutations in four complementation groups (designated *hhr*) (Table II). It is not yet clear whether any of the *hhr* mutations are alleles of the *MOP2* gene, mutations in which lead to decreased H^+-ATPase activity (J. Haber, personal communication).

Goffeau and co-workers (1990) have reported the discovery of proteolipids contained within purified plasma membrane ATPase preparations from *S. cerevisiae*, *S. pombe*, and *Neurospora*. One of these proteolipids is phosphorylated by a zinc-sensitive protein kinase that also phosphorylates the plasma membrane ATPase (Kolarov *et al.*, 1988). Goffeau *et al.* (1990) postulate that the proteolipids associated with the plasma membrane ATPase may represent an additional subunit(s) although their function remains unknown. Vai *et al.* (1986) reported the association of a 115-kDa glycoprotein with the H^+-ATPase during biochemical purification steps, raising the possibility that the glycoprotein might be a subunit of the H^+-ATPase or may at least affect its activity *in vitro*. However, Serrano *et al.* (1991) have shown that a mutant lacking the glycoprotein contains H^+-ATPase activity that is normal.

8. Regulation of the Plasma Membrane ATPase

Eraso *et al.* (1987) have shown that physiological conditions that increase the need for proton pumping, such as decreased intracellular pH and glucose metabolism, result in increased H^+-ATPase activity. The amount of enzyme present in the membrane, however, does not change significantly under these conditions. Even when *PMA1* containing its native promoter is placed on a multicopy plasmid in yeast, the apparent copy number of plasmids containing *PMA1* was reduced compared to those containing vector sequences alone. These results suggested that expression of *PMA1* is tightly controlled. The basis of *PMA1* regulation is beginning to unfold with the discovery that it is controlled at least in part by *TUF1*, also known as *RAP1* (Huet *et al.*, 1985), a gene encoding a transcriptional activator that controls the expression of many genes, including those required for translation, transcription, glycolysis, cell differentiation, and, with *PMA1*, nutrient uptake and intracellular pH (Capieaux *et al.*, 1989). Capieaux *et al.* suggest that the proton pump plays an integral part in the regulation of yeast growth control by being regulated by RAP1.

As mentioned above, site(s) for glucose metabolism regulation of the proton pump have been genetically identified. Serrano (1989) pointed out that glucose metabolism leads to the synthesis of hexose-phosphates, which results in a decrease in intracellular pH. Serrano also hypothesized that this may be the first intracellular messenger in the signal transduction that leads to H^+-ATPase activation via phosphorylation (Kolarov *et al.*, 1988) of the ATPase. This hypothesis is supported by the observation that in *cdc25* cells, which are conditionally defective for adenylate cyclase activity (Broek *et al.*, 1987), the H^+-ATPase is not activated by glucose at the nonpermissive temperature (Portillo and Mazon, 1986). The demonstration by Koralov *et al.* (1988) that the H^+-ATPase is phosphorylated by protein kinase and that acid phosphatase decreases ATPase activity, further supports the likelihood that the H^+-ATPase is regulated by phosphorylation.

Chang and Slayman (1991) demonstrated that phosphorylation of the H^+-ATPase is indeed a mechanism by which the enzyme is regulated. These investigators used phosphopeptide mapping to show that, although the enzyme is phosphorylated during its maturation, at least one site becomes phosphorylated only upon final localization to the plasma membrane. Furthermore, this site is dephosphorylated during glucose starvation and rephosphorylated within three minutes following the restoration of glucose. Thus, the relationship between glucose activation of the H^+-ATPase and site-specific phosphorylation has been firmly established.

The degree to which the H^+-ATPase controls cell growth was demonstrated by Portillo and Serrano (1989) through their analysis of a series of

TABLE II
Ion-Related Mutants

Mutation[a]	Ion	Organism	Gene product[b]	Effect/Phenotype	Ref.[c]
amt	NH_4^+	*S. cerevisiae*	Unknown	Confers decreased NH_4^+ (methylamine) uptake	Roon *et al.* (1977)
ca1	Ca^{2+}	*S. cerevisiae*	Unknown	Confers temperature-sensitive growth in calcium-poor medium	Ohya *et al.* (1984)
csg2	Ca^{2+}	*S. cerevisiae*	Putative plasma membrane Ca^{2+} transporter	Confers sensitivity to high Ca^{2+} ($>$ 1 m*M*)	T. Dunn and T. Beeler (unpublished)
csg1, *scg3*, *csg4*	Ca^{2+}	*S. cerevisiae*	Unknown	Confer sensitivity to high Ca^{2+} ($>$1 m*M*)	T. Dunn and T. Beeler (unpublished)
cls1–cls18	Ca^{2+}	*S. cerevisiae*	Unknown[d]	Confer defects in rate and/or extent of Ca^{2+} uptake	Ohya *et al.* (1986a)
cta3	Ca^{2+}	*S. pombe*	Ca^{2+}-ATPase	Confers defective Ca^{2+} uptake and sensitivity to high concentrations of Ca^{2+}	Goffeau *et al.* (1990)
hhr1–hhr4	H^+?	*S. cerevisiae*	Unknown	Increase resistance to hygromycin	J. Haber (personal communication)
HOL1-1	Cations	*S. cerevisiae*	Putative plasma membrane cation transporter	Confers uptake of cations, including histidinol; also confers hypersensitivity of some cations	Gaber *et al.* (1990), M. Wright and R. F. Gaber (unpublished results)

Gene	Ion	Organism	Product	Effect	Reference
hol2, hol3	Cations	S. cerevisiae	Unknown	Confer increased uptake of cations, including histidinol; also confer hypersensitivity of some cations	Gaber et al. (1988)
kdm2–kdm7	K^+?	S. cerevisiae	Unknown	Confer increased K^+ requirement	Dubois and Grenson (1979)
mep1, mep2	NH_4^+	S. cerevisiae	Unknown	Confer defects in NH_4^+ uptake	J. Haber (personal communication)
MOP1	H^+	S. cerevisiae	Unknown	Decreases H^+-ATPase activity	Boguslawski (1985)
pbs1	?	S. cerevisiae	Unknown	Confers resistance to polymyxin B in Trk$^+$ cells	G. Boguslawski and R. F. Gaber (unpublished observations)
PBS2	K^+ or H^+	S. cerevisiae	Kinase	Required for resistance to polymyxin B in Trk$^+$ cells	Boguslawski and Polazzi (1987) G. Boguslawski, M. Vidal, and R. F. Gaber (unpublished observations)
pho84	PO_4^{2-}	S. cerevisiae	Inorganic phosphate transporter	Confers loss of one component of phosphate uptake	Ueda et al. (1975), Bun-Ya et al. (1991)
pmal	H^+	S. cerevisiae and S. pombe	Plasma membrane ATPase	Alters intracellular pH and membrane potential	See Table I for references

(continued)

319

TABLE II
(Continued)

Mutation[a]	Ion	Organism	Gene product[b]	Effect/Phenotype	Ref.[c]
pma2	H^+	S. cerevisiae and S. pombe	Plasma membrane ATPase	Confers sensitivity to low pH	Schlesser et al. (1988)
pmr1	Ca^{2+}	S. cerevisiae	Ca^{2+}-ATPase	Confers increased Ca^{2+} requirement for growth; hypersensitivity to trifluoperazine	Rudolph et al. (1989)
pmr2 (ena1)	Na^+	S. cerevisiae	Na^+-ATPase	Confers sensitivity to high pH (>pH 5.0), Na^+ and Li^+	Rudolph et al. (1989) Haro et al. (1991)
RDP103-1 HXT1-1	K^+	S. cerevisiae	Allelic to HXT1; mutant allele encodes a K^+/glucose cotransporter	Confers increased K^+ uptake in trk1 trk2 cells	C. Ko, T. Herman, and R. F. Gaber (unpublished)
rpd102-1	K^+	S. cerevisiae	Unknown	Confers increased K^+ uptake in trk1 trk2 cells	C. Ko and R. F. Gaber (unpublished)
RDP104-1	K^+	S. cerevisiae	Mutant allele encodes a putative K^+ cotransporter	Confers increased K^+ uptake in trk1 trk2 cells	C. Ko and R. F. Gaber (unpublished)
RPD4-1	K^+	S. cerevisiae	Mutant allele encodes putative K^+ transporter	Suppresses low pH sensitivity and K^+ transport defect of trk1 trk2 cells	M. Sherrer and R. F. Gaber (unpublished)

Gene	Ion	Organism	Product	Phenotype	Reference
RPD5-1 HXT3-1[e]	K^+	*S. cerevisiae*	Mutant allele encodes putative K^+/glucose co-transporter	Confers increased K^+ uptake in *trk1 trk2* cells	C. Ko, T. Herman, B. Kennedy, C. Lin, and R. F. Gaber (unpublished)
rpd6-1	K^+	*S. cerevisiae*	Unknown	Confers ability of *trk1 trk2* cells to grow below pH 4.5	M. Sherrer and R. F. Gaber (unpublished)
rpd1, rpd3	K^+	*S. cerevisiae*	Putative transcriptional regulators	Increase expression of *TRK2*	Vidal *et al.* (1990, 1991) Vidal and Gaber (1991)
TEA1	TEA, K^+?	*S. cerevisiae*	Unknown	Polymorphic dominant allele; confers resistance to tetraethylammonium	J. Anderson and R. F. Gaber (unpublished)
tea2	K^+?	*S. cerevisiae*	Unknown	Polymorphic recessive allele; confers resistance to tetraethylammonium	J. Anderson and R. F. Gaber (unpublished)
trk1	K^+	*S. cerevisiae, S. uvarum, K. lactis*	High-affinity K^+ transporter	Confers defect in high-affinity K^+ uptake	Ramos *et al.* (1985), Gaber *et al.* (1988), Anderson *et al.* (1991), A. Brunner, R. Ortega, and R. F. Gaber (unpublished)
trk2	K^+	*S. cerevisiae*	Low-affinity K^+ transporter	Confers defect in low-affinity K^+ uptake	Ko *et al.* (1990), Ko and Gaber (1991)

(continued)

TABLE II
(Continued)

Mutation[a]	Ion	Organism	Gene product[b]	Effect/Phenotype	Ref.[c]
TRK2[D]	K[+]	S. cerevisiae	5′-nontranslated region of low-affinity K[+] transporter gene	Increases expression of TRK2	Vidal et al. (1990)
TFP1/ vma1	Ca[2+]	S. cerevisiae	69-kDa subunit of vacuolar H[+]-ATPase	TFP1 allele confers resistance to trifluoperazine; vma1 allele confers inability to acidify vacuole; slow growth	Kane et al. (1990), Hirata et al. (1990, No. 23)
vat2	Ca[2+]	S. cerevisiae	60-kDa subunit of vacuolar H[+]-ATPase	Confers slow growth	Yamashiro et al. (1990)
vma3	Ca[2+]	S. cerevisiae	19-kDa subunit of vacuolar H[+]-ATPase	Confers inability to acidify vacuoles and undergo fluid-phase endocytosis	Umemoto et al. (1990)
VDAC	Anions	S. cerevisiae	Voltage-gated mitochondrial anion channel	Null alleles confer decrease in ability to grow on nonfermentable carbon sources at 37°C	Mihara and Sato (1985), Forte et al. (1987a)

[a] Unless otherwise specified, alleles in upper case letters denote dominant mutations and those in lower case denote recessive mutations.
[b] In cases where the functional identiy is unknown, information gained by DNA sequence analysis of the gene is provided (if available).
[c] The first report of the isolation and molecular analysis of the gene and/or it mutant phenotype(s) is referenced.
[d] cls24 is allelic to cdc24 (Ohya et al., 1986a).
[e] RKT3 alleles were renamed HXT3 due to the similarity of this gene (85% amino acid identity) to HXT1.

pmal mutants that exhibited different levels of plasma membrane ATPase activity (Table I). The mutations in their study proved to be more useful than mutations that completely abolish activity because correlations could be established between ATPase activity and the effect on intracellular pH and growth rate. They showed that *in vitro* ATPase activities of six mutants correlated directly with proton transport *in vivo*. Furthermore, the *in vitro* ATPase activity present in the mutants, which ranged from 10 to 70% of wild type, also correlated completely with the growth rate of the cells. This established quite convincingly that the plasma membrane ATPase is "strictly rate limiting for yeast growth" (Portillo and Serrano, 1989).

Lentzen *et al.* (1987) reported that the activity of the H^+-ATPase oscillates during progression through the cell cycle. Taken together with the observation that the plasma membrane ATPase seems to be rate limiting for growth, one might expect to find *pmal* mutations that affect the cell cycle. Indeed, McCusker *et al.* (1987) reported that one of their hygromycin-resistant mutants showed an unusual alteration of the cell cycle. When *pmal-101* cells were grown in synthetic medium the mother cell would enlarge and develop a large number of buds, each of which contained a nucleus. They suggested that this phenotype may be due to the inability of the *pmal-101* mutant to "keep pace" with the cell cycle. Further investigation into the relationship between the alteration of H^+-ATPase function caused by this mutation and the cell cycle may add significantly to our knowledge of the critical control points between cell growth and the mitotic cell cycle.

9. Physiology of Mutants with Reduced Expression of Plasma Membrane ATPase

Mutants that have reduced plasma membrane ATPase activity show decreased capability for proton efflux, slow growth, and lowered intracellular pH (McCusker *et al.*, 1987; Ulaszewski *et al.*, 1987; Vallejo and Serrano, 1989). Combined with the observation that the catalytic activity of the ATPase in wild-type cells is stimulated by low pH medium, these results underscore the significance of a primary regulatory mechanism that alters proton pumping activity depending on the intracellular pH needs of the cell. In drawing conclusions about the effect of reducing the activity of the plasma membrane ATPase on the cell, it is essential to understand the nature of the mutation(s) leading to decreased activity. Two examples illustrate this point: (1) Although McCusker *et al.* (1987) found that several of their ATPase-deficient mutants exhibited low pH sensitivity that could be suppressed by 0.1 M K^+, the low pH-sensitive phenotype of *pmal* mutants constructed by Vallejo and Serrano (1989) were not K^+ suppressible. It is now known that the K^+-suppressible mutants of McCusker *et al.*

altered the structure of the enzyme (Perlin *et al.*, 1989), whereas the mutants of Vallejo and Serrano simply expressed lower levels of the authentic enzyme. (2) Cells containing the *pma1-101* mutation can lead to the formation of a large mother cell surrounded by many nucleated bud cells (McCusker *et al.*, 1987). In contrast, Vallejo and Serrano (1989) observed an unusual but entirely different morphology among cells that consistently express equally low levels of the enzyme.

10. Expression of *PMA1* in Fibroblasts

Expression of the H^+-ATPase from *S. cerevisiae* in mouse and monkey fibroblasts has been used to provide strong evidence that an increase in intracellular pH is not merely coincident with the activation of cell proliferation but plays a crucial role in this process. Fibroblasts expressing the yeast proton pump exhibit properties typical of transformed cells, including growth to high density, "rounding up," growth in soft agar, and tumor formation in nude mice (Perona and Serrano, 1988; Perona *et al.*, 1990). Previous strategies to increase the intracellular pH of animal cells have relied on the use of alkaline media or ammonia (Epel and Dube, 1987), both of which produce long-term toxicities and cannot be used to study growth regulation in intact organisms. With the expression of the yeast H^+-ATPase in mammalian cells, not only has a specific pH-altering mechanism been incorporated in these cells, but now transgenic animals can be generated using this approach to examine the effects of increased intracellular pH in the whole animal. With the additional specificity offered by placing *PMA1* under the control of tissue-specific promoters, investigations into the role of intracellular pH in growth control can now be carried out with extraordinary precision.

11. Identification of *PMA1* Genes in Other Organisms

In addition to yeast, *PMA1* cognate genes have been identified in other organisms, including *Neurospora* (Addison, 1986; Hager *et al.*, 1986), tobacco (Boutry *et al.*, 1989), *Arabidopsis* (Harper *et al.*, 1989, 1990; Pardo and Serrano, 1989), and parasitic protozoans (Meade *et al.*, 1987). Regardless of the source, these enzymes show a high degree of amino acid sequence conservation.

12. Genetic Identification and Molecular Analysis of *PMA2* in Yeasts

A second gene encoding an H^+-ATPase has been discovered in *S. cerevisiae* (Schlesser *et al.*, 1988) and *S. pombe* (A. Goffeau, personal communication). Remnants of a second *PMA* gene in *Neurospora* have also been found but this gene appears to be nonfunctional (C. W. Slayman, unpublished results). The *S. cerevisiae PMA2* gene shares nearly 90% amino acid sequence identity with *PMA1*. The principal differences are located at the amino and carboxy termini, where they exhibit almost complete divergence. Since structural models of the yeast H$^+$-ATPases have varied quite widely with regard to the number of transmembrane domains (Serrano *et al.*, 1986; Goffeau *et al.*, 1989; Serrano, 1989, 1990), comparative analyses of the *PMA1* and *PMA2* genes of both *S. cerevisiae* and *S. pombe* may serve to derive more accurate structural models since these proteins are likely to be structurally highly conserved.

Unlike *PMA1*, *PMA2* in *S. cerevisiae* is not an essential gene (Schlesser *et al.*, 1988). Thus far, the only phenotype observed to be clearly associated with *PMA2* is an increase in resistance to alkaline conditions conferred by the deletion of *PMA2* from cells harboring the *pma1-1* mutation (Schlesser *et al.*, 1988). This phenotype suggested that the function of *PMA2* is similar to *PMA1*. Indeed, Ghislain and Goffeau (1991) have shown that the *PMA2* gene in *S. pombe* is a functional proton pump. They placed the *PMA2* gene under control of the *ADH* promoter and found that this construct could rescue the lethality of a *pma1* null mutant. Thus, the two H$^+$-ATPases in *S. pombe* are functionally interchangeable under these circumstances. The reason why the *PMA2* genes in *S. cerevisiae* and *S. pombe* are apparently expressed at so low a level in wild-type cells (neither rescue *pma1* null mutations when expressed from their own promoters) remains unknown. Genetic redundancy and disparate functional relationships among the members of an ion transport "family" is a recurring theme in yeast (see Section II,B,2,h).

Several laboratories have observed genetic "cross-talk" between *PMA1* and *PMA2* (Goffeau *et al.*, 1990; J. Haber, personal communication). In attempts to isolate intragenic revertants of *pma1* mutations, many of the phenotypically Pma$^+$ cells were found to be true revertants that arose from gene conversion between the mutant *pma1* sequence and the *PMA2* gene. The structural similarity of *PMA1* and *PMA2* could prove advantageous in molecular genetic analysis since there is the potential for recombination between *PMA1* and *PMA2* genes, generating chimeric H$^+$-ATPases.

B. K$^+$ Transport

The extrusion of protons via the proton pump necessitates the uptake of a counter ion in order to achieve overall electrical neutrality. This is accomplished most efficiently by the uptake of K$^+$ (Conway and Brady, 1950).

We now know that such K^+/H^+ antiport activity is indirectly coupled since the primary vehicles for H^+ efflux and K^+ transport are proteins encoded by separate genes. Thus, the energy for the uptake of K^+ does not originate directly from ATP but from the electrical gradient established across the yeast membrane by the proton pump. Consistent with this conclusion is the observation that the extrusion of protons and the uptake of potassium ions occurs with a stoichiometry of 1 : 1 (Ogino et al., 1983). The electrical potential across the plasma membrane of yeast has been estimated at −180 mV (inside negative) (Borst-Pauwels, 1981; Peña et al., 1987), a potential that is sufficient to generate and maintain a 1,000-fold concentration gradient of potassium. This accounts for the high intracellular potassium concentration (130–170 mM) (Ogino et al., 1983) and the ability of yeast cells to grow on micromolar levels of the ion (Rodriguez-Navarro and Ramos, 1984).

Potassium transport is, however, quite complex. Rodriguez-Navarro and Ramos (1984) described a dual affinity for the uptake of potassium in S. cerevisiae. They reported that the high-affinity (K_m = 24 μM), high-velocity (V_{max} = 34 nmol mg^{-1} min^{-1}) transport activity develops from cells exhibiting low-affinity (K_m = 2 mM), low-velocity (V_{max} = 7 nmol mg^{-1} min^{-1}) activity. Rodriguez-Navarro and Ramos found that high-affinity uptake reached maximum activity approximately 4 hr after potassium in the growth medium was reduced to micromolar levels but concluded that protein synthesis was not required.

Ramos et al. (1985) isolated the first mutant (trk1) defective in the high-affinity component of K^+ uptake. Deletion of TRK1 from haploid yeast cells abolished high-affinity K^+ uptake (Gaber et al., 1988), demonstrating that the dual affinity for K^+ does not occur by interconversion of a single system. Rather, the different affinities reflect the activity of functionally distinct transporters.

1. The *TRK* Class of K^+ Transporters

Genetic screening for mutations that confer an increase in the minimum concentration of potassium required to support growth led to the identification of the genes encoding the high-affinity (Ramos et al., 1985; Gaber et al., 1988) and low-affinity (Ko et al., 1990) K^+ transporters. DNA sequence analysis revealed that the genes encoding the high-affinity (Gaber et al., 1988) and low-affinity (Ko and Gaber, 1991) transporters are sufficiently related to define a specific class of K^+ transporter designated TRK, for Transport of K^+.

a. Genetic Identification of TRK1   In mutant hunts designed to identify the gene(s) in S. cerevisiae encoding the high-affinity K^+ transporter, heat-killing enrichment schemes were used to enrich for mutants that require increased levels of potassium in the medium to support growth (Ramos *et al.*, 1985; Gaber *et al.*, 1988). Although mutations in several complementation groups were obtained in the later study, only mutations in *trk1* showed obvious defects in K^+ uptake (Rodriguez-Navarro and Ramos, 1984; Gaber *et al.*, 1988). Curiously, *kdm2* and *kdm4*, two mutants that require increased potassium for growth (Gaber *et al.*, 1988), are unable to grow on complete synthetic medium lacking tryptophan but grow well on a minimal medium (R. F. Gaber, unpublished). Thus, mutations other than those conferring a defect in transport can cause an increased requirement for the ion.

b. Molecular Analysis of the Saccharomyces cerevisiae and Saccharomyces uvarum TRK1 Genes The wild-type *TRK1* gene from *S. cerevisiae* was cloned by exploiting its ability to suppress the increased K^+ requirement of *trk1* cells (Gaber *et al.*, 1988). DNA sequence analysis indicated that *TRK1* encodes a membrane protein containing 12 putative transmembrane domains (M1–M12). Antibodies raised against the transporter were used to show that TRK1 is localized to the plasma membrane. Several structural aspects of TRK1 are noteworthy. First, although it contains a region that closely resembles a consensus sequence for the class of nucleotide binding domains described by Higgins *et al.* (1986), this region is not essential for transport of K^+ (J. Anderson and R. F. Gaber, unpublished results). Second, located between putative transmembrane domains M3 and M4 is an exceptionally large, 650-amino acid hydrophilic region. Mutations constructed *in vitro*, consisting of small (four amino acids) insertions, have been constructed throughout *TRK1*. Six of these, located in the large hydrophilic region, apparently have little or no affect on K^+ transport since these cells retain the ability to grow on minimal concentrations of potassium (J. Anderson and R. F. Gaber, unpublished results). Thus, whatever its relationship to K^+ transport, this domain can accommodate minor structural alterations. Insertion mutations that confer a Trk^- phenotype were found to reside within or adjacent to putative transmembrane domains.

The genomes of most *Saccharomyces* species contain sequences highly related to the *S. cerevisiae TRK1* gene (Anderson *et al.*, 1991). However, in Southern blot hybridization experiments using the *S. cerevisiae TRK1* gene (c-*TRK1*) as a probe, the endogenous *TRK1* sequence in *Saccharomyces uvarum* (u-*TRK1*) was detected only under low-stringency conditions. u-*TRK1* was cloned and found to confer high-affinity potassium uptake in *S. cerevisiae* (Anderson *et al.*, 1991). DNA sequence analysis revealed that the inferred amino acid sequences of u-TRK1 and c-TRK1 are 78% identical (86% similar). The putative membrane-spanning domains are the most highly conserved regions in these proteins (95% identical). The degree of sequence conservation observed between either high-affinity transporter and TRK2, the low-affinity transporter, is significantly reduced (see Section II,B,1,c), suggesting that these regions directly affect affinity for K^+.

Saccharomyces cerevisiae cells containing a *trk1* null allele, generated by deleting most of the *TRK1* open reading frame (*trk1Δ*), are viable but have lost high-affinity K^+ uptake (Gaber et al., 1988). Wild-type cells growing in the presence of typical concentrations of ammonium (5% NH_4SO_4) require at least 30–50 μM K^+ to support growth; *trk1Δ* cells require 3–5 mM K^+ in the presence of ammonium. The viability of *trk1Δ* cells and their increased requirement for K^+ in the growth medium revealed that low-affinity uptake is mediated by a K^+ transporter that is functionally distinct from TRK1.

c. Genetic Identification of the Low-Affinity K^+ Transporter in Saccharomyces cerevisiae Deletion of the high-affinity K^+ transporter from *S. cerevisiae* not only exposed the activity of the low-affinity K^+ transporter, but also generated cells in which the genetic identification of the low-affinity transporter was possible. Two independent approaches resulted in the genetic identification of this second K^+ transporter. First, using *trk1Δ* cells and a variation of the heat-killing enrichment scheme mentioned above, 42 independent mutants were isolated, each of which required 50–100 mM KCl to allow growth (Ko et al., 1990). Genetic analysis indicated that each of these mutants harbored a recessive mutation at the same locus, designated *TRK2*. Growth curves showed that the concentration of potassium required to produce half-maximal growth of *trk1* cells is 3–5 mM (in medium containing ammonium), while this value increases to approximately 25 mM for *trk1 trk2* cells (Ko and Gaber, 1991).

A second genetic scheme that gave rise to mutations at *TRK2* was based on the selection for Trk$^+$ pseudorevertants of a *trk1Δ* mutant. By selecting for suppressor mutants in *trk1Δ* cells that allow growth on medium containing low levels of potassium (0.1 mM), dominant mutations at *TRK2* were obtained. Molecular analysis of these mutations (designated *TRK2^D*) revealed that they consisted of substitutions at any of three adjacent sites located 101–103 nucleotides upstream from the *TRK2* open reading frame (Vidal et al., 1992). Expression of β-galactosidase activity from wild-type (*TRK2::lacZ*) and mutant (*TRK2^D::lacZ*) fusions confirmed that the *TRK2^D* mutations confer an increased expression of *TRK2*. The increased expression of *TRK2^D* alleles leads to an increase in the rate of K^+ uptake (Vidal et al., 1990) sufficient to allow *trk1Δ TRK2^D* cells to grow on medium containing only 0.1 mM potassium. The need for *TRK2* repression remains unknown although preliminary evidence indicates that *TRK2* may be required for full thermotolerance following heat shock (C. Ko and R. F. Gaber, unpublished). Selection for mutations that increase *TRK2* expression also gave rise to recessive mutations, unlinked to *TRK2* (described in Section II,B,1,g).

d. Cloning and Molecular Analysis of TRK2   Attempts to clone the *TRK2* and *HOL1* (see Section II,H) genes from existing libraries failed due to a toxic effect these genes have in *Escherichia coli*. We developed a strategy for cloning genes directly from yeast to yeast that avoids the initial propagation of the plasmids in *E. coli* (Gaber *et al.*, 1990). This method is an efficient way to clone dominant mutations in general. It has proven successful for the cloning of genes such as *HOL1* (Gaber *et al.*, 1990), *TRK2* (Ko and Gaber, 1991), and the K^+/glucose transporters described in Section II,B,3.

DNA sequence analysis of *TRK2* revealed that, like TRK1, TRK2 is predicted to contain 12 membrane-spanning domains (M1–M12) with a large hydrophilic region between M3 and M4. The low-affinity K^+ transporter is significantly smaller than the high-affinity transporter (101 vs 141 kDa). Nevertheless, the inferred amino acid sequences of the two transporters are 55% identical with the putative transmembrane domains exhibiting the highest degree of sequence conservation (66–100% identical). Thus, speaking only of the putative transmembrane domains, there is greater sequence divergence between the high- and low-affinity transporters of *S. cerevisiae* than there is between the *S. cerevisiae* and *S. uvarum* high-affinity transporters. The large hydrophilic domains between M3 and M4 in all three transporters share little sequence similarity.

Functional comparisons of the low- and high-affinity K^+ transporters were facilitated by the disruption of either *TRK1* or *TRK2*. *TRK1 trk2Δ* cells exhibit wild-type high-affinity K^+ uptake and *trk1Δ TRK2* cells exhibit low-affinity uptake (Ko *et al.*, 1990). Thus, TRK1 can mask defects in TRK2 and can function normally in cells deleted for the low-affinity transporter.

e. TRK1 and TRK2 Are Nonessential   Cells in which both the high- and low-affinity K^+ transporters have been disrupted are viable, revealing the existence of other K^+ transport mechanisms (Ko and Gaber, 1991). Deletion of *TRK2* in cells from which *TRK1* has already been deleted increases the potassium requirement by an additional 5- to 10-fold (see above). A second, related phenotype of *trk1Δ trk2Δ* cells is their hypersensitivity to low pH. Unlike wild-type cells, which can grow in medium well below pH 3.0, *trk1Δ trk2Δ* cells fail to grow on medium below pH 4.5 even in the presence of 100 m*M* KCl. The low pH-sensitive phenotype can be suppressed by very high concentrations of potassium (>400 m*M*) but not sodium or sorbitol. The phenotypes associated with deletion of the TRK transporters revealed important functional aspects of the K^+ transporters that remain in these cells. First, the non-TRK transporters function so inefficiently that, even under permissive conditions, normal growth of $trk\Delta$ $trk2\Delta$ cells occurs only when the extracellular concentration of K^+ is approximately equal to the internal concentration (Ko and Gaber, 1991). Since the uptake of K^+ is severely impaired in low-pH medium, the non-TRK transporters could either be K^+/H^+ antiporters, which would be inhibited in H^+ extrusion in the face of a large inward-facing H^+ gradient, or they could be cation/solute symporters with K^+ transport being competitively inhibited by protons. The latter possibility has been further supported by genetic analysis of mutations that increase the efficiency of some of the non-TRK K^+ transporters (see Section II,B,3).

Another phenotype of $trk1\Delta$ $trk2\Delta$ cells suggests that one of the non-TRK K^+ transporters could be a yeast K^+ channel. Compared to $TRK1$ $trk2\Delta$ or $TRK1$ $TRK2$ cells, $trk1\Delta$ $trk2\Delta$ cells are hypersensitive to the K^+ channel-blocking agent 4-aminopyridine (4AP) (C. Ko and R. F. Gaber, unpublished), indicating that $trk1\Delta$ $trk2\Delta$ cells rely on 4AP-sensitive proteins for K^+ uptake. Whether or not K^+ channels in yeast can allow sufficient uptake of the ion to allow growth should be revealed by genetic and molecular analysis of mutations in $trk1\Delta$ $trk2\Delta$ cells that give rise to 4AP resistance.

f. TRK2 Confers Sensitivity to Tetraethylammonium   Aside from their different apparent affinities for K^+, TRK1 and TRK2 show a functional difference in their ability to confer tetraethylammonium (TEA) sensitivity to *S. cerevisiae* cells. Yeast cells can be sensitive to as little as 1 mM TEA if, and only if, they harbor a functional $TRK2$ gene (C. Ko, J. Anderson, and R. F. Gaber, unpublished results). $TRK1$ $trk2\Delta$ and $trk1\Delta$ $trk2\Delta$ cells are completely resistant to TEA (>1 M tested). Possibly, TEA depends on its uptake by TRK2. If so, this represents a powerful genetic tool that would allow direct selection for $TRK2$ mutations that confer decreased uptake of TEA while maintaining the ability to transport K^+. Such mutations could establish a genetic map of potential binding/transport sites for TEA and K^+. Other genes conferring TEA resistance to $TRK2$ cells ($TEA1$ and $tea2$; see Table II) have also been identified (J. Anderson and R. F. Gaber, unpublished results).

g. Mutations in Genes Regulating the Low-Affinity K^+ Transporter   The selection for increased expression of $TRK2$ described above also resulted in the identification of recessive mutations at $RPD1$ and $RPD3$ (Vidal *et al.*, 1990). Epistasis tests demonstrated that the ability of $trk1$ $rpd1$ and $trk1\Delta$ $rpd3$ cells to grow on medium containing minimal concentrations of potassium (0.1 mM) depends on *TRK2*: *trk1 trk2 rpd1* and *trk1 trk2 rpd3* cells are unable to grow on this medium. Subsequent molecular analysis of *TRK2* expression confirmed that *RPD1* and *RPD3* are repressors of the low-affinity transporter (Vidal and Gaber, 1991; Vidal *et al.*, 1991). As expected from an increase in the amount of the transporter present in the membrane, *rpd1* and *rpd3* mutations increase the rate of K^+ uptake but do not affect the K_m for K^+ transport (Vidal *et al.*, 1990; R. F. Gaber, unpublished results).

h. Why Two TRK Genes? Using the *TRK1* and *TRK2* genes from *S. cerevisiae* as molecular probes, it was discovered that both of these genes are present in the genomes of eight *Saccharomyces* species examined (J. Anderson, C. Ko, and R. F. Gaber, unpublished results). The role of TRK2 in the cell is still unknown despite the functional differences it displays compared to TRK1 (see above). In contrast, hybridization experiments suggest that only a single *TRK*-related gene exists in *Kluyveromyces lactis* (J. Anderson, C. Ko, and R. F. Gaber, unpublished results). DNA sequence analysis of the *K. lactis TRK* gene, cloned by its ability to hybridize to the *TRK1* gene from *S. cerevisiae* (R. Ortega, A. Brunner, and R. F. Gaber, unpublished results), is currently underway. Further data on the physiological and structural differences between TRK K^+ transporters in these two yeasts may help us to understand the basis of the TRK redundancy in *Saccharomyces*.

2. K^+ Efflux

The genetic basis for K^+ efflux from yeast remains unknown. Peña's group has shown that the efflux mechanism is energy dependent and functions normally in cells from which the high-affinity transporter (TRK1) has been deleted, demonstrating that uptake and efflux do not depend on the same transporter. An energy-dependent efflux activity has also been shown to occur in cells that harbor mutations in both TRK1 and TRK2 (Ko *et al.*, 1990; Ko and Gaber, 1991), indicating that the transporter responsible for K^+ efflux is independent of both of these proteins.

3. K^+ Transport by Sugar Transporters

The deletion of *TRK1* and *TRK2* revealed the existence of other K^+ transporter(s) and generated the appropriate cell type with which to investigate their function and determine their genetic identification. The author's laboratory is currently investigating the genetic and molecular basis of suppressor mutations in *trk1Δ trk2Δ* cells that restore K^+ transport (designated *RPD*). *RPD* mutations have been obtained in two genetic selection schemes. First, when suppressor mutations in *trk1Δ* *trk2Δ* cells were selected by their ability to confer growth on lower levels of potassium (7 m*M*, pH 5.9), dominant mutations in three genes (*RPD103*, *RPD104*, and *RPD5*) and recessive mutations in another (*rpd102*) were obtained. In contrast, selection for suppression of the low-pH sensitive phenotype (pH 3.0, 100 m*M* KCl) of *trk1Δ* *trk2Δ* cells resulted in dominant mutations in two genes (*RPD103* and *RPD4*) and recessive mutations in a third (*rpd6*). Thus, although the increased potassium requirement and the low pH-sensitive phenotypes are related (described above), overlapping sets of genes were identified among the suppressor mutants obtained from the two selection schemes.

Direct selection for suppressor mutations that functionally bypass the *trk1Δ* *trk2Δ* potassium transport defect has provided new information about the function of other transport proteins in yeast. DNA sequence analysis of *RPD103* and *RPD5* revealed that they encode highly related proteins (86% amino acid identity) that contain 12 putative transmembrane domains. *RPD103* shares 62% identity with the galactose transporter encoded by *GAL2* (Szkutnicka *et al.*, 1989a) and the hexose transporter encoded by *HXT2* (Kruckeberg and Bisson, 1990) and is allelic to the recently identified glucose transporter gene *HXT1* (Lewis and Bisson, 1991). Thus, the *RPD103-1* and *RPD5-1* suppressor alleles encode mutant glucose transporters capable of transporting K^+ ions, apparently coupled to uptake of the sugar. Because these genes encode sugar transporters, some of which have been already identified and named we refer to the K^+-transporting alleles of *RPD103* and *RPD5* as *HXT1-1* and *HXT3-1*, respectively.

The hypothesis that *HXT1-1* and *HXT3-1* are sugar-dependent K^+ transporters is further supported by the observation that cells containing these mutations can only suppress the *trk1Δ* *trk2Δ* phenotype if glucose is the carbon source. *HXT1-1* *trk1Δ* *trk2Δ* cells, for example, are unable to grow on glycerol/ethanol medium supplemented with 7 m*M* KCl but grow normally on this medium if the potassium concentration is increased to 100 m*M* KCl (C. Ko and R. F. Gaber, unpublished observations). Furthermore, uptake of ^{86}Rb in *trk1Δ* *trk2Δ* *HXT1-1* cells can be driven by the uptake of 2-deoxyglucose, indicating that glucose transport and not glucose metabolism is required for K^+ uptake by these mutant glucose transporters.

DNA sequence analysis of the wild-type and suppressor forms of *HXT1* and *HXT3* revealed that the suppressor mutations reside within or immediately adjacent to different putative membrane-spanning domains (C. Ko, T. Herman, and R. F. Gaber, unpublished results). These results explain the dominant nature of the suppressors and suggest that K^+ uptake may occur as a result of an increase in the permeability of the membrane to ions when the transporters are in the act of transporting glucose.

4. K⁺ Transport and Drug Resistance

pbs1 mutations confer resistance of *S. cerevisiae* cells to polymyxin B (Boguslawski, 1985). A correlation between K⁺ transport and polymyxin B resistance has been established by comparing the polymyxin B resistance of K⁺ transport-proficient and K⁺ transport-deficient cells that harbor the *pbs1* resistance mutation (G. Boguslawski, M. Vidal, and R. F. Gaber, unpublished results). *trk1 pbs1* cells exhibit significantly reduced resistance to polymyxin B compared to *TRK1 pbs1* cells. Resistance does not rely on a particular transporter but on a minimum capacity for K⁺ uptake: *trk1Δ pbs1* cells in which *TRK2* is overexpressed exhibit polymyxin B resistance (G. Boguslawski, M. Vidal, and R. F. Gaber, unpublished results). Thus the K⁺ transport capacity required for polymyxin B resistance can be provided by *TRK1* or overexpression of *TRK2* but not by the normal expression level of *TRK2* in *trk1Δ* cells. Since the rate of H⁺ extrusion depends, in large part, on the uptake of K⁺ as a counter ion (Ogino *et al.*, 1983), the activity of the plasma membrane H⁺-ATPase rather than K⁺ uptake may be directly responsible for conferring polymyxin B resistance. In support of this hypothesis, mutations in *PMA1* that lead to cesium resistance (M. Gustin, D. Conklin, and M. Culbertson, personal communication), presumably by decreasing the membrane potential, also confer resistance to polymyxin B (G. Boguslawski, unpublished results). This hypothesis could be further supported if mutations known to decrease the membrane potential (Perlin *et al.*, 1989) are found to confer resistance to polymyxin B.

5. K⁺ Transport by a Plant K⁺ Channel in Yeast

Identification of the first non-animal ion channel gene was recently accomplished through the isolation and sequence analysis of a cDNA from *Arabadopsis thaliana* that fully suppressed the K⁺ transport defect in *S. cerevisiae* cells delated for *TRK1* and *TRK2* (Anderson *et al.*, 1992). This K⁺ channel cDNA from plants, designated *KAT1*, is capable of encoding a protein that contains several of the hallmarks of the *Shaker* family of K⁺ channels including six amino terminal transmembrane domains, a putative voltage gating domain (S4) and the highly conserved H5 region thought to form the pore lining. Pharmacological evidence further supports the likelihood that *KAT1* encodes a K⁺ channel: TEA and barium ion, known

K$^+$ channel inhibitors, inhibit the function of KAT1 in the *trk1Δ trk2Δ* yeast recipient. Our observation that a K$^+$ channel will allow the growth of yeast cells on medium containing as little as 50 μm potassium reveals that K$^+$ uptake in yeast can, at least in this case, rely on only two biophysical parameters: the membrane potential and a K$^+$-specific port of entry.

The author has recently become aware of another putative K$^+$ channel cDNA (designated *AKT1*) that was isolated from *Arabidopsis* by its ability to confer K$^+$ uptake in *S. cerevisiae* (Sentenac *et al.*, 1992). The amino acid sequences of *KAT1* and *AKT1* share extensive identity but are clearly not allelic, demonstrating the existence of multigene K$^+$ channel families in higher plants.

C. Ca^{2+} Transport

1. Calcium-Related Mutants

Ca^{2+} transport across the yeast plasma membrane was demonstrated by Eilam (1982) to be inhibited when the extracellular pH is increased, suggesting the existence of Ca^{2+}/H$^+$-antiporter activity. Morphological analysis of the first Ca^{2+}-dependent mutant isolated, *cal1-1* (Ohya *et al.*, 1984), revealed that calcium plays a crucial role in bud development and nuclear division. When *cal1-1* cells were placed in Ca^{2+}-poor medium they exhibited temperature-sensitive growth and arrested as large cells with a tiny bud. Nuclear staining indicated taht the nuclei in these cells had arrested at the G$_2$ stage. Ohya *et al.* (1986a) isolated Ca^{2+}-sensitive mutants (*cls*) that fell into 18 complementation groups. Type I (*cls5, cls6, cls13–cls18*) showed increased initial rates of Ca^{2+} uptake and increased intracellular Ca^{2+} content; type II (*cls4*) showed normal initial rates of Ca^{2+} uptake and normal intracellular Ca^{2+} content; type III (*cls1-cls3*) showed normal initial rates of Ca^{2+} uptake but increased intracellular Ca^{2+} content; and type IV (*cls8-cls11*) showed increased initial rates of Ca^{2+} uptake but normal intracellular Ca^{2+} content. The type IV mutants were also unable to grow on nonfermentable carbon sources, suggesting the presence of Ca^{2+}-related respiratory defects.

cls4 and *cdc24* were shown to be alleles of the same gene (Ohya *et al.*, 1986b). The presence of 100 m*M* Ca^{2+} in the growth medium is sufficient to arrest bud development but not nuclear division in *cls4* (*cdc24*) cells. Anand and Prasad (1987) subsequently showed that the rate of Ca^{2+} uptake is increased in *cdc24* and *cdc7* mutant cells following cell cycle arrest at nonpermissive temperature, a result that further attests to the importance of Ca^{2+} to bud development in yeast.

The first evidence that mobilized Ca^{2+} plays an essential role in yeast was obtained by Iida *et al.* (1990). They showed that a transient increase in intracellular Ca^{2+} occurs in **a** cells only during the late phase of the mating response to α cells when the **a** cells elongate. The availability of sufficient Ca^{2+} in the growth medium is essential for the viability of **a** cells during this stage. The observation that an increase in intracellular Ca^{2+} depends on extracellular Ca^{2+} demonstrated that the mobilized ion is recruited across the plasma membrane and not the vacuole.

In recent unpublished experiments T. Dunn and T. Beeler isolated approximately 80 independent mutants that are sensitive to 100 mM Ca^{2+}. Of these, 36 are specifically sensitive to Ca^{2+} and not Sr^{2+} or Ba^{2+}. The Ca^{2+}-specific group consists of at least four complementation groups (*csg1–csg4*). During the development of Ca^{2+} transport assays to be used to further screen their *csg* mutants, T. Dunn and T. Beeler (unpublished observations) discovered that no net accumulation of Ca^{2+} occurs when cells are grown in low Ca^{2+} medium (<1 mM). However, in high Ca^{2+} medium (1–100 mM) yeast cells accumulate Ca^{2+} to a maximum concentration of 300 nmol/mg protein. They also observed that although yeast cells tolerate high levels of Ca^{2+} in medium containing glucose, Ca^{2+} rapidly kills cells at 37°C in the absence of glucose.

2. Putative Plasma Membrane Ca^{2+} Transporters

PMR2 encodes a putative P-type ATPase that was identified by two independent strategies: (1) hybridization to a probe which encodes the phosphorylation site found in all of the P-type ATPases, and (2) complementation of the sporulation defect of diploids homozygous for a mutation (*ssc1*) that conditionally altered secretion (Rudolph *et al.*, 1989). DNA sequence analysis of *PMR2* revealed that it encodes a protein containing 10 conserved regions found in all P-type ion pumps and its predicted membrane topology is similar in many respects to Ca^{2+}-ATPases (Brandl *et al.*, 1986; Shull and Greeb, 1988). The PMR2 protein also contains a region near its carboxy terminus that shares significant sequence similarity with the calmodulin-binding domains of all known plasma membrane Ca^{2+}-ATPases (James *et al.*, 1988; Verma *et al.*, 1988).

The *PMR2* locus consists of at least five highly related *PMR2* genes repeated in tandem (H. Rudolph, A. Antebi, and G. Fink, personal communication). The first two *PMR2* genes of this repeat show only 13 amino acid differences; 12 of these are located near the amino termini. These changes could reflect different affinities for Ca^{2+} since Brandl *et al.* (1986) reported that the amino termini of Ca^{2+}-ATPases may dictate ion specificity. Given what is known of the sequences of the first two *PMR2* genes of the locus, and given the highly conserved restriction site patterns of each of the member genes, was considered that this complex locus might encode a family of plasma membrane Ca^{2+}-ATPases regulated by calmodulin (H. Rudolph, A. Antebi, and G. Fink, personal communication).

G. Fink and H. Rudolph (personal communication) have found that a phosphorylated plasma membrane protein of the appropriate size is present in wild-type cells but missing from cells from which the entire *PMR2* cluster has been deleted. Surprisingly, none of the *PMR2* genes is essential. *pmr2Δ* mutants, in which all of the *PMR2* genes have been deleted, are pH sensitive: they grow well only below pH 5.0 and this sensitivity is exacerbated by the presence of auxotrophic mutations. Although the significance of these observations is still unclear, because deletion of the *PMR2* cluster confers negative phenotypes the isolation and analysis of suppressor mutations should lead to an increased understanding of the function of the *PMR2* genes.

A startling discovery concerning the function of the P-type ATPases encoded by the *PMR2* locus was recently reported by Haro, Garcideblas and Rodriguez (Haro *et al.*, 1991). In a search for *S. cerevisiae* strains that exhibited different degrees of Li^+ and Na^+ sensitivity, they found that one line contained a single locus, designated *ENA1*, that conferred sensitivity. The wild-type *ENA1* locus was cloned by its ability to suppress the Li^+ and Na^+ sensitivity of their sensitive strain. DNA sequence analysis revealed that *ENA1* is allelic to *PMR2*. Furthermore, disruption of *ENA1* (*PMR2*) led to decreased ability to efflux sodium and lithium from the mutant and sensitivity to these ions. The ability of ENA1 to pump sodium has not been shown to depend on calcium. Thus, although the sequence similarity of *PMR2* to specific Ca^{2+}-ATPase motifs and to calmodulin-binding proteins suggested that this complex locus encoded multiple Ca^{2+}-ATPases, the most recent evidence more strongly suggests that this locus encodes a sodium pump. This would better explain why the deletion of the entire locus did not confer an increased requirement for calcium.

The *CSG2* gene is likely to encode a membrane protein capable of regulating Ca^{2+} transport (T. Dunn and T. Beeler, personal communication). The inferred CSG2 protein sequence contains a consensus Ca^{2+} binding loop and several putative transmembrane domains. *csg2* null alleles confer increased accumulation of Ca^{2+} and a wild-type copy of *CSG2* is required for growth on high Ca^{2+} medium (>3 mM).

Nakajima *et al.* (1991) have determined that the jellyfish protein aegorin, which emits light on binding Ca^{2+}, can be expressed at sufficiently high levels in *S. cerevisiae* to allow its use as an *in vivo* reagent to measure changes in Ca^{2+} concentrations. This development may prove highly significant, not only for the relevant physiological determinations but also for screening the effects of various mutants on the rates and extent of Ca^{2+} transport.

D. Mg^{2+} Transport

Little is known about the genetic basis of Mg^{2+} uptake by yeast cells. In 1982 Cooper reported that "as yet no genetic studies have been reported for Mg^{2+} transport." Unfortunately, a decade later this remains true. Genetic redundancy and the ability of Mg^{2+} to be transported by Ca^{2+} transporters may be responsible for the apparent lack of progress in the genetic identification of the Mg^{2+} transporter. For a review of the physiology of Mg^{2+} transport the reader is referred elsewhere (Borst-Pauwels, 1981).

E. NH_4^+ Transport

Bogoñez et al. (1983) first described the NH_4^+ uptake system in yeast as a single-component energy-driven system when uptake is quantitated at lower pH values. Unlike many of the ions typically investigated, NH_4^+ undergoes an interconversion with NH_3. Bogoñez et al. (1983) demonstrated the occurrence of at least one additional component in the transport kinetics due to passive NH_3 diffusion across the plasma membrane, significantly complicating transport studies.

More recently, Peña et al. (1987) reported NH_4^+ to be coupled indirectly to the proton pump. They suggested that the electrical potential is directly responsible for the 200-fold accumulation of NH_4^+ against its concentration gradient. Both Peña et al. (1987) and Egbosimba and Slaughter (1987) reported that the rate and extent of NH_4^+ uptake was maximal when cells were provided with carbon sources that maximally activated the proton pump. Peña et al. (1987) used the ratio of NH_4^+ and the pH gradient to conclude that the electrical potential across the plasma membrane of NH_4^+-transporting cells was approximately 180 mV (negative inside). This allowed them to conclude that the NH_4^+ transporter probably functions much like the K^+ transporter(s), i.e., uptake of NH_4^+ may be driven by the membrane potential. Ammonium is not an essential ion since yeast cells can utilize a wide variety of nitrogen sources. Perhaps because of this obstacle, genetic analysis of the NH_4^+ transporter lags behind that of many other ions. Roon et al. (1977) described a mutant, amt, that exhibited decreased NH_4^+ (methylamine) uptake. Two other complementation groups, *mep1* and *mep2*, were described by Dubois and Grenson (1979) as having defects in NH_4^+ uptake. Whether or not *amt*, *mep1*, or *mep2* represent mutations in the structural genes that encode the NH_4^+ transporter(s) remains to be determined.

F. Phosphate Transport

Mutations in *PHO84* (originally designated *PHOT;* Ueda and Oshima, 1975 no. 227) were shown to be defective in the uptake of inorganic phosphate (Ueda and Oshima, 1975). Bun-Ya *et al.* (1991) showed that *PHO84* encodes a membrane protein that shares significant amino acid sequence identity with the superfamily of sugar transporters (Celenza *et al.*, 1988; Szkutnicka *et al.*, 1989). PHO84 is predicted to contain 12 transmembrane domains and harbors many of the sequence characteristics that are highly conserved in the sugar transporter family (Szkutnicka *et al.*, 1989), including duplicated GRR sequences in membrane-spanning domains 2 and 8, the PESPR sequence in the membrane-spanning domain 6, and the PETK sequence in membrane-spanning domain 12. PHO84 also contains 11 of 13 sites that are completely conserved among the sugar transporters (Kruckeberg and Bisson, 1990). Perhaps more surprising than its similarity to sugar transporters is its lack of sequence similarity to the putative phosphate transporter encoded by the *Neurospora crassa pho-4$^+$* gene, which also encodes a protein likely to contain 12 transmembrane domains (Mann *et al.*, 1989). Northern blot hybridization experiments demonstrated that *PHO84* is transcriptionally regulated by the concentration of extracellular phosphate (Bun-Ya *et al.*, 1991).

Disruption of *PHO84* is not lethal (Bun-Ya *et al.*, 1991). Although haploid cells carrying the *pho84* null allele show significantly impaired uptake of phosphate, the viability of these cells confirmed the existence of an additional phosphate transporter. Perhaps *pho84* cells can be used for the direct selection of mutations that increase uptake through the remaining phosphate transporter(s), thus permitting their identification.

G. Plasma Membrane Ion Channels

With the notable exceptions of the K^+ channel activity exhibited by the H^+-ATPase (Ramirez *et al.*, 1989) and the voltage-gated anion channel of mitochondria (described in Section IV,A) the genetics of ion channels in yeast is only in its infancy. Nevertheless, with the ability to mutate or delete genes encoding other ion transporters such as those responsible for scavenging K^+ from the medium (Gaber *et al.*, 1988; Ko and Gaber, 1991), new ion transport activities may become amenable to genetic analysis.

1. The K^+-Selective Channel

Gustin *et al.* (1986) used the patch-clamp technique in the first demonstration of ion channels in the yeast plasma membrane. Using on-cell, whole-cell, and outside-out patch clamps they showed that a K^+-specific current of approximately 20 pS was induced on depolarization of the membrane and they further showed that known K^+ channel inhibitors, including tetraethylammonium and $BaCl_2$, block this current. A more extensive analysis of this channel will almost certainly depend on its genetic identification since Gustin *et al.* (1986) estimated the presence of only 7 to 10 channels per cell based on their whole-cell vs unitary K^+ conductance measurements.

2. The Mechanosensitive Ion Channel

Gustin *et al.* (1988) used the patch-clamp technique to identify an ion channel activity that conducts both cations and anions through the plasma membrane in a stress-dependent manner. The investigators showed that stretching of the membrane greatly increased the activity of this mechanosensitive (MS) channel. Its sensitivity to gadolinium and its resistance to the tetraethylammonium demonstrated the independence of the MS channel from the K^+ channel present in the same patches of membrane. Although the MS channel can pass both cations and anions, it displays clear selectivities among both monovalent and divalent cations and anions. Gustin *et al.* suggested possible roles for the MS channel, including osmoregulation via salt conductance, signal transduction via Ca^{2+} uptake, and the control of budding via Ca^{2+} entry and subsequent signaling.

3. Ion Channels Are Formed in the Plasma Membrane by Yeast Killer Toxin

Biochemical studies of the effect of killer toxin, encoded by the M1 double-stranded RNA of K1 killer yeast virus, have shown that sensitive cells undergo rapid ion leakage from the plasma membrane (Skipper and Bussey, 1977; de la Peña *et al.*, 1981). Using the patch-clamp technique, Martinac *et al.* (1990) confirmed the previous prediction (Bostian *et al.*, 1984) that toxin-induced ion leakage results from the activity of an ion channel formed by killer toxin in the plasma membrane of sensitive yeast cells. These investigators also showed that killer toxin was able to form ion channels when incorporated into liposomes, strongly implicating the toxin as an active agent that has evolved "to kill sensitive cells by draining them electrically and energetically."

H. *HOL1*, a Promiscuous Cation Transporter

Genetic selection often leads to the discovery of previously unknown cellular functions. *HOL1* mutations, selected on the basis of their ability to take up the histidine biosynthetic precursor histidinol, confer hypersensitivity to cations (Gaber *et al.*, 1990). It was shown that *HOL1* mutants exhibit cation hypersensitivity due to the novel ability of these cells to take up sodium (^{22}Na), histidinol ([^{14}C]histidinol), and presumably the other toxic ions tested (Li^+, Cs^+, and guanidinium ion) under growth conditions. Unlike animal cells, yeast cells take up little if any Na^+ under normal growth conditions (Borst-Pauwels, 1981; Gaber *et al.*, 1990) and can grow in medium containing less than 2 μM Na^+ (Rodriguez-Navarro and Ramos, 1984; Ramos *et al.*, 1985). Thus the toxicity of Na^+ in *HOL1* mutants could be due to their ability to decrease the membrane potential or due to a chemical effect within the cytoplasm. Experiments measuring Na^+ flux confirmed that this ion is taken up by actively growing *HOL1* cells but not by cells containing the wild-type (*hol1*) or null (*hol1Δ*) alleles (Gaber *et al.*, 1990).

Experiments have shown that increasing the Ca^{2+} concentration in the medium inhibits histidinol-dependent growth of *HOL1* cells. Ca^{2+} uptake experiments revealed that *HOL1* mutants exhibited a sixfold increase in the rate of Ca^{2+} uptake (M. Wright and R. F. Gaber, unpublished), demonstrating that the Ca^{2+} inhibition of histidinol-dependent growth is likely due to competition between Ca^{2+} and histidinol (itself a divalent cation at low pH). Thus, dominant mutations in the *hol1* gene appear to convert it into a general transporter of cations.

DNA sequence analysis of the wild-type *hol1* gene indicated that it encodes a protein containing 11 putative membrane-spanning domains (M. Wright and R. F. Gaber, unpublished). Comparisons with sequences in the available databases showed no sequence similarity to other proteins. Molecular analysis of seven independent mutant *HOL1* alleles indicated that V509F (four occurrences) or L510F (three occurrences) substitutions confer the ability of this protein to act as a novel ion transporter. These mutations reside at the midpoint of putative transmembrane domain 10, suggesting that steric hindrance between adjacent membrane-spanning domains might be the molecular basis of the enhanced rate, and the more promiscuous nature of ion transport observed in *HOL1* cells.

The construction of yeast strains from which both the high- and low-affinity K^+ transporters have been deleted has afforded a unique opportunity to assess further the function of the mutant *HOL1* transporter. *HOL1* mutations significantly suppress the increased K^+ requirement of *trk1Δ trk2Δ* cells, indicating that in addition to the ion species mentioned above the mutant HOL1 protein confers the uptake of K^+. Unfortunately, because of the absence of phenotypes associated with *hol1Δ* null allele, the function of the wild-type protein remains unknown. Two other genes have been implicated in histidinol/ion transport. By selecting for histidinol uptake by cells from which the *hol1* gene has been deleted, recessive mutations at two unlinked loci (*hol2* and *hol3*) were obtained (R. F. Gaber, unpublished results). These mutations exhibit not only the same cation hypersensitivities of *HOL1* cells but are also sensitive to potassium when present in concentrations above 20 mM. An intriguing possibility is that *HOL2* or *HOL3* might encode potassium channels responsible for K^+ efflux.

III. Vacuolar Ion Transport

The principal function of the yeast vacuole include controlling the homeostasis of intracellular pools of calcium and metabolites (Ohsumi and Anraku, 1981; Halachmi and Eilam, 1989). Other functions include facilitating the degradation of metabolites and proteins by generating a subcellular compartment that allows selective sequestration and activation of the appropriate hydrolytic enzymes (Weimken *et al.*, 1979). The driving force for these functions is the electrochemical gradient generated by the vacuolar H^+-ATPase, a proton pump that acidifies the vacuole. The low internal pH of the vacuole activates vacuolar enzymes and presumably mediates the release of ligands from receptors. Although the structure of the vacuolar H^+-ATPase has been highly conserved (yeast, bacterial, and archaebacterial enzymes show immunological cross-reactivity (Konishi, 1990), surprisingly, the vacuolar H^+-ATPase is not an essential enzyme in yeast (see this section).

The proton gradient across the vacuolar membrane is used to drive the uptake of 10 species of amino acids from the cytoplasm. The transport of these amino acid is mediated by at least seven independent H^+/amino acid antiporters, the function of which depends on the potential energy inherent in the proton gradient across this membrane (Sato *et al.*, 1984a). Interestingly, the uptake of arginine into the vacuole is mediated by a high-affinity arginine/histidine antiporter, indicating that the uptake of this amino acid is one step further removed from, but nevertheless dependent on, the vacuolar H^+-ATPase (Sato et al., 1984b).

Transport of Ca^{2+} into the vacuole is ATP dependent and appears to be mediated by a Ca^{2+}/H^+ antiporter that exhibits a K_m for Ca^{2+} of approximately 0.1 mM (Ohsumi and Anraku, 1983; Okorokov et al., 1985). The ability of the vacuole to concentrate Ca^{2+} and thus regulate Ca^{2+} homeostasis is further facilitated by the generation of Ca^{2+}–polyphosphate complexes within the organelle (Ohsumi et al., 1988).

The observation that the yeast vacuole contains approximately 30% of the K^+ of the cell (Ohsumi et al., 1988) is consistent with the existence of a cation channel protein in the membrane of this organelle. Such a channel, showing little selectivity between monovalent cations but great selectivity of cations over anions, was first described by Anraku and colleagues (Wada et al., 1987; Tanifuji et al., 1988) but the function of this activity in vitro depended on nonphysiologically high concentrations of calcium. Bertl and Slayman (1990) showed that vacuolar channel activity, likely to be synonomous with that described above, is greatly enhanced in vitro by reducing agents. They were able to detect activity of the channel in the presence of as little as 1 μM calcium after treatment with dithiothreitol (DTT) or 2-mercaptoethanol (2-ME) and suggested that the redox state of –SH groups in the channel could mediate channel regulation in vivo.

A. The Vacuolar H^+-ATPase

The vacuolar H^+-ATPase was biochemically purified and characterized by Uchida et al. (1985). They described three major subunits (89, 65, and 19.5 kDa) and provided evidence that the vacuolar H^+-ATPase is significantly different from the plasma membrane ATPase and the mitochondrial F_0F_1 H^+-ATPase. The two larger subunits were postulated to be peripheral membrane subunits and the small subunit was thought to be an integral membrane protein (Uchida et al., 1985; Kane et al., 1989). The first 12 amino acids of the 2 larger subunits are removed in the mature protein (Kane et al., 1990; Yamashiro et al., 1990).

1. TFP1 (VMA1)

The gene encoding the 69-kDa catalytic subunit of the vacuolar ATPase (equivalent to the 89-kDa subunit of Uchida et al.) was independently cloned and analyzed by two groups: as TFP1 by Kane et al. (1990) and as VMA1 by Hirata et al. (1990). The inferred amino acid sequence of TFP1 was found to be similar to the 70-kDa subunits of the *Neurospora* vacuolar ATPase and the H^+-ATPases from mitochondria, chloroplasts, and *E. coli*. However, the similarities matched only the N-terminal and C-terminal thirds of TFP1. It was suggested by Hirata *et al.* and elegantly demonstrated by Kane *et al.* that the middle third of the TFP1 protein undergoes a novel modification: posttranslational splicing. *TFP1* is nonessential (Shih *et al.*, 1988; Hirata *et al.*, 1990) and dominant mutations at *TFP1*, selected by their ability to confer trifluoperazine resistance, can lead to Ca^{2+} sensitivity.

2. *VAT2/VMA2*

VAT2 encodes the 60-kDa subunit of the vacuolar H^+-ATPase (Yamashiro *et al.*, 1990). Although the precise function of the VAT2 protein has not yet been identified, disruption of VAT2 is not lethal but confers slow growth, sensitivity of the cells to high pH (7.0), and the inability to acidify the vacuole. Vacuoles isolated from *vat2Δ* cells showed no ATPase activity, indicating that the vacuolar proton pump is the major ATPase in the membrane of this organelle (Yamashiro *et al.*, 1990).

3. *VMA3*

VMA3, the gene encoding the 19-kDa subunit of the vacuolar H^+-ATPase, was cloned and analyzed by Umemoto *et al.* (1990). The inferred 160-amino acid peptide contains hydrophobic segments thought to be transmembrane domains. Cells from which *VMA3* has been deleted are viable but show no vacuolar membrane H^+-ATPase activity. *vma3Δ* cells show severely impaired vacuolar biogenesis, are unable to assemble the large subunits of this enzyme, and are completely unable to endocytose the fluid-phase marker lucifer yellow.

B. K^+ Transport

Ohsumi *et al.* (1988) used $CuCl_2$ (100 mM) to mediate the selective release of small molecules from the cytoplasm without affecting the concentration of these molecules in the vacuole. Their results indicated that 30% of intracellular potassium resides within the vacuole. Thus, although there is an electrical potential across the vacuole (internal positive) generated by the vacuolar H^+-ATPase, there does not appear to be a significant difference in the concentration of potassium on either side of this membrane. K^+ transport across the vacuolar membrane has been documented by Kitamoto *et al.* (1988) in studies that show a 1 : 1 stoichiometry of arginine efflux coupled to K^+ influx. Although the occurrence of vacuolar membrane channels capable of transporting potassium ions (discussed above) may account for the permeability to K^+, the genetic identification of this protein(s) remains unknown.

IV. Other Organelles

A. Mitochondrial Voltage-Dependent Anion Channel

The mitochondrial voltage-dependent anion channel (VDAC) is located in the outer mitochondrial membrane (Mannella and Colombini, 1984) and allows the transport of molecules, including ions, into the intermembrane space. Unlike the high molecular weight channels that form narrow aqueous pores, the VDAC is relatively small (approximately 30 kDa) (Mihara and Sato, 1985; Forte et al., 1987a,b) and forms a 3-nm pore through which a variety of molecules, including ions, can pass. Although somewhat promiscuous, this channel exhibits selectivity for certain anions, Cl^- being the most permeable (Colombini, 1979). VDAC structural genes from S. cerevisiae (Mihara and Sato, 1985; Forte et al., 1987b) and Neurospora (Kleene et al., 1987) have been cloned and sequenced. In S. cerevisiae the VDAC is a 283-amino acid protein. A structural model, based on computer-assisted identification of alternating hydrophilic and hydrophobic residues, has been developed in which approximately 11 or 12 β sheets form a single layer between the hydrophobic surface of the membrane and a 3-nm pore generated by the channel.

Cells containing null mutations in the VDAC gene are viable but fail to grow on nonfermentable carbon sources at 37°C or only after a significant delay (Dihanich et al., 1987; Forte et al., 1987a,b; Benz et al., 1989). Why the absence of such a significant mitochondrial structure is not lethal is not yet understood but clearly the cells must be able to utilize an alternative route across the outer mitochondrial membrane. Because cells containing a null mutation of the VDAC gene are viable, electrophysiological analysis of mutants was directly testable by introducing specific mutant VDAC alleles generated by site-specific mutagenesis into recipients containing a null mutation, purifying the mutant channel and reconstituting it in vitro (Blachly-Dyson et al., 1989, 1990).

Selectivity of mutant and wild-type VDACs was quantitated by measuring the reversal potential. The results of these experiments yielded two classes of mutations: those that had a minimal effect on the selectivity of the channel and those that had a significant and equivalent effect on selectivity. The former class is thought to represent residues that reside on either side of the membrane and thus would not be expected to have a significant effect on selectivity. The latter class was observed not only to affect selectivity, but to alter it in the direction predicted by a model in which the side chain of the mutant amino acid resided within the lining of the channel that faced the pore. Although the electrophysiological analysis of mutant channels was limited to measuring the effect of mutations when the channels were in the open (Blachly-Dyson *et al.*, 1990), the results of these experiments strongly support a model consisting of a channel that contains primarily β sheets spanning the membrane. However, mutations, in the amino-terminal end of the protein, predicted to form an amphipathic helix, also affect selectivity of the channel, indicating that this domain, too, forms part of the polar pathway through the mitochondrial membrane. This domain has been postulated to be part of the gating mechanism of the channel (M. Colombini, personal communication).

The entry and exit of metabolites into the mitochondrion is likely to be mediated by redundant transport systems since cells from which the VDAC has been deleted are viable. VDAC-deleted cells offer the appropriate genetic background in which to identify such alternate pathways. The inability of such cells to grow on a nonfermentable carbon source at 37°C should allow the direct selection of suppressor mutations to facilitate this identification.

B. P-Type Ca²⁺-ATPases

Several genes encoding P-type Ca^{2+}-ATPases have been identified in *S. cerevisiae* and *S. pombe*. In *S. cerevisiae*, the same strategies used to identify *PMR2*, a putative plasma membrane Ca^{2+}-ATPase (described above), also led to the identification of *PMR2* (Rudolph *et al.*, 1989). DNA sequence analysis of *PMR1* indicated that it encodes a protein highly reminiscent of other Ca^{2+}-ATPases (Brandl *et al.*, 1986; Shull and Greeb, 1988).

1. *PMR1*

PMR1 is capable of encoding a protein of 104 kDa and contains a putative Ca^{2+}-binding site in its amino terminus similar to the E–F Ca^{2+}-binding pocket of calmodulin (Kretsinger, 1980). The *PMR1* protein has sequence similarity with two known Ca^{2+}-ATPases within its fifth and sixth putative transmembrane domains. Clarke *et al.* (1989) reported that in the sarcoplasmic reticulum Ca^{2+}-ATPase contains high-affinity Ca^{2+}-binding sites. Thus the genetic evidence favors the hypothesis that *PMR1* encodes a P-type Ca^{2+}-ATPase.

A null mutation of *PMR1* is viable and confers phenotypes that are consistent with an alteration of Ca^{2+} homeostasis. *pmr1-1::LEU2* cells require higher concentrations of Ca^{2+} in the medium to support growth and they are hypersensitive to EGTA, the Ca^{2+} ionophore A23187, and the anti-calmodulin drug trifluoperazine (Rudolph *et al.*, 1989).

Although *pmr1* cells exhibit a higher requirement for Ca^{2+}, *PMR1* does not appear to encode a Ca^{2+}-ATPase responsible for uptake of the ion directly, nor does it appear to encode a plasma membrane-localized protein. Rather, *PMR1* appears to play a role in the secretory pathway. In *pmr1* cells heterologous proteins exhibit a higher rate of secretion and secreted glycoproteins, such as invertase, appear not to be fully glycosylated (Rudolph *et al.*, 1989). Also, whereas the null allele of *PMR1* suppresses the conditional lethality of a number of different secretory mutants, including *ypt1-1* (Rudolph *et al.*, 1989), *sec13-1*, and *sec19-1*, overexpression of *PMR1* leads to an exacerbation of the growth defects of these mutants (H. Rudolph, A. Antebi, and G. Fink, personal communication). Finally, PMR1 protein has been localized, using an epitope-tagged but functional derivative of *PMR1*, to the Golgi or post-Golgi secretory vesicles.

Rudolph *et al.* suggested that the Ca^{2+}-ATPase encoded by *PMR1* is required for the normal flow of secreted proteins through the Golgi, exposing them to a rate-limiting step that precedes and/or includes the recognition and glycosylation steps that include the addition of high-mannose sugars (Rudolph *et al.*, 1989). In *pmr1-1::LEU2* cells, perhaps because of improper sequestration of Ca^{2+}, traffic through this rate-limiting step is bypassed. The result appears to be faster secretion of proteins that are not fully glycosylated.

2. CTA3

M. Ghislain and A. Goffeau have cloned and sequenced the *CTA3* gene from *S. pombe* and deduced that it encodes a 1037-amino acid protein that is only 15% identical to the *S. pombe* H^+-ATPase and only somewhat similar to *PMR1* and *PMR2* (Goffeau *et al.*, 1990). *CTA3* is nonessential but *cta3* null mutants are hypersensitive to sodium acetate (at pH 5.0) and sensitive to high concentrations of Ca^{2+} in the growth medium. Ca^{2+} uptake in *cta3* mutants is severely impaired compared to wild-type cells although why the *cta3* mutant is sensitive to Ca^{2+} in the growth medium when its uptake is impaired is not yet understood.

The inferred protein sequence of CTA3 contains a region similar to the phospholambin binding sites of the endoplasmic reticulum-localized Ca^{2+}-ATPases in mammalian systems (James *et al.*, 1989). Thus, *CTA3* may encode the fungal equivalent of these calcium pumps. If so, one scenario that would explain the phenotypes of *cta3* cells is that they are defective in transporting Ca^{2+} into the lumen of the endoplasmic reticulum and this defect results in negative feedback to the plasma membrane Ca^{2+}-ATPases, thus decreasing the uptake of the ion by the cell.

V. Concluding Remarks

A. The Genetics of Redundancy

In the last 5 years the combined approaches of genetics and molecular biology have led to the identification of most of the essential ion transport systems in yeast. We now know the identity of genes encoding transporters for H^+, K^+, Na^+, and Ca^{2+} and P_i. Hopefully, genetic identification of the transporters for Mg^{2+}, SO_4^-, and NH_4^+ will follow shortly. Genetic redundancy is clearly the rule for ion transporters in yeast despite the fact that this eukaryote is single celled and free living. The genes encoding (at least) two H^+ pumps, two K^+ transporters, and perhaps six to eight Na^+ pumps have been identified and in most cases their sequences determined. Why such redundancy has evolved in a unicellular organism is enigmatic. At the very least new interpretations of previous studies can now be made knowing that multiple transporters have to be accounted for. Even better is the present opportunity for physiological studies of ion transport to be accomplished in genetically "simplified" strains.

B. More Genetics?

Genetic studies will continue to yield significant contributions to our understanding of yeast ion transport. This prediction is a safe one, based the diversity of genetic approaches that can be undertaken with yeast. In this regard, it is worth emphasizing two approaches detailed in this article that represent opposite extremes of genetic analysis. Focused mutational analysis of the plasma membrane ATPase has allowed the assignment of specific functions to specific regions of the enzyme. In contrast, selection for suppressor mutations that increase K^+ uptake in transport-deficient cells allowed the identification of previously unknown transporter genes.

Observations from this laboratory, that a class of structurally related glucose transporters can be mutationally altered to become efficient K^+ transporters underscore the power of suppressor analysis in the investigation of many problems in cell biology, can be productive in the transport field as well. By genetically deleting a function such as ion transport and requiring the cell to restore this function through extragenic suppressors, we can rely on the organism not only to accomplish for us what would otherwise be tedious, but also, if the selections are properly designed, to sift and winnow the countless mutational possibilities. Thus, genetic selection has proven to be not only efficient but (unexpectedly) revealing as well. Moreover, the molecular analysis of these sugar/K^+ transporters will likely serve to enhance an understanding of the structure and function of these proteins in ways we had not anticipated. For example, the analysis of sufficient numbers of independent mutations in these genes will permit us to generate a genetic map of the amino acids directly involved in K^+ transport.

C. Molecular and Genetic Approaches to Ion Transport: The Limitations of Dancing Alone

We are now at the point where molecular genetic techniques can be employed to confer previously unimagined precision in the manipulation of transporter structure. In our quest to determine structure–function relationships we must be able to make sense out of the functional alterations that result from mutations. The problem we face is the increasing need for collaboration between geneticists and molecular biologists on the one hand, and biochemists and physiologists on the other. Although it has permitted significant advances in our understanding of ion transport, molecular genetics cannot supplant rigorous physiological and biochemical analysis. To the latter group I offer, on behalf of my fellow geneticists, our newly developed tools in the hope that the marriage between molecular genetics and physiology/biochemistry may be harmonious and productive.

Acknowledgments

I wish to thank T. Beeler, T. Dunn, A. Goffeau, J. Haber, D. Hoeppner, and R. Serrano for critical reading of segments of the manuscript. This work was supported by grants from the National Institutes of Health (GM45739) and National Science Foundation (DCB-8711346 and DCB-8657150) to R.F.G.

References

Addison, R. (1986). *J. Biol. Chem.* **267,** 14896–14907.

Anand, S., and Prasad, R. (1987). *Biochem. Int.* **14,** 963–970.

Anderson, J. A., Best, L. A., and Gaber, R. F. (1991). *Gene* **99,** 39–46.

Anderson, J. A., Huprikar, S. S., Kochian, L. V., Lucas, W. J., and Gaber, R. F. (1992). *Proc. Natl. Acad. Sci. U.S.A.* **89,** 3736–3740.

Benz, R., Schmid, A., and Dihanich, M. (1989). *J. Bioenerg. Biomembr.* **21,** 439–450.

Bertl, A., and Slayman, C. L. (1990). *Proc. Natl. Acad. Sci. U.S.A.* **87,** 7824–7828.

Blachly-Dyson, E., Peng, S. Z., Colombini, M., and Forte, M. (1989). *J. Bioenerg. Biomembr.* **21,** 471–483.

Blachly-Dyson, E., Peng, S., Colombini, M., and Forte, M. (1990). *Science* **247,** 1233–1236.

Bogañez, E., Machado, A., and Satrustegui, J. (1983). *Biochim. Biophys. Acta* **733,** 234–241.

Boguslawski, G. (1985). *Mol. Gen. Genet.* **199,** 401–405.

Boguslawski, G., and Polazzi, J. O. (1987). *Proc. Natl. Acad. Sci. U.S.A.* **84,** 5848–5852.

Borst-Pauwels, G. W. F. H. (1981) *Biochim. Biophys. Acta* **650,** 88–127.

Bostian, K., Elliott, Q. A., Bussey, H., Burn, V., Smith, A., and Tipper, D. J. (1984). *Cell* (*Cambridge, Mass.*) **36,** 741–751.

Boutry, M., Foury, F., and Goffeau, A. (1977). *Biochim. Biophys. Acta* **464,** 602–612.

Boutry, M., Michelet, B., and Goffeau, A. (1989). *Biochem. Biophys. Res. Commun.* **162,** 567–574.

Brandl, C. J., Green, N. M., Korczak, B., and MacLennan, D. H. (1986). *Cell* (*Cambridge, Mass.*) **44,** 597–607.

Broek, D., *et al.* (1987) *Cell* (*Cambridge, Mass.*) **48,** 789–799.

Bun-Ya, M., Nishimura, M., Harashima, S., and Oshima, Y. (1991). *Mol. Cell. Biol.* **11,** 3229–3238.

Cantley, L. C., Cantley, L. G., and Josephson, L. (1978). *J. Biol. Chem.* **253,** 7361–7368.

Capieaux, E., Vignais, M. L., Sentenac, A., and Goffeau, A. (1989). *J. Biol. Chem.* **264,** 7437–7446.

Calenza, J. L., Marshall-Carlson, L., and Carlson, M. (1988). *Proc. Natl. Acad. Sci. U.S.A.* **85,** 2130–2134.

Chang, A. and Slayman, C. W. (1991). *J. Cell Biol.* **115,** 289–295.

Cid, A., and Serrano, R. (1988). *J. Biol. Chem.* **263,** 14134–14139.

Cid, A., Perona, R., and Serrano, R. (1987). *Curr. Genet.* **12,** 105–110.

Clarke, E. M., Loo, W. W., Inesi, G., and MacLennan, D. H. (1989). *Nature* (*London*) **339,** 476–478.

Colombini, M. (1979). *Nature* (*London*) **279,** 643–645.

Conway, E. J., and Brady, T. G. (1950). *Biochem. J.* **47,** 360–369.

Conway, E. J., and Downey, M. (1951). *Biochem. J.* **47,** 355–360.

Cooper, T. G. (1982). *In* "The Molecular Biology of the Yeast Saccharomyces: Metabolism and Gene Expression" (J. N. Strathern, E. W. Jones, and J. R. Broath, eds.) Cold Spring Harbor Lab., pp. 39–99. Cold Spring Harbor, New York.

de la Peña, P., Barros, F., Gascon, S., Lazo, P. S., and Ramos, S. (1981). *J. Biol. Chem.* **256,** 10420–10425.

Dihanich, M., Suda, K., and Schatz, G. (1987). *EMBO J.* **6,** 723–728.

Dubois, E., and Grenson, M. (1979). *Mol. Gen. Genet.* **175,** 67.

Egbosimba, E. E., and Slaughter, J. C. (1987). *J. Gen. Microbiol.* **133,** 375–379.

Eilam, Y. (1982). *Microbios* **35,** 99–110.

Epel, D., and Dube, F. (1987). *In* "Control of Animal Cell Proliferation" (A. L. Boynton and H. L. Leffert, eds.) Academic Press, San Diego, California.

Eraso, P., Cid, A., and Serrano, R. (1987). *FEBS Lett.* **224,** 193–197.

Forte, M., Adelsberger-Mangan, D., and Colombini, M. (1987a). *J. Membr. Biol.* **99,** 65–72.

Forte, M., Guy, H. R., and Mannella, C. (1987b). *J. Bioenerg. Biomembr.* **19,** 341–350.

Foury, F., and Goffeau, A. (1975). *J. Biol. Chem.* **250,** 2354–2362.

Foury, F., Boutry, M., and Goffeau, A. (1977). *J. Biol. Chem.* **252,** 4577–4583.

Gaber, R. F., Styles, C. A., and Fink, G. R. (1988). *Mol. Cell. Biol.* **8,** 2848–2859.

Gaber, R. F., Kielland-Brandt, M. C., and Fink, G. R. (1990). *Mol. Cell. Biol.* **10,** 643–652.

Ghislain, M., and Goffeau, A. (1991). *J. Biol. Chem.* (in press).

Ghislain, M., Schlesser, A., and Goffeau, A. (1987). *J. Biol. Chem.* **262,** 17549–17555.

Goffeau, A., and de Meis, L. (1990). *J. Biol. Chem.* **265,** 15503–15505.

Goffeau, A., and Green, N. M. (1990). *In* "Monovalent Cations in Biological Systems."

Goffeau, A., Coddington, A., and Schlesser, A. (1989). "Molecular Biology of the Fission Yeast." Academic Press, San Diego, California.

Goffeau, A., Ghislain, M., Navarre, C., Purnelle, B., and Supply, P. (1990). *Biochim. Biophys. Acta* **1018,** 200–202.

Gustin, M., Martinac, B., Saimi, Y., Culbertson, M. R., and Kung, C. (1986). *Science* **233,** 1195–1197.

Gustin, M., Zhou, X.-L., Martinac, B., and Kung, C. (1988). *Science* **242,** 762–765.

Hager, K. M., Mandala, S. M., Davenport, J. W., Speicher, D. W., Benz, E. J. J., and Slayman, C. W. (1986). *Proc. Natl. Acad. Sci. U.S.A.* **83,** 7693–7697.

Halachmi, D., and Eilam, Y. (1989). *FEBS Lett.* **256,** 55–61.

Haro, R., Garciadeblas, B., and Rodriguez, N. A. (1991). *FEBS Lett.* **291,** 189–191.

Harper, J. F., Surowy, T. K., and Sussman, M. R. (1989). *Proc. Natl. Acad. Sci. U.S.A.* **86,** 1234–1238.

Harper, J. F., Manney, L., Dewitt, N. D., Yoo, M. H., and Sussman, M. R. (1990). *J. Biol. Chem.* **265,** 13601–13608.

Higgins, C. F., *et al.* (1986). *Nature (London)* **323,** 448–450.

Hirata, R., Ohsumk, Y., Nakano, A., Kawasaki, H., Suzuki, K., and Anraku, Y. (1990). *J. Biol. Chem.* **265,** 6726–6733.

Huang, W.-H., and Askari, A. (1984). *J. Biol. Chem.* **259,** 13287–13291.

Huet, J., *et al.* (1985). *EMBO J.* **4,** 3539–3547.

Iida, H., Yagawa, Y., and Anraku, Y. (1990). *J. Biol. Chem.* **265,** 13391–13399.

James, P., Maeda, M., Fischer, R., Verma, A. K., Krebs, J., Penniston, J. T., and Carafoli, E. (1988). *J. Biol. Chem.* **263,** 2905–2910.

James, P., Inui, M., Tada, M., Chiesi, M., and Carafoli, E. (1989). *Nature (London)* **342,** 90–92.

Kane, P. M., Yamashiro, C. T., and Stevens, T. H. (1989). *J. Biol. Chem.* **264,** 19236–19244.

Kane, P. M., Yamashiro, C. T., Wolczyk, D. F., Neff, N., Goebl, M., and Stevens, T. H. (1990). *Science* **250,** 651–657.

Kitamoto, K., Yoshizawa, K., Ohsumi, Y., and Anraku, Y. (1988). *J. Bacteriol.* **170,** 2683–2686.

Kleene, R., Pfanner, N., Pfaller, R., Link, T. A., Sebald, W., Neupert, K. W., and Tropshug, M. (1987). *EMBO J.* **6,** 2627–2633.

Ko, C. H., and Gaber, R. F. (1991). *Mol. Cell. Biol.* **11,** 4266–4273.

Ko, C. H., Buckley, A. M., and Gaber, R. F. (1990). *Genetics* **125,** 305–312.

Kolarov, J., Kulpa, J., Baijot, M., and Goffeau, A. (1988). *J. Biol. Chem.* **263,** 10613–10619.

Konishi, J., Denda, K., Oshima, T., Wakagi, T., Uchida, E., Ohsumi, Y., Anraku, Y., Matsumoto, T., Wakabayashi, T., Mukohata, Y., *et al.* (1990). *J. Biochem. (Tokyo)* **108,** 554–559.

Kretsinger, R. H. (1980). *CRC Crit. Rev. Biochem.* **8**, 119–174.

Kruckeberg, A. L., and Bisson, L. F. (1990). *Mol. Cell. Biol.* **10**, 5903–5913.

Lentzen, H., Arreguin, M., Kappeli, O., Fiechter, A., and Fuhrmann, G. F. (1987). *J. Biotechnol.* **6**, 281–291.

Lewis, D. A., and Bisson, L. F. (1991). *Mol. Cell. Biol.* **11**, 3804–3813.

Mann, B. J., Bowman, B. J., Grotelueschen, J., and Metzenberg, R. L. (1989). *Gene* **83**, 281–289.

Mannella, C. A., and Colombini, M. (1984). *Biochim. Biophys. Acta* **774**, 206–214.

Martinac, B., Zhu, H., Kubalski, A., Zhou, X., Culbertson, M., Bussey, H., and Kung, C. (1990). *Proc. Natl. Acad. Sci. U.S.A.* **87**, 6228–6232.

McCusker, J. H., Perlin, D. S., and Haber, J. E. (1987). *Mol. Cell. Biol.* **7**, 4082–4088.

Meade, J. C., Shaw, J., Lemaster, S., Gallageher, G., and Stringer, J. R. (1987). *Mol. Cell. Biol.* **7**, 3937–3946.

Mihara, K., and Sato, R. (1985). *EMBO J.* **4**, 769–774.

Nakajima, S. J., Iida, H., Tsuji, F. I., and Anraku, Y. (1991). *Biochem. Biophys. Res. Commun.* **174**, 115–122.

Nakamoto, R. K., Rao, R., and Slayman, C. W. (1989). *Ann. N. Y. Acad. Sci.* **574.**

Nakamoto, R. K., Rao, R., and Slayman, C. W. (1991). *J. Biol. Chem.* **266**, 7940–7949.

Novick, P., Ferro, S., and Schekman, R. (1981). *Cell (Cambridge, Mass.)* **25**, 461.

Orino, T., Den, H. J. A., and Shulman, R. G. (1983). *Proc. Natl. Acad. Sci. U.S.A.* **80**, 5185–5189.

Ohsumi, Y., and Anraku, Y. (1981). *J. Biol. Chem.* **256**, 2079–2082.

Ohsumi, Y., and Anraku, Y. (1983). *J. Biol. Chem.* **258**, 5614–5617.

Ohsumi, Y., Kitamoto, K., and Anraku, Y. (1988). *J. Bacteriol.* **170**, 2676–2682.

Ohta, T., Nagano, K., and Yoshida, M. (1986). *Proc. Natl. Acad. Sci. U.S.A.* **83**, 2071–2075.

Ohya, Y., Ohsumi, Y., and Anraku, Y. (1984). *Mol. Gen. Genet.* **193**, 389–394.

Ohya, Y., Miyamoto, S., Ohsumi, Y., and Anraku, Y. (1986a). *J. Bacteriol.* **165**, 28–33.

Ohya, Y., Ohsumi, Y., and Anraku, Y. (1986b). *J. Gen. Microbiol.*

Okorokov, L. A., Kulakovskaya, T. V., Lichko, L. P., and Polorotova, E. V. (1985). *FEBS Lett.* **192**, 303–306.

Ovchinnikov, Y., Dzhandzhugazyan, K. N., Lutsenko, S. V., Mustayev, A. A., Modyanov, N. N., and Dzhandzugazyan, K. D. K. N. (1987). *FEBS Lett.* **217**, 111–116.

Pardo, J. M., and Serrano, R. (1989). *J. Biol. Chem.* **264**, 8557–8562.

Peña, A., Pardo, J. P., and Ramirez, J. (1987). *Arch. Biochem. Biophys.* **253**, 431–438.

Perlin, D. S., Brown, C. L., and Haber, J. E. (1988). *J. Biol. Chem.* **263**, 18118–18122.

Perlin, D. S., Harris, S. L., Seto-Young, D., and Haber, J. E. (1989). *J. Biol. Chem.* **264**, 21857–21864.

Perona, R., and Serrano, R. (1988). *Nature (London)* **334**, 438–440.

Perona, R., Portillo, F., Giraldez, F., and Serrano, R. (1990). *Mol. Cell. Biol.* **10**, 4110–4115.

Pick, U. (1982). *J. Biol. Chem.* **257**, 6111–6119.

Portillo, F., and Mazon, M. J. (1986). *J. Bacteriol.* **168**, 1254–1257.

Portillo, F., and Serrano, R. (1988). *EMBO J.* **7**, 1793–1798.

Portillo, F., and Serrano, R. (1989). *Eur. J. Biochem.* **186**, 501–507.

Portillo, F., de Larrinoa, I. F., and Serrano, R. (1989). *FEBS Lett.* **247**, 381–385.

Ramirez, J. A., Vacata, V., McCusker, J. H., Haber, J. E., Mortimer, R. K., Owen, W. G., and Lecar, H. (1989). *Proc. Natl. Acad. Sci. U.S.A.* **86**, 7866–7870.

Ramos, J., Contreras, P., and Rodriguez-Navarro, A. (1985). *Arch. Microbiol.* **143**, 88–93.

Rodriguez-Navarro, A., and Ramos, J. (1984). *J. Bacteriol.* **159**, 940–945.

Roon, R. J., Meyer, G. M., and Larimore, F. S. (1977). *Mol. Gen. Genet.* **158**, 185.

Roon, R. J., Larimore, F. S., Meyer, G. M., and Kreisle, R. A. (1978). *Arch. Biochem. Biophys.* **185**, 142–150.

Rudolph, H. K., et al. (1989). Cell (Cambridge, Mass.) 58, 133–145.

Sato, T., Ohsumi, Y., and Anraku, Y. (1984a). J. Biol. Chem. 259, 11505–11508.

Sato, T., Ohsumi, Y., and Anraku, Y. (1984b). J. Biol. Chem. 259, 11509–11511.

Schlesser, A., Ulaszewski, S., Ghislain, M., and Goffeau, A. (1988). J. Biol. Chem. 263, 19480–19487.

Sentenac, H., Bonnaud, N., Minet, M., Lacroute, F., Salmon, J.-M., Gaymard, F., Grignon, C. (1992). Science 256, 663–665.

Serrano, R. (1983). FEBS Lett. 156, 11–14.

Serrano, R. (1984). Curr. Top. Cell. Regul. 23, 87–126.

Serrano, R. (1989). Annu. Rev. Plant Physiol. Plant Mol. Biol. 40, 61–94.

Serrano, R. (1990). In "The Plant Plasma Membrane" (C. Larsson and I. M. Moller, eds.). Springer-Verlag, Berlin.

Serrano, R., and Portillo, F. (1990). Biochim. Biophys. Acta 1018, 195–199.

Serrano, R., C., K.-B. M., and Fink, G. R. (1986). Nature (London) 319, 689–693.

Serrano, R., et al. (1991). Biochim. Biophys. Acta 1062, 157–164.

Shih, C. K., Wagner, R., Feinstein, S., Kanik, E. C., and Neff, N. (1988). Mol. Cell. Biol. 8, 3094–3103.

Shull, G. E., and Greeb, J. (1988). J. Biol. Chem. 263, 8646–8657.

Shull, G. E., Schwartz, A., and Lingrel, J. B. (1985). Nature (London) 316, 691–695.

Skipper, N., and Bussey, H. (1977). J. Bacteriol. 129, 668–677.

Sussman, M. R., Strickler, J. E., Hager, K. M., and Slayman, C. W. (1987). J. Biol. Chem. 262, 4569–4573.

Szkutnicka, K., Tschopp, J. F., Andrews, L., and Cirilli, V. P. (1989). J. Bacteriol. 171, 4486–4493.

Tanifuji, M., Sato, M., Wada, Y., Anraku, Y., and Kasai, M. (1988). J. Membr. Biol. 106, 47–55.

Uchida, E., Ohsumi, Y., and Anraku, Y. (1985). J. Biol. Chem. 260, 1090–1095.

Ueda, Y., and Oshima, Y. (1975). Mol. Gen. Genet. 136, 255.

Ulaszewski, S., Grenson, M., and Goffeau, A. (1983). Eur. J. Biochem. 130, 235–239.

Ulaszewski, S., Coddington, A., and Goffeau, A. (1986). Curr. Genet. 10, 359–364.

Ulaszewski, S., Balzi, E., and Goffeau, A. (1987a). Mol. Gen. Genet. 207, 38–46.

Ulaszewski, S., Van, H. J. C., Dufour, J. P., Kulpa, J., Nieuwenhuis, B., and Goffeau, A. (1987b). J. Biol. Chem. 262, 223–228.

Umemoto, N., Yoshihisa, T., Hirata, R., and Anraku, Y. (1990). J. Biol. Chem. 265, 18447–18453.

Vai, M., Popolo, L., and Alberghina, L. (1986). FEBS Lett. 206, 135–141.

Vallejo, C. G., and Serrano, R. (1989). Yeast 5, 307–319.

Van Dyck, L., Petretski, J. H., Wolosker, H., Rodrigues, Jr., G., Schlesser, A., Ghislain, M., Goffeau, A. (1990). Eur. J. Biochem. 194, 785–790.

Verma, A. K., et al. (1988). J. Biol. Chem. 263, 14152–14159.

Vidal, M., and Gaber, R. F. (1991). Mol. Cell. Biol. 11, 6317–6327.

Vidal, M., Buckley, A. M., Hilger, F., and Gaber, R. F. (1990). Genetics 125, 313–320.

Vidal, M., Strich, R., Esposito, R. E., and Gaber, R. F. (1991). Mol. Cell. Biol. 11, 6306–6316.

Vidal, M., Buckley, A. M., Yohn, C. and Gaber, R. F. (1992). Proc. Natl. Acad. Sci. U.S.A. Submitted for publication.

Vilsen, B., Andersen, J. P., Clarke, D. M., and MacLennan, D. H. (1989). J. Biol. Chem. 264, 21024–21030.

Wada, Y., Ohsumi, Y., Tanifuji, M., Kasai, M., and Anraku, Y. (1987). J. Biol. Chem. 262, 17260–17263.

Walderhaug, M. O., Post, R. L., Saccomani, G., Leonard, R. T., and Briskin, D. P. (1985). *J. Biol. Chem.* **260,** 2852–3859.
Weimken, A., Schellenberg, M., and Urech, K. (1979). *Arch. Microbiol.* **123,** 23–35.
Yamashiro, C. T., Kane, P. M., Wolczyk, D. F., Preston, R. A., and Stevens, T. H. (1990). *Mol. Cell. Biol.* **10,** 3737.

Index

355

Z